W9-DFR-186

Carbon Sequestration in Forest Ecosystems

Carbon Sequestration in Forest Ecosystems

By

Klaus Lorenz
School of Environment and Natural Resources
Ohio State University
USA

Rattan Lal
School of Environment and Natural Resources
Ohio State University
USA

Klaus Lorenz
Carbon Management
and Sequestration Center
School of Environment
and Natural Resources
The Ohio State University
Columbus, Ohio
USA
lorenz.59@osu.edu

Rattan Lal
Carbon Management
and Sequestration Center
School of Environment
and Natural Resources
The Ohio State University
Columbus, Ohio
USA
lal.1@osu.edu

ISBN 978-90-481-3265-2 e-ISBN 978-90-481-3266-9
DOI 10.1007/978-90-481-3266-9
Springer Dordrecht Heidelberg London New York

Library of Congress Control Number: 2009935335

Cover picture: Sequoia sempervirens (D. Don) Endl., Mariposa Grove, Yosemite National Park, California, USA (Nicola Lorenz)

Printed on acid-free paper

Springer is part of Springer Science+Business Media (www.springer.com)

Foreword

Carbon (C) sequestration in forest ecosystems has become an important issue both in the political discussion about abrupt climate change (ACC) and forest ecosystem research. This book is the first to synthesize information on relevant processes, factors, and causes of C turnover in forest ecosystems and the technical and economic potential of C sequestration. Accordingly, the authors are able to fill an important gap between the needs of global environmental policy and local forest management. In fact, the book collates valuable knowledge which is necessary to define a sustainable and adaptive forest management in terms of both slowing-down ACC and preparing forests to potential scenarios of a future climate.

Notably soil organic matter (SOM) may act as a powerful sink for atmospheric C in the long-term. On the other hand, soils can also become a source of C when environmental conditions are subject to a change (e.g., in the long-term, when climate changes during soil formation, in mid-term when a forest is clear-felled, or in the short-term such as after rewetting of the soil following an extreme and extended drought). All source-sink functions of forest soils are related to biotic processes since litter production, decomposition, and humus synthesis are controlled by a large number of autotrophic or heterotrophic organisms that interact in the ecosystem. Furthermore, the amount and quality of SOM is closely related to biogeochemical cycles of other elements. Notably the availability of nitrogen (N) plays a key role in the SOM dynamics. Soil N may be affected by natural soil formation, but also by human activities (e.g., atmospheric N deposition or cultivation of N-fixing species). Such examples underline that forest managers (as silviculturists in the classical sense) have to go beyond their traditional concepts of sustainable forestry. Up-to-now their approaches have been focusing mostly on controlling growth of trees and stands (e.g., by species selection and regulating stand structure). However, under the auspices of C sequestration they have to consider likewise the 'belowground forest' by looking on site-specific root distribution and turnover, humus formation, microbial activity etc. To cope with this challenging task forest managers have to integrate modern knowledge resulting from basic soil science and forest ecology into the management plans. The well-documented history of forest use in Central Europe may be helpful in demonstrating how detrimental an improper management of SOM can be: former practices like litter raking and fuelwood coppicing have led to soils severely depleted of SOM and nutrient reserves. Thus, forests in many regions suffer from those extractive management practices. The potential of SOM to provide basic services (e.g., buffers and filters in the water and element cycling) has not yet been restored.

In many respects the topics and structure of this book is highly meritorious. The book will contribute significantly to academic teaching, and stimulating the dialog among different groups in policy, science, and practical land management!

April 26, 2009

K.H. Feger
Professor of Forest Soils and Nutrition
Dresden University of Technology (Germany)
Tharandt (Germany)

Foreword

This comprehensive work by Klaus Lorenz and Rattan Lal on carbon sequestration provides a significant biology focused contribution to discussions in school rooms, university lecture halls and Parliamentary debating chambers around the world – how can forest ecosystems help mitigate climate change?

While debate will continue for some time about the relationship between climate change and the activities of civilisation, this book firmly establishes that the potential benefits of forests in terms of carbon sequestration will not be fully realised if these ecosystems are not carefully managed.

Clearly, well before humans evolved, forests ecosystems played an important role in the development of the global environment. Today, we depend on forest natural resources, ecosystem services and benefits from forests and fossil fuels that have evolved from forests established thousands of centuries ago. In modern times, global forest cover has greatly diminished and society is relying on the remaining forests to continue to provide the full range of environmental services and benefits upon which our survival depends. The significance of forests in maintaining a habitable planet has never been more in the spotlight, nor have we ever been as dependent on the use fossil fuels. So, although the area of new forests established is increasing, the importance of the remaining existing forests in maintaining the global carbon cycle cannot be understated.

By bringing current knowledge on carbon sequestration in forest ecosystems together in one place, this book will advance our ability to manage the remaining forest resources and safeguard their continuing contribution to the global carbon cycle. This will ensure that the small amounts of carbon locked up in forests on a day by day, week by week, month by month, year by year, century by century, millennium by millennium basis continue to accumulate, ensuring that forests support human kind.

This book makes a valuable contribution to the collective knowledge of students, scientists and policy makers, which will, in turn, guide efforts to manage the world's remaining forests and new forests in the millennia to come. It provides a series of questions and identifies knowledge gaps that will encourage further debate and inquiry, leading to the identification of policy and management practices that will see the realisation of the full potential of forest ecosystems to sequester carbon. These questions and our commitment to filling the knowledge gaps identified provide a useful benchmark against which progress in our science around forest ecosystem carbon sequestration can be measured. In time, that progress will be judged by future scientists and, ultimately, observers of human history. Society stands to gain much from books like this. It is our collective responsibility to consider very carefully the complexity and connectivity of forests to the global carbon cycle and put in place measures to ensure that the remaining forest ecosystems prosper.

April 2009

Peter Clinton
Scion (formerly New Zealand Forest Research Institute)
Christchurch, New Zealand

Preface

Forest ecosystems cover large parts of the terrestrial land surface and are major components of the terrestrial carbon (C) cycle. Most important, forest ecosystems accumulate organic compounds with long C residence times in vegetation, detritus and, in particular, the soil by the process of C sequestration. Trees, the major components of forests, absorb large amounts of atmospheric carbon dioxide (CO_2) by photosynthesis, and forests return an almost equal amount to the atmosphere by auto- and heterotrophic respiration. However, a small fraction of C remaining in forests continuously accumulates in vegetation, detritus, and soil. Thus, undisturbed forest ecosystems are important global C sinks.

The forest ecosystem service of C sequestration is central to the well-being of the human society and to the well-being of planet Earth. However, abrupt climate change (ACC) threatens the C sink in forests as a consequence of burning of fossil fuels and land use changes, effectively disposing increasing amounts of CO_2 in the atmosphere. Thus, atmospheric CO_2 concentrations and temperatures are increasing, and precipitation regimes are altered which all may impact C sequestration processes in forest ecosystems. Recent ACC has had limited consequences for the forest C sink compared to human activities such as deforestation for agriculture. However, future ACC as result of increasing fossil fuel emissions may turn forests into a source for atmospheric CO_2 which will further exacerbate ACC impacts on forests by a positive feedback. Thus, the ultimate solution for ACC is the de-carbonization of the global economy. Until effective technological measures are implemented, C sequestration in forest ecosystems can help to slow-down ACC. Also, sustainable and adaptive forest management can better prepare forests for future ACC change. Sustainable and adaptive forest management practices must be implemented to ensure that future forests absorb C despite ongoing perturbations by ACC. International agreements on climate change must appreciate the role of forest ecosystems for ACC mitigation. Future international climate agreements will, in particular, address the importance of reducing deforestation and forest degradation (REDD). Important for C sequestration in forest ecosystems is the reduction in tropical deforestation, and the protection of the large amounts of C stored in peatland and old-growth forests.

However, there is a lack of reference and text books for graduate and undergraduate students interested in understanding basic processes of C dynamics in forest ecosystems and the underlying factors and causes which determine the technical and economic potential of C sequestration. This book provides the information on processes, factors and causes influencing C dynamics in forest ecosystems. It illustrates the topic with appropriate examples from around the world, and lists a set of questions at the end of each chapter to stimulate thinking and promote academic dialogue. Each chapter provides up-to-date references on the current issues, and summarizes the current understanding while identifying the knowledge gaps for future research.

This book is the first to describe the effects of ACC on the various processes by which forests exchange C with the environment. Exchanges of C with the atmosphere and surrounding ecosystems occur through photosynthesis, respiration, and fluxes of carbon monoxide (CO), methane (CH_4), biogenic volatile organic compounds (BVOCs), dissolved inorganic carbon (DIC), dissolved organic carbon (DOC), and particulate carbon (PC). The discussion of effects of ACC on forest ecosystem C sequestration processes is based on a broad review of current literature on the possible impacts of increasing atmospheric CO_2 concentrations, temperature and altered precipitation regimes on ecosystem processes. Carbon sequestration is defined as the increase in the amount of C bound in organic compounds with long C residence times in vegetation, detritus and soil. Major nutrient and water limitations on C sequestration in forest ecosystems are also described. Finally, the future roles of forests as bioenergy source and for ACC mitigation are discussed. Focus of the book is C sequestration in existing forests and not in those established by afforestation and reforestation or in the forest products sector. Thus, this book is valuable source of information intended for use by graduate and undergraduate students, scientists, forest managers and policy makers.

May 1, 2009

Klaus Lorenz
Rattan Lal
Columbus, OH, USA

Contents

List of Figures

List of Tables

Chapter 1
Introduction

Forest ecosystems cover the largest part of ice-free land surface among all terrestrial ecosystems. Trees, the main component of forest ecosystems, contain the largest stock or absolute quantity of the living forest biomass. The total forest biomass is about 677 petagram (Pg), and trees constitute 80% of the world's biomass (Kindermann et al. 2008). Forest ecosystems absorb large amounts of CO_2 from the atmosphere via photosynthesis, and return a large part of the fixed carbon (C) back to the atmosphere through auto- and heterotrophic respirations. However, a small fraction of assimilated C is stored in above- and belowground biomass, litter, and soil. About half of the terrestrial C sink is located in forests (Canadell et al. 2007; Fig. 1.2). Based on FAO statistics, about 234 Pg C are stored aboveground in forests, 62 Pg C belowground, 41 Pg C in dead wood, 23 Pg C in litter, and 398 Pg C in forest soils (Kindermann et al. 2008). Forest C data are, however, highly uncertain as, for example, up to 691 Pg C may be stored in forest plant biomass and up to 968 Pg C in forest soils to 1-m depth (Fig. 1.2). Yet, more C is stored in forests than in the atmospheric pool which is estimated to contain about 817 Pg C. In particular, pristine, undisturbed, old-growth forests accumulate large amounts of C and are, therefore, important components of the terrestrial C cycle. Historically, the conversion of forest ecosystems to other land uses (e.g., agricultural and urban) and forest degradation have been major threats to the forest C stock. However, with unprecedented increases in atmospheric CO_2 emissions from burning of fossil fuel and deforestation accompanied by unprecedented global population growth, direct and indirect human-induced pressures on the C stock of forests are dramatically increasing. Specifically, global climate change may weaken the C uptake by forest ecosystems and render forests into a C source which will then have a positive feedback on the global climate. While C sequestration in forest ecosystems cannot stop increases in atmospheric CO_2 originating from fossil fuel combustion, enhancing and strengthening C-fluxes into stable forest C pools can offset anthropogenic CO_2 emissions and minimize risks of abrupt climate change (ACC). Thus, C sequestration, the transfer and secure storage of atmospheric CO_2 into long-lived C pools such as forest ecosystems, buys time for the development and implementation of low-C technologies and de-carbonization of the global economy.

The introductory chapter defines terms related to C sequestration in forest ecosystems. A brief overview of the long-term development of forest ecosystems is presented with a focus on forest ecosystems in Europe and North America.

K. Lorenz and R. Lal, *Carbon Sequestration in Forest Ecosystems*,
DOI 10.1007/978-90-481-3266-9_1,

The global C cycle, ACC and the importance of C sequestration in forest ecosystems are also discussed.

1.1 Forest Ecosystems

The term 'ecosystem' was proposed by A. R. Chapman in the early 1930s and first used in print by A. R. Tansley (Tansley 1935; Willis 1997). Although not clearly formulated, however, the basic concept of an ecosystem is not new, since the Greek philosopher Theophrastus (371–c. 287 BC) was aware of the importance of climate in plant distribution and the 'sympathetic relationships' between the life cycles of plants and the season. Thus, an 'ecosystem' is defined as a "unit of biological organization made up of all of the organisms in a given area interacting with the physical environment so that a flow of energy leads to characteristic trophic structure and material cycles within the system" (Odum 1967). In terrestrial ecosystem studies, ecosystem boundaries can be determined by a watershed (i.e., a topographically defined area such that all precipitation falling into the area leaves through a single stream) or by a stand (i.e., an area of sufficient homogeneity with regards to vegetation, soils, topography, microclimate and disturbance history) (Aber and Melillo 2001). Key to the ecosystem function is the transfer of energy or C from producers (e.g., trees) to consumers (e.g., animals) and decomposers (e.g., microorganisms) (Lindeman 1942). The flow of energy strongly interacts with the flow of nutrient elements and water (Ovington 1962).

Definitions for 'forest' can be grouped into administrative or legal units, a land cover or a land use (Lund 1999). A working definition of 'forest' was used for the Global Forest Resource Assessment 2005 (FAO 2006). Accordingly, 'forest' is a land spanning more than 0.5 hectare (ha) with trees taller than 5 m and a canopy cover of more than 10%, or trees able to reach these thresholds in situ. A forest is determined both by the presence of trees and the absence of other land uses. This definition includes areas with bamboo (*Bambuseae* Kunth ex Dumorth.) and palms (*Arecaceae* Schultz Schultzenstein) provided that height and canopy cover criteria are met; forest roads, firebreaks and other small open areas; forest in national parks, nature reserves and other protected areas such as those of specific scientific, historical, cultural or spiritual interest; windbreaks, shelterbelts and corridors of trees with an area of >0.5 ha and width of >20 m; plantations primarily used for forestry or protective purposes.

As forests are comprised of trees, many forest definitions also define 'trees' (Lund 1999). For example, 'trees' can be defined as large, long-lived (i.e., perennial) woody plants that attain a height of at least 6 m at maturity in a given locality and have a single main self-supporting stem which gives off spreading branches, twigs and foliage to make a crown (Seth 2004). More comprehensively, 'trees' are woody perennials with a single main stem or, in the case of coppice, with several stems, having a more or less definite crown (FAO 2001). This definition also includes bamboos, palms and other woody plants meeting the above criterion.

Thus, trees consist of roots, stem(s), branches, twigs and leaves (Kozlowski et al. 1991). Tree stems consist mainly of support and transport tissues (xylem and phloem). Wood consists of xylem cells, and bark is made of phloem and other tissues external to the vascular cambium. Trees can be classified in several ways. For example, 'evergreen trees' remain green in the dormant season as all leaves do not drop simultaneously and the trees are never leafless (Seth 2004). Otherwise, 'deciduous trees' shed all the leaves at the end of the growing season. Cone-bearing trees are called 'coniferous trees' or 'conifers', and all non-cone bearing but flower bearing trees are called 'flowering trees' or 'broad-leaved trees'. Conifers have needle-shaped leaves whereas flowering trees have broad or flattened leaves. Trees in which seeds are borne naked are called 'gymnosperms' and those in which seeds are enclosed within an ovary/fruit wall are called 'angiosperms'. The angiosperm trees are further classified into dicotyledonous or dicot trees if they have two cotyledons (i.e., embryonic first leaves of a seedling) in their seeds, or monocotyledons or monocot trees if they have only one cotyledon in their seeds (Seth 2004).

In summary, a forest ecosystem includes all living components of the forest, and extends vertically upward into the atmospheric layer enveloping forest canopies and downward to the lowest soil layers affected by roots and biotic processes (Waring and Running 2007). Forest ecosystems are open systems and exchange C, energy and materials with other systems including adjacent forests, aquatic ecosystems and the atmosphere. Thus, a forest ecosystem is never in equilibrium.

1.2 Historic Development of Forest Ecosystems

Long-term changes in forest ecosystems occur over thousands and millions of years, and cause plant migration, speciation, and evolution (Barnes et al. 1998). Natural forests occur in all regions capable of sustaining tree growth, at altitudes up to the tree line, except where natural fire frequency and other disturbances are too high, or where the environment has been altered by human activity. Of importance for the current distribution of forest ecosystems are direct and indirect consequences of Quaternary climate variations which were characterized by cold periods with intermittent warm periods (Schulze et al. 2005). During the cold periods, enormous inland ice masses developed on both hemispheres and forests were eliminated but some trees survived in refuges. However, since the end of the last glaciation in the last Pleistocene (about 10,000 years before present), the ice masses retreated as the climate warmed.

During the present Holocene Epoch many tree species migrated back to earlier positions, and depending on requirements of species for a site, the speed of migration and the position of the cold period refuges distinct forest communities developed everywhere. In the United States, for example, mixed conifer-hardwood or pure hardwood replaced boreal forests over much of the east while pine forests established in the southeast (Perry et al. 2008). In western North America, the community composition developed into a combination of western hemlock (*Tsuga heterophylla* (Raf.) Sarg.), red cedar (*Thuja plicata* Donn ex D.Don.), and Douglas fir (*Pseudotsuga menziesii* (Mirb.)) since withdrawal of the most recent glaciers. Glacial-interglacial cycles, otherwise, determine the extent of the Amazon rainforests and its community composition (Mayle et al. 2000).

The natural development of forest communities in the current interglacial is altered by human perturbations. Almost half of the world's forest has been converted to farms, pasture, and other uses over the past 8,000 years (Bryant et al. 1997). Tens of thousands of years ago, hunter-gatherers used fire to reduce fuels and manage wildlife and plants (Bowman et al. 2009). On a larger scale, forests were cleared by fire as Neolithic agriculture started in south-central Europe. In the following millennia, forest clearance spread to northern regions (Schulze et al. 2005). Aside from clearing, natural forest vegetation was also changed in large areas around settlements by grazing, trampling, tree felling and other uses of trees. The following medieval forest clearing resulted in pre-industrial extensive forest management in central Europe. At the onset of industrialization around 1800, these thinned and degraded forests were planted with coniferous trees instead of the once dominant deciduous trees. In the 18th century, forest management was developed as science in Europe (Bravo et al. 2008). Principles from biology, ecology, and economics were applied in forest management to the regeneration, density control, use and conservation of forests (Helms 1998). During the 19th and 20th centuries, forest management was initiated in European countries (Bravo et al. 2008). The aim of forest management was to produce commercial timber as quickly and as frequently, but as sustainably as possible. However, long-distance transport of strong acids from industrial processes increasingly acidified the upper soil layers, and caused significant damage in large areas of coniferous forests since mid 1970s. Furthermore, increasing N influx alters the forest site conditions and vegetation in many European regions independent of forest management (Schulze et al. 2005).

Similar to the developments in Europe, vast areas of forest were cleared worldwide over the past two to three centuries (Martin 2008). For example, in the conterminous U.S. forest land was first cleared in the east, and then abandoned as settlers migrated westwards (Clawson 1979). The main reasons for forest clearing were cereals and cotton (*Gossypium hirsutum*) production in North America, whereas cattle pastures and establishing plantations of sugarcane (*Saccharum* L.), tea (*Camellia sinensis* (L.) Kuntze), coffee (*Coffea* L.), rubber trees (*Hevea brasiliensis* Müll.Arg.) and oil palm (*Elaeis* Jacq.) were main reasons for forest clearance in Latin America, the Carribean, Africa and Asia (Martin 2008). Thus, up to 110 million hectare of forest were lost globally in the past 300 years, primarily due to agricultural expansion and timber extraction (Foley et al. 2005). Forests modified by selective logging and other human interventions and forest plantations have replaced primary forests and now cover about 2.5 billion hectare (FAO 2006).

The initiation of forest management for timber production typically depletes the forest biomass C stock. For example, previously unmanaged tropical forests lose between 30% and 70% of their C stock upon conversion for timber production, and C stocks of forest in Europe managed for timber production lie between 100 and 120 megagram (Mg) C ha^{-1} whereas unmanaged forests in national parks contain between 140 and 300 Mg C ha^{-1} (Mollicone et al. 2007). About 1.5 billion hectare are still covered by primary forests, and, in particular, Brazil, the Russian Federation, Canada and the USA contain large areas of primary forests (FAO 2006). Just one fifth of the world's original forest cover remains in large tracts of relatively undisturbed forest (Bryant et al. 1997). Today, forests remain only where people cannot farm sustainably because of difficult market access, poor soils, slope or lack of water and the want of even meager economic returns (Martin 2008).

Aside from changes in forest area, many land-use practices can degrade forest ecosystems in terms of productivity, biomass, stand structure, and species composition (Foley et al. 2005). Introduction of pests and pathogens, changing fire-fuel loads, changing patterns and frequency of ignition sources, and changing local meteorological conditions can also degrade forest ecosystems. However, scenarios of global change raise concerns about alterations in forest ecosystem goods and services. For example, the distribution of a number of typical tree species in the Mediterranean region is likely to decrease under climate change scenarios (Schröter et al. 2005). Otherwise, afforestation is predicted to cause a net increase in forest soil C in Europe despite losses caused by the projected warming. Furthermore, climate change may cause an increased forest growth especially in northern Europe. However, forest management is predicted to have a greater influence on wood production in Europe than climate change. Globally, boreal and temperate forest ecosystems are predicted to shift considerably polewards in the Northern Hemisphere, and a substantial degradation of tropical forest vegetation is indicated by climate projections (Alo and Wang 2008). However, over most of the globe net primary production (NPP) and growing season leaf are index (LAI) are predicted to increase.

1.3 The Global Carbon Cycle and Climate Change

Carbon is the core element for life on Earth (Roston 2008). Life contributes to the regulation of the C content of the atmosphere whereas geological forces predominate over geological timescales. Most important, the Earth's temperature and the C content of the atmosphere are correlated on geological time scales. The global C cycle describes the biogeochemical cycling of C among the atmosphere, biosphere, hydrosphere, pedosphere and geosphere on Earth. The C cycle processes take place over hours to millions of years, and a long-term and a short-term C cycle can be distinguished (Berner 2003). The long-term C cycle, in particular, describes the exchange of C among the rocks and the surficial system consisting of the ocean, atmosphere, biosphere and soil. This cycle is the main controller of the atmospheric CO_2 concentration over geological timescale (>100,000 years), and can be represented by simplified equations (Eqs. (1.1) and (1.2); Berner 2003).

$$CO_2 + CaSiO_3 \leftrightarrow CaCO_3 + SiO_2 \quad (1.1)$$

$$CO_2 + H_2O \leftrightarrow CH_2O + O_2 \quad (1.2)$$

Carbon dioxide (CO_2), methane (CH_4), carbon monoxide (CO) and non-methane hydrocarbons are major carboniferous gases in the atmosphere, but only CO_2 is relevant from the perspective of the C cycle (Houghton 2007). Any process, activity or mechanism that removes carboniferous greenhouse gases (GHG), aerosols or their precursors from the atmosphere is a C sink (IPCC 2007). Several GHGs (e.g., CO_2, CH_4 and non-methane hydrocarbons) are constituents of the atmosphere that absorb and emit radiation at specific wavelengths within the spectrum of thermal infrared radiation emitted by the Earths's surface, the atmosphere itself, and by clouds. However, GHGs differ in their radiative forcing on the efficiency in heating the atmosphere. For example, the greenhouse warming potential (GWP) of CH_4 is about 23 times that of CO_2 on a century timescale (Forster et al. 2007). Furthermore, tropospheric ozone (O_3) has the third strongest positive radiative forcing on climate after the long-lived GHGs CO_2 and the combined forcing of CH_4, non-methane hydrocarbons and nitrous oxide (N_2O).

On land, CO_2 is taken up from the atmospheric pool during the weathering of Ca-and Mg-silicates (Eq. (1.1)). Under pre-industrial conditions, an estimated 0.2 Pg C year^{-1} was absorbed from the atmospheric pool through weathering (Denman et al. 2007). The dissolved weathering products (e.g., Ca^{2+}, Mg^{2+} and HCO_3^-) are transported to the ocean and precipitated in sediments as Ca- and Mg-carbonates, and C is thus stored for geological times as carbonate rocks reach ages between 10^6 and 10^9 years (Holmen 2000). During deep ocean burial, the carbonates are transformed and decomposed at low temperatures and pressures during diagenesis, and at higher temperatures and pressures during metamorphism. By both processes, some CO_2 is released into the oceans and the atmosphere (Eq. (1.1)).

By lateral and vertical movements of plates of the Earth's oceanic crust (subduction), carbonate-C in sediments is delivered to the mantle-C and some is released mostly as CO_2 to the atmosphere by volcanism or undersea vents. Aside from weathering, atmospheric CO_2 is also absorbed on land by photosynthesis (pre-industrial net flux of +0.4 Pg C, Denman et al. 2007), and some C is buried as organic matter (OM) in sediments (Eq. (1.2)). Surface sediments contain about 150 Pg C (Denman et al. 2007). Similar to carbonate C, sediment OM may be deeply buried by subduction and CO_2 released from mantle C through volcanism. Otherwise, buried OM is eventually transformed by diagenesis and metamorphism into kerogen, oil, gas and coal. The stored C in lignite (brown coal) may reach ages of 10^3 to 10^5 years, and ages of 10^6 to 10^9 years in hard coal (Holmen 2000). The pre-industrial fossil fuel pool contained about 3,700 Pg C (Denman et al. 2007). Aside through volcanism and undersea vents, C is also released into the atmosphere by natural oxidative weathering processes of OM exposed on the continents through erosion. However, by burning of fossil fuels humans have accelerated the rate of OM oxidation by a factor of about 100 and greatly perturbed and short-circuited the long-term C cycle (Berner 2003).

The short-term C cycle is of greater importance than the long-term C cycle with respect to C sequestration in forest ecosystems (Fig. 1.1). This cycle controls the atmospheric concentrations of both CO_2 and CH_4 through continuous flows of large amounts of C among the oceans, the terrestrial biosphere and the atmosphere

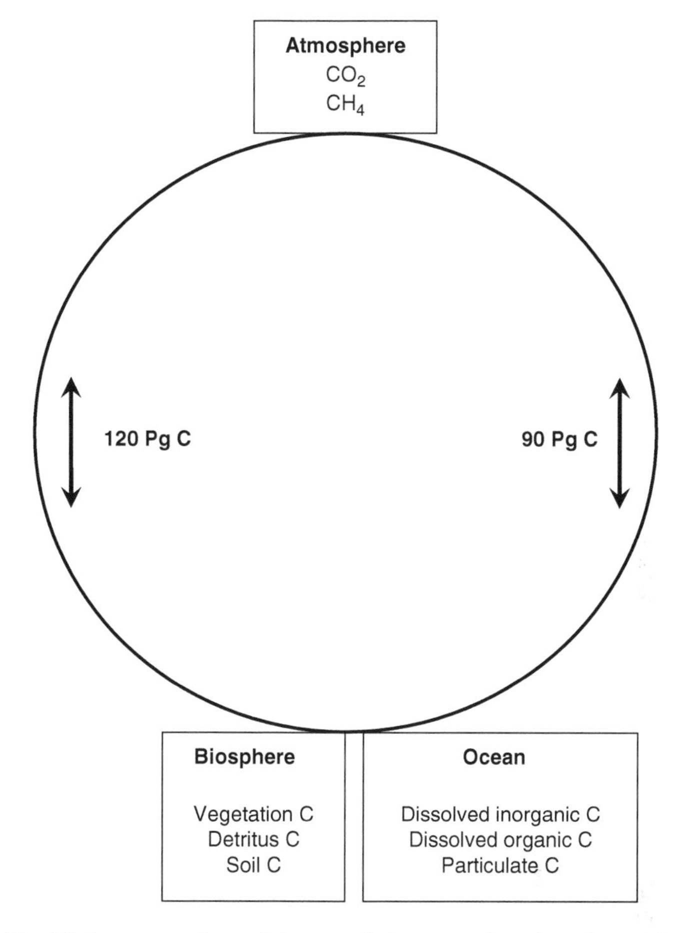

Fig. 1.1 Simplified representations of the natural short-term C cycle and natural annual flux between the atmosphere and the biosphere, and between the atmosphere and the ocean (Denman et al. 2007)

(Denman et al. 2007). Atmospheric C fixed during photosynthesis is returned by plant, microbial and animal respiration, and released to the atmosphere as CO_2 under aerobic, and as CH_4 under anaerobic conditions. On annual timescales, vegetation fires can also be significant sources of CO_2 and CH_4 but subsequent vegetation re-growth captures much of the CO_2 on a decadal time scale.

Similar to exchange processes between the land surface and its vegetation, CO_2 is continuously exchanged between the atmosphere and the ocean. HCO_3^- and CO_3^{2-} (dissolved inorganic C (DIC)) are formed by the reaction of CO_2 with surface layers in the ocean (Fig. 1.1). In winter, DIC sinks at high latitudes with cold waters to deeper ocean depths. This downward transport is roughly balanced by a distributed diffuse upward transport of DIC primarily into warm surface waters.

Phytoplankton growth in the ocean also takes up CO_2 through photosynthesis, and some of this assimilated C sinks as dead organisms and particle to deeper ocean layers (called biological pump) or is transformed into dissolved organic C (DOC). During sinking, most particulate C is respired and eventually re-circulated to the surface as DIC. However, some particles reach abyssal depths and deep ocean sediments, some of which is re-suspended and other is buried and connected to the long-term C cycle described above. The exchange of CO_2 between the atmosphere and the ocean is regulated by solubility and biological pumps which maintain the vertical gradient between surface and deep ocean layers (Denman et al. 2007).

The natural or unperturbed pre-industrial C fluxes have been estimated to be 120 Pg C year^{-1} for the exchange between the terrestrial biosphere and the atmosphere, and 90 Pg C year^{-1} for the exchange between the oceans and the atmosphere (Denman et al. 2007). The pre-industrial pool sizes have been estimated to be 597 Pg C for the atmosphere, 2,300 Pg C for the terrestrial biosphere, and 900 and 37,100 Pg C for surface and intermediate, and deep ocean, respectively. The natural fluxes among these pools have been approximately in balance over longer time periods. However, human activities have severely perturbed the natural global C cycle by starting the anthropogenic greenhouse area. In particular, since the onset of the industrial revolution in 1800 (Steffen et al. 2007), CO_2 is added to the atmospheric pool from hundreds of millions of year old geological pools by burning fossil fuels (coal, oil, gas) and by cement production (i.e., heating limestone). Since 1950 the perturbation of the global C cycle has accelerated as human enterprise has experienced a remarkable explosion (Steffen et al. 2007). Thus, the Earth has left its natural geological epoch, the Holocene, and entered the Anthropocene in which humans and societies have become a global geophysical force. The increases in atmospheric CO_2 concentrations associated with the progression of the Anthropocene impact forest ecosystems from the molecule to the ecosystem level (Valladares 2008). The processes affected include the regulation of gene expression, photosynthetic C oxidation/reduction at the cellular level, and water-use efficiency, secondary compounds, and C:N ratio at the leaf level. At the whole plant level, increasing CO_2 concentrations may impact resource acquisition, the ability to reproduce (fecundity), germination and phenology. Associated increases in aridity and temperature also impact processes in forest ecosystems form the gene to the whole plant level.

In addition to fossil fuel burning and cement production, deforestation and agricultural development add CO_2 to the atmosphere from decadal to centuries old pools in the terrestrial biosphere. Deforestation can be defined as clear cutting and conversion of the forest to other land uses such as cattle pasture, crop agriculture, and urban and suburban areas (Asner et al. 2005). It is also hypothesized that forest clearance and biomass burning 8,000 years ago in Eurasia contributed to the anthropogenic greenhouse era by causing an anomalous increase in atmospheric CO_2 (Ruddiman 2003; Brook 2009). Presently, the net land-atmosphere and ocean-atmosphere CO_2 fluxes are not balanced, and different from zero, and measurable changes in the C pools occurred since pre-industrial times (~10,000 years ago). For example, 140 Pg C have been lost from the terrestrial biosphere through land-use change. Primarily deforestation was responsible for 20% of anthropogenic CO_2

emissions during the 1990s and about 80% resulting from fossil fuel burning (Denman et al. 2007). In 2007, China emitted 21% of the global CO_2 emissions followed by the U.S. which was responsible for 19% of the global CO_2 emissions (Guan et al. 2009).

Due to burning of fossil fuels, cement production, deforestation and agricultural development, the atmospheric CO_2 concentration increased from the pre-industrial level of about 280 parts per million (ppm) to a global monthly mean level of 385 ppm in 2008, and is increasing at the rate of about 2 ppm $year^{-1}$ (Tans 2009). Global surface temperatures are also increasing but none of the natural processes such as solar variability, El Niño-Southern Oscillation (ENSO) or volcanic eruptions can account for the overall warming trend in global surface temperatures from 1905 to 2005 (IPCC 2007; Lean and Rind 2008). For at least the past 1,700 years, recent increases in the Northern Hemisphere surface temperature are likely anomalous. However, conclusions are less definitive for the Southern Hemisphere because proxy data for the region are only sparsely available (Mann et al. 2008). Increases in temperature and aridity associated with increases in atmospheric CO_2 concentrations impact competition among forest plants, improvement of conditions for a plant due to the presence of another (facilitation), reproductive success and diversity at the forest plant community level (Valladares 2008). Furthermore, animal-microbial-plant interactions, and mass and energy fluxes at the forest ecosystem level are affected by increases in temperature and aridity.

Increasing atmospheric abundance of GHGs (including CO_2 and tropospheric aerosols) has been identified as the source of recent global surface warming or the abrupt climate change or ACC (Allen et al. 2006). An increase in GHGs causes a change in Earth's energy balance or radiative forcing (Shine and Sturges 2007). Non-CO_2 GHGs have contributed about 1 W m^{-2} to radiative forcing since pre-industrial times but the largest single contributor to radiative forcing is CO_2, contributing about 1.66 W m^{-2} (IPCC 2007). Thus, ACC is caused by increases in atmospheric CO_2 concentrations originating primarily from fossil fuel burning and land use change or deforestation (Solomon et al. 2009). Increases in GHGs since the pre-industrial era has most likely committed the world to a warming of 2.4°C above the pre-industrial surface temperatures (Ramanathan and Feng 2008). This human-caused climate change causes pronounced worldwide changes within ecosystems (Rosenzweig et al. 2008). In addition, global climate change and ENSO are closely linked. Thus, more extreme ENSO events have been observed during the industrial than under pre-industrial era, and this connection may synergistically stress natural ecosystems (Gergis and Fowler 2009). Projected global average surface warming by the end of the twenty-first century (2090–2099) relative to 1980–1999 ranges from +1.1°C to +6.4°C (IPCC 2007). However, it is not clear if changes in the global climate happen gradually or suddenly (Broecker 1987).

At a global mean temperature change of +3–4°C and +3–5°C above present (1980–1999) the Amazon rainforest and the boreal forest, respectively, are projected to reach a critical threshold at which a tiny perturbation may result in forest dieback (Lenton et al. 2008). Dieback of the Amazon rainforest, in particular, is defined as the conversion of at least half of the current area into raingreen forest, savannah or grassland (Kriegler et al. 2009). Dieback of the boreal forest, on the other hand, is defined as transition to open woodlands or grasslands. However, the current knowledge base about critical thresholds for forest dieback in the Amazon and boreal region is poor (Kriegler et al. 2009). Across a broad range of forests around the world, ACC-induced drought and heat stress have the potential to cause forest dieback defined as tree mortality noticeably above usual mortality levels (Allen 2009). However, information gaps and scientific uncertainties limit the conclusion that ACC-induced forest dieback is an escalating global phenomenon.

Anthropogenic increase in the atmospheric CO_2 abundance is accelerating, and principal drivers have been identified as the dominant contributors (Raupach et al. 2008). Specifically, the CO_2 emission growth rate from fossil fuel emissions increased from 1.3% year^{-1} in the 1990s to 3.3% year^{-1} in 2000–2006 (Canadell et al. 2007). However, despite a reduced growth in the global gross domestic product following the economic crisis, the atmospheric CO_2 concentration still increased by 2.3 ppm in 2008, more than in 2007 (Pep Canadell, Director Global Carbon Project, cited in Spiegel Online, http://www.spiegel.de/wissenschaft/natur/0,1518,614208,00.html). Because of the economic slowdown, the growth rate of global fossil fuel emissions should also have been lower in 2008. Whether this increase in atmospheric CO_2 abundance indicates a decrease in C sink activity of forests and oceans is a topic of further discussion. To avert a dangerous degree of ACC, the concentrations of atmospheric CO_2 must be stabilized by mitigation actions. Avoiding ACC is more easily achievable and more effective by commencing mitigation actions sooner (Vaughan et al. 2009). However, the designed level of stabilization remains a debatable issue (Leigh Mascarelli 2009; Mann 2009). Maximum warming targets considered acceptable are discussed by scientists and policy-makers. For example, temperatures of 1.7°C or 2°C above pre-industrial levels have been proposed (Hansen 2005; European Council 2007). Ironically, the ACC may persist for thousands of years even as CO_2 emissions were to stop (Solomon et al. 2009). Furthermore, even modest increases in global mean temperature above the levels of circa 1990 could commit the climate system to the risk of very large impacts on multiple-century time scales (Smith et al. 2009). For example, by using the most complex class of Earth System Model (ESM), a coupled climate-carbon cycle general circulation model (GCM), Lowe et al. (2009) demonstrated that only very low rates of temperature reduction follow even massive reductions in emissions. Otherwise, the increase in atmospheric CO_2 abundance can be slowed down through forest management, and, in particular, avoided tropical deforestation, reforestation and afforestation of temperate and tropical forests, and establishing plantations on non-forested land (Pacala and Socolow 2004). Sequestration of C in the soil, in particular, is an important option to stabilize the atmospheric abundance of GHGs (Lal 2004; Barker et al. 2007).

The global C cycle is not entirely understood (Leigh Mascarelli 2009). Specifically, all sources and sinks are not completely known. For example, the term 'residual land sink' (also called the missing sink or fugitive CO_2) was introduced in the context of deforestation dominating over forest regrowth vs. the observed net uptake of CO_2 by the land biosphere (Denman et al. 2007). One of the major uncertainties in the global C cycle are reasons for renewed growth of atmospheric CH_4 since the beginning of 2007 after a decade with little change (Rigby et al. 2008). Due to weakness of the observational network for measuring atmospheric CO_2 abundance, the seasonal, interannual and longer-term variability of C fluxes has not been satisfactorily quantified in key locations, in particular the vast expanses of continental Asia, the tropics of South America, Africa and Southeast Asia, and in the Southern Ocean (Heimann 2009). Thus, the sustainability of C sinks in the boreal forest and the Arctic tundra, the fate of the vast amounts of C stored in thawing permafrost such as Yedoma sediments, and the C balance of the remaining tropical forests are unknown. Forests have been exposed to environmental changes over geological and historical periods of time but it is unclear whether forest ecosystems can cope with the current ACC (Valladares 2008). More specifically, it is not certain whether natural C sinks in forests will strengthen through increase in uptake of CO_2 as the planet warms (Kintisch 2009).

1.4 Carbon Sequestration

The rate of increase in atmospheric CO_2 concentration can be reduced through the process of C sequestration. The term 'carbon sequestration' is defined as the uptake of C containing substances, in particular CO_2, into a long-lived reservoir (IPCC 2007). It is a natural process. Thus, the net flux of –1 Pg C year^{-1} from the atmosphere to vegetation, detritus and soil, and the net flux of –1.6 Pg C year^{-1} from the atmosphere to the ocean is C sequestration (Denman et al. 2007). More specifically, 'carbon sequestration' can be defined as the transfer and secure storage of atmospheric CO_2 into other long-lived pools that would otherwise be emitted or remain in the atmosphere (Lal 2008). These pools are located in the ocean, biosphere, pedosphere and geosphere. Most important for the short-term C cycle in forest ecosystems is the exchange with the atmospheric CO_2 pool. Thus, C sequestration in forest ecosystems occurs primarily by uptake of atmospheric CO_2 during tree photosynthesis and the subsequent transfer of some fixed C into vegetation, detritus and soil pools for secure C storage.

Aboveground but more important belowground C inputs from vegetation and detritus C pools are the main C sources for sequestration in the soil organic C (SOC) pool in the soil profile. The average ratio of C pool in soils relative to vegetation ranges from about 5:1 in boreal, 2:1 in temperate, to 1:1 in tropical forests (Jarvis et al. 2005). The efficiency in C sequestration differs among the 100,000 tree species as they vary widley in properties that drive C sequestration such as growth, mortality, decomposition and their dependency on climate (Purves and Pacala 2008). In total, estimates of the C uptake vary from between 0.49 and 0.7 Pg C year^{-1} for the boreal, to 0.37 Pg C year^{-1} for the temperate, and between 0.72 and 1.3 Pg C year^{-1} for the tropical forest biome (Fig. 1.2; see Chapter 4). However, biome data are lacking specifically with regards to forest stands of all species at all stages

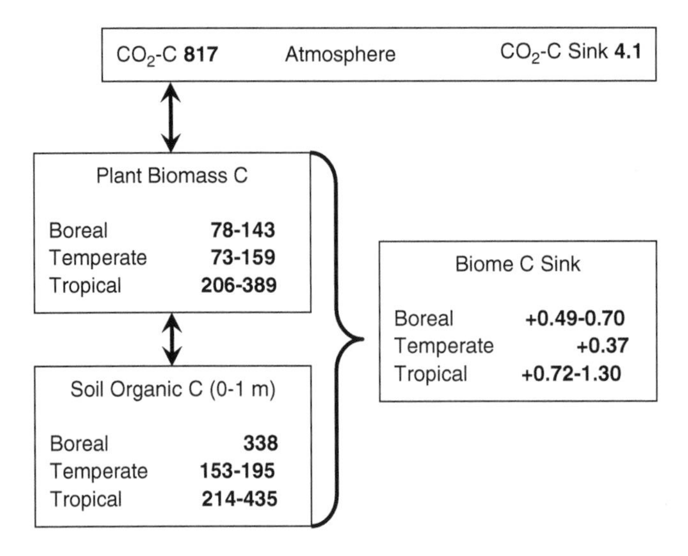

Fig. 1.2 Estimates for plant biomass C and soil organic C pools (Pg C) to 1-m depth for boreal, temperate and tropical forest biomes in relation to the atmospheric CO_2 pool (Pg C), and annual net changes (Pg C $year^{-1}$) (atmospheric pool in 2008 based on global average atmospheric C pool of 805 Pg in 2005, increasing by 4.1 Pg $year^{-1}$; Houghton 2007; Canadell et al. 2007; References for forest biome C pools and sinks see Chapter 4)

in the life cycle from regeneration to harvest, and the impacts of disturbances and effects of climate change. Thus, estimates of C sequestration in global forest biomes and their net C budget are uncertain (Jarvis et al. 2005).

Atmospheric C can be securely stored through binding in inorganic and organic compounds. The C sequestered in forests is primarily bound as organic compounds in vegetation, detritus and soil. However, C may also be sequestered in soil as carbonates (Chapter 2). Organic compounds in vegetation and, in particular, in trees and herbaceous plants may also be sequestered by occlusion within phytoliths which are silicified features resulting from biomineralization within plants (Parr and Sullivan 2005). Biomineralization in some plants, such as the tropical iroko tree (*Milicia excelsa* (Welw.) C. C. Berg), results in the accumulation of Ca-carbonate (Cailleau et al. 2004). Nonetheless, C sequestration in forest ecosystems implies primarily the transfer of C into OM, because soil inorganic C is less dynamic and less effective in the storage of atmospheric CO_2 (Schlesinger 2006). Thus, C sequestration in forest ecosystem in the following Chapters refers to sequestration of atmospheric CO_2 in organic compounds which results in increasing pools of organic C in forest vegetation, detritus and soil.

Sequestration of C implies that the net change in vegetation, detritus and soil C pools of a forest area in a specified time interval is positive. Thus, aside from C uptake processes, efflux processes such as C loss through respiration, leaching and erosion must also be considered (Chapter 2). Soil respiration, in particular, plays a major role in determining the C sequestration potential of forests (Valentini et al. 2000).

Furthermore, natural disturbances such as fire and pest outbreaks, and human disturbances such as land-use change and harvest can also contribute to C losses from a forest ecosystem (Chapter 3). The net changes in forest ecosystem C pools are also affected by increases in atmospheric CO_2 concentrations and temperature, and altered precipitation regimes associated with the ACC (Chapters 2 and 4). Thus, for a quantitative comparison of C sequestration among different forest ecosystems and for reducing uncertainties in C accounting in current and future international agreements it is necessary to measure, estimate or model the net C pool, flux and its direction for each reservoir (Chapter 6). A comprehensive analysis of the role of forest ecosystem C contribution to the C balance must be based on inventories of C pools and their changes in time, direct C flux measurements, and process-based biochemical models that derive net ecosystem production (NEP) from estimates of gross primary production (GPP) and ecosystem respiration (Chapter 2 and 6; Bombelli et al. 2009).

The secure storage of C in a pool or reservoir is an essential pre-requisite to C sequestration. This implies that C inputs especially result in increases in stable vegetation, detritus and soil C pools. A measure of the stability of a pool implies its turnover rate or the residence time. In ecosystem studies, the turnover rate is the fraction of material in a component that enters or leaves in a specified time interval (Aber and Melillo 2001). The residence time, on the other hand, is the inverse of the turnover rate. The C residence time can be approximated by calculating the ratio of C pool size divided by the corresponding C flux (Zhou and Luo 2008). However, not all pools and fluxes in forest ecosystems, in particular, belowground can be easily measured. Another method to estimate C residence time takes advantage of the drastic increase in radiocarbon (^{14}C) in the atmosphere by nuclear bomb tests in the 1960s. This signal has been transferred to vegetation and soil, and can now be used as a tracer to estimate C residence times. Recently, inverse analysis has also been used to estimate the C residence times (Zhou and Luo 2008).

The stability of specific compounds in vegetation, detritus and soil can be assessed by the analyses of biomarkers. The latter are organic compounds with a defined structure indicative of their producer or origin such as plants, fungi, bacteria, animals, fire, or anthropogenic (Amelung et al. 2008). Compound-specific stable isotope analyses allow both tracking back the source of a molecule and its turnover time or mean residence time (MRT). Fumigating a forest with $^{13}CO_2$ or adding ^{13}C labeled substrates to a forest soil can, thus, be used to study mechanisms and rates of soil organic matter (SOM) genesis and transformation on a decadal to centennial scale. However, for compounds with turnover times greater than hundred to thousands of years compound-specific ^{14}C ages need to be estimated.

Continued direct input of organic compounds with long C residence times results in their accumulation in forest ecosystems. However, compounds with long residence times can also be generated through metabolic and decomposition processes by modification of organic compounds with shorter C residence times (Chapter 2). The major input into forest ecosystems with inherent long C residence times up to millennia is the black C (BC) (char) from biomass burning (Table 1.1; Kuzyakov et al. 2009). Increases in BC inputs can, thus, directly contribute to

Table 1.1 Residence times of bulk organic matter, organic compounds and biomarkers in the soil-plant system (Modified from Kuzyakov et al. 2009; Amelung et al. 2008; Lorenz et al. 2007; Nieder et al. 2003)

Organic matter/chemical compound	Residence time
I Plant residues	
Leaf litter	Months to years
Root litter	Years
Bark	Decades to centuries
Wood	Decades to centuries
Soil organic matter (SOM)	Years to centuries
Available SOM	Years to decades
Stable SOM	Millenia
Black C (BC)	Decades to millenia
II Organic compounds	
Cellulose	Years to decades
Lignin	Years to decades
Lipids	Decades
Proteins	Decades
III Biomarker	
Lignin-derived phenols	Years to decades
Aliphatic structures	Years to centuries
Carbohydrates	Hours to decades
Proteins	Decades
Phospholipid fatty acids	Decades to centuries
Amino sugars	Years to decades

increases in stable forest C pools. Whether bio(macro)molecules from plants, microorganisms and fauna can also directly contribute to the stable C pool due to inherent long residence times is a researchable topic (Amelung et al. 2008; Lorenz et al. 2007; Marschner et al. 2008). Stabilization of C in soils probably depends on the integration of C from any precursor substance into new microbially derived molecules (Chabbi and Rumpel 2009). The importance of phytolith organic C and carbonates from biomineralization in forest vegetation as a source for the stable C is also a debatable topic.

Tree leaves and fine roots can live for several months, whereas the residence time of C in wood can be centuries (Zhou and Luo 2008). Typical mean ages of organic C in plant detritus (exudates, leaves, roots, stems) based on radiocarbon analysis range from days to centuries (Trumbore and Czimczik 2008). Mean ages for microbial C range from days to years and those of organic C in microbial products (exudates, cell walls) from years to decades. Thus, trees store C for a shorter period of time than bulk, available and stable SOM, and BC originating from biomass burning (Table 1.1). Estimates for MRT of aliphatic structures of various origin and microbial-derived phospholipid fatty acids can be centuries. However, no biomarker MRT exceed several centuries (Amelung et al. 2008). Otherwise, long C residence times up to millennia have been reported for chemically and physically separated SOM fractions (e.g., Eusterhues et al. 2007; von Lützow et al. 2007). Physically stabilized or isolated SOC fractions can be old due to weak processes such as aggregation and sorption, and strong processes such as mineral surface interactions (Trumbore and Czimczik 2008). Thus, aside from increasing the input of BC, C sequestration in forest ecosystems can be strengthened by the transfer of atmospheric CO_2 into C pools with long MRT in vegetation, detritus and soil.

Sequestration of atmospheric CO_2 must particularly occur in stable C pools in forest soil profiles given the size of this pool and its long MRT (Lukac et al. 2009). However, stabilization and destabilization mechanisms of SOM are not clearly understood and, in particular, quantification of their importance for C sequestration is difficult (Smernik and Skjemstad 2009). Furthermore, the long-term C sequestration rates in soil profiles in boreal, temperate and tropical forests are likely to be small and estimated to be 0.008–0.117, 0.007–0.120 and 0.023–0.025 Mg C ha^{-1} $year^{-1}$, respectively (Schlesinger 1990). Rates of SOC sequestration may be higher by afforestation on agricultural and degraded soils (Kimble et al. 2003; Laganière et al. 2009). Yet, estimates for C sequestration in tree biomass in pristine, undisturbed, old-growth forests range from 0.4 Mg C ha^{-1} $year^{-1}$ in boreal and temperate to 0.49 Mg C ha^{-1} $year^{-1}$ in tropical forests (Luyssaert et al. 2008; Lewis et al. 2009). Tropical forests can sequester larger amounts of C annually than temperate forests whereas boreal forests sequester the smallest amounts of C among global forest biomes (Bonan 2008). The current additional SOC sequestration potential appears to be tiny against the amount of C that can be potentially lost in the future (Reichstein 2007). Thus, the vulnerability of SOC to potential losses requires special attention.

The changes in SOC pools over a millennial scale are small (~0.02 Mg C ha^{-1} $year^{-1}$), and 100 to 500 times slower than changes in litter pools (~2–10 Mg C ha^{-1} $year^{-1}$) which occur on time scales of months to years (Trumbore and Czimczik 2008). Otherwise, changes in SOC pools occur at intermediate rates (~0.1–10 Mg C ha^{-1} $year^{-1}$) on decadal to centennial time scales. These changes may be caused by alterations in quantity, age, and quality of plant litter inputs, shifts in community composition, spatial distribution, and function of soil fauna and microorganisms, alterations in weak stabilization processes such as aggregate formation, and changes in mineral surfaces (Trumbore and Czimczik 2008). Thus, one of the long-term goals of C sequestration in forest ecosystems is to increase the SOM storage directly through management of OM–mineral interactions.

The ACC affects C exchange, transfer and stabilization processes in forest ecosystems (Chapter 2). How the vegetation, detritus and soil C pools respond to ACC is, however, a debatable issue. In particular, the temperature response of SOM decomposition is uncertain. However, differences in the thermal stability of specific organic compounds in detritus and soil have been observed by molecular-level analysis during a forest soil warming experiment. Specifically, lignin degradation was accelerated but leaf-cuticle-derived compounds were increasingly sequestered in the heated forest soil (Feng et al. 2008). Thus, increases in input of 'heat-proof C compounds' from forest vegetation may help to slow-down increases in atmospheric

CO_2 concentration as less respiratory losses may occur from decomposition in a future warmer climate (Prescott 2008).

In summary, C sequestration in forest ecosystems occurs primarily when

1. **The total pool of organic C in forest vegetation, detritus and soil in a specified forest area increases in a specified time interval through absorption of atmospheric CO_2, and, in particular,**
2. **The pool of organic compounds with long C residence times in forest vegetation, detritus and soil increases over time.**

Thus, C sequestration in this book refers to the sequestration of stable organic compounds in an existing forest ecosystem. Discussions about the C sequestration through increase in the forest area by afforestation and reforestation activities can be found elsewhere (e.g., Nabuurs et al. 2007; Bravo et al. 2008; Laganière et al. 2009).

In the following Chapter 2, C influx- and efflux processes, C turnover processes in forests and C sequestration in vegetation, detritus and soil pools in forest ecosystems are described with a focus on processes occurring in trees and soils. Aside the major CO_2 exchange through processes associated with photosynthesis and respiration, the exchange of CO, CH_4, biogenic volatile organic compounds (BVOC), DIC, DOC and PC are also presented as they contribute to the net ecosystem C balance (NECB) in forests. The potential effects of increasing atmospheric CO_2 concentrations, increasing temperatures and altered precipitation regimes on the different processes and C pools are also discussed. In Chapter 3, the effects of some common major natural disturbances on the C balance of forest stands are compared with those of some minor disturbances. Subsequently, C dynamics and sequestration processes are characterized during the natural succession cycle of forest stand development. This is followed by a comparison of different forest management activities and their effects on the C balance. The chapter concludes with some examples of the effects of disturbances in peatlands, and in forests affected by mining activities and urbanization on C sequestration in forests. In the following Chapter 4, C influx- and efflux processes, C turnover and C pools for the major global forest biomes (i.e., the boreal, temperate and tropical forest biome) are compared, and how climate change may affect them. Furthermore, a few examples for perturbations in C exchange processes associated with natural disturbances and human alterations of the C balance in each forest biome are given. In Chapter 5, the role of nitrogen (N) and phosphorous (P) on C sequestration in vegetation, detritus and soil in forest ecosystems are briefly discussed. Some major effects of water on C sequestration in forest ecosystems are also discussed in this Chapter. The final chapter (Chapter 6) begins with a discussion of the importance of forest residues, trees and dedicated forest biomass plantations for bioenergy production. Finally, the role of forest ecosystems in current international agreements on climate change is compared with their role as significant component in any future agreement on climate change. This section contains also a discussion about methods of forest C monitoring and accounting. The importance of the protection of the large and most vulnerable global forest ecosystem C pools in the tropics and in peatland forests are also highlighted with a special emphasis on the major role old-growth forests play for C sequestration.

1.5 Review Questions

1. Describe the major components of a forest ecosystem and the C exchange processes among them.
2. Which natural and human drivers determine the current global distribution of forest ecosystems?
3. The global C cycle is not entirely understood. What are the major knowledge gaps?
4. Which other GHGs determine the global surface temperature? What processes determine Earth's temperature over millions of years?
5. Aside their effects on GHG concentrations, how do forests regulate local, regional and global surface temperature?
6. The term C sequestration is often referring to geological C sequestration. Contrast and compare advantages and disadvantages of this technology with terrestrial C sequestration to slow down anthropogenic increases in atmospheric CO_2.
7. The anthropogenic increase in atmospheric CO_2 concentration continues but the C sink in forest ecosystems apparently weakens. What processes may occur when forests turn from a C sink into a C source?
8. Why are the C sequestration rates in forest soils so low?
9. Contrast and compare the MRT of C in forest ecosystem pools with those in other ecosystems.
10. Planting of trees outside of forest appears to have benefits for the global C balance. Should tree planting be promoted? What may be major advantages and drawbacks of a dense tree cover outside of forests?

References

Aber JD, Melillo JM (2001) Terrestrial ecosystems. Academic, San Diego, CA

Allen CD (2009) Climate-induced forest dieback: an escalating global phenomenon? Unasylva 60:43–49

Allen MR, Gillett NP, Kettleborough JA, Hegerl G, Schnur R, Stott PA, Boer G, Covey C, Delworth TL, Jones GS, Mitchell JFB, Barnett TP (2006) Quantifying anthropogenic influence on recent near-surface temperature change. Surv Geophys 27:491–544

Alo CA, Wang G (2008) Potential future changes of the terrestrial ecosystem based on climate projections by eight general circulation models. J Geophys Res 113:G01004. doi:10.1029/2007JG000528

Amelung W, Brodowski S, Sandhage-Hofmann A, Bol R (2008) Combining biomarker with stable isotope analyses for assessing the transformation and turnover of soil organic matter. Adv Agron 100:155–250

Asner GP, Knapp DE, Broadbent EN, Oliveira PJC, Keller M, Silva JN (2005) Selective logging in the Brazilian Amazon. Science 310:480–482

Barker T, Bashmakov I, Alharthi A, Amann M, Cifuentes L, Drexhage J, Duan M, Edenhofer O, Flannery B, Grubb M, Hoogwijk M, Ibitoye FI, Jepma CJ, Pizer WA, Yamaji K (2007) Mitigation from a cross-sectoral perspective. In: Metz B, Davidson OR, Bosch PR, Dave R, Meyer LA (eds) Climate change 2007: mitigation. Contribution of working group III to the fourth assessment report of the intergovernmental panel on climate change. Cambridge University Press, Cambridge/New York, pp 620–690

Barnes BV, Zak DR, Denton SR, Spurr SH (1998) Forest ecology. Wiley, New York

Berner RA (2003) The long-term carbon cycle, fossil fuels and atmospheric composition. Nature 426:323–326

Bombelli A, Henry M, Castaldi S, Adu-Bredu S, Arneth A, de Grandcourt A, Grieco E, Kutsch WL, Lehsten V, Rasile A, Reichstein M, Tansey K, Weber U, Valentini R (2009) The sub-Saharan Africa carbon balance, an overview. Biogeosci Discuss 6:2085–2123

Bonan GB (2008) Forests and climate change: forcings, feedbacks, and the climate benefits of forests. Science 320:1444–1449

Bowman DMJS, Balch JK, Artaxo P, Bond WJ, Carlson JM, Cochrane MA, D'Antonio CMD, DeFries R, Doyle JC, Harrison SP, Johnston FH, Keeley JE, Krawchuk MA, Kull CA, Marston JB, Moritz MA, Prentice IC, Roos CI, Scott AC, Swetnam TW, van Der Werf GR, Pyne SP (2009) Fire in the Earth system. Science 324:481–484

Bravo F, del Río M, Bravo-Oviedo A, Del Peso C, Montero G (2008) Forest management strategies and carbon sequestration. In: Bravo F, LeMay V, Jandl G, von Gadow K (eds) Managing forest ecosystems: the challenge of climate change. Springer, New York, pp 179–194

Broecker WS (1987) Unpleasant surprises in the greenhouse? Nature 328:123–126

Brook EJ (2009) Atmospheric carbon footprints? Nat Geosci 2:170–172

Bryant D, Nielsen D, Tangley L (1997) The last frontier forests: ecosystems and economics on the edge. World Resources Institute, Washington, DC

Cailleau G, Braissant O, Verrecchia EP (2004) Biomineralization in plants as a long-term carbon sink. Naturwissenschaften 91:191–194

Canadell JG, Le Quéré C, Raupach MR, Field CB, Buitenhuis ET, Ciais P, Conway TJ, Gillett NP, Houghton RA, Marland G (2007) Contributions to accelerating atmospheric CO_2 growth from economic activity, carbon intensity, and efficiency of natural sinks. Proc Natl Acad Sci USA 104:18866–18870

Chabbi A, Rumpel C (2009) Organic matter dynamics in agro-ecosystems – the knowledge gaps. Eur J Soil Sci 60:153–157

Clawson M (1979) Forests in the long sweep of American history. Science 204:1168–1174

Denman KL, Brasseur G, Chidthaisong A, Ciais P, Cox PM, Dickinson RE, Hauglustaine D, Heinze C, Holland E, Jacob D, Lohmann U, Ramachandran S, da Silva Dias PL, Wofsy SC, Zhang X (2007) Couplings between changes in the climate system and biogeochemistry. In: Intergovernmental panel on climate change (ed) Climate Change 2007: the physical science basis, Chapter 7. Cambridge University Press, Cambridge

European Council (2007) Limiting global climate change to 2°C – the way ahead for 2020 and beyond. Communication from the Commissions to the Council, the European Parliament, the European Economic and Social Committee and the Committee of the Region

Eusterhues K, Rumpel C, Kögel-Knabner I (2007) Composition and radiocarbon age of HF-resistant soil organic matter in a Podzol and a Cambisol. Org Geochem 38:1356–1372

FAO (Food and Agricultural Organization of the United Nations) (2001) Global forest resources assessment 2000-main report. FAO Forestry paper 140. FAO, Rome

FAO (Food and Agricultural Organization of the United Nations) (2006) Global forest resources assessment 2005. Progress towards sustainable forest management. FAO Forestry paper 147. FAO, Rome

Feng X, Simpson AJ, Wilson KP, Williams DD, Simpson MJ (2008) Increased cuticular carbon sequestration and lignin oxidation in response to soil warming. Nat Geosci 1:836–839

Foley JA, DeFries R, Asner GP, Barford C, Bonan G, Carpenter SR, Chapin FS, Coe MT, Daily GC, Gibbs HK, Helkowski JH, Holloway T, Howard EA, Kucharik CJ, Monfreda C, Patz JA, Prentice IC, Ramankutty N, Snyder PK (2005) Global consequences of land use. Science 309:570–574

Forster P, Ramaswamy V, Artaxo P, Berntsen T, Betts R, Fahey DW, Haywood J, Lean J, Lowe DC, Myhre G, Nganga J, Prinn R, Raga G, Schulz M, Van Dorland R (2007) Changes in atmospheric constituents and in radiative forcing. In: Intergovernmental panel on climate change (ed) Climate Change 2007: the physical science basis, Chapter 2. Cambridge University Press, Cambridge

Gergis JL, Fowler AM (2009) A history of ENSO events since A.D. 1525: implications for future climate change. Clim Change 92:343–387
Guan D, Peters GP, Weber CL, Hubacek K (2009) Journey to world top emitter: an analysis of the driving forces of China's recent CO_2 emissions surge. Geophys Res Lett 36:L04709. doi:10.1029/2008GL036540
Hansen J (2005) A slippery slope: how much global warming constitutes 'dangerous anthropogenic interference'? Clim Change 68:269–279
Heimann M (2009) Searching out the sinks. Nat Geosci 2:3–4
Helms JA (ed) (1998) The dictionary of forestry. Society of American Forestry, Bethesda, MD
Holmen K (2000) The global carbon cycle. In: Jacobsen MC, Charlston RJ, Rohde H, Orians GH (eds) Earth system science. Academic, Amsterdam, pp 282–321
Houghton RA (2007) Balancing the global carbon budget. Annu Rev Earth Planet Sci 35:313–347
IPCC (2007) Summary for policymakers. In: Solomon S, Qin D, Manning M, Chen Z, Marquis M, Averyt KB, Tignor M, Miller HL (eds) Climate change 2007: the physical science basis. Contribution of working group I to the fourth assessment report of the intergovernmental panel on climate change. Cambridge University Press, Cambridge
Jarvis PG, Ibrom A, Linder S (2005) 'Carbon forestry': managing forests to conserve carbon. In: Griffiths H, Jarvis PG (eds) The carbon balance of forest biomes. Taylor & Francis, Oxon, UK, pp 331–349
Kimble JM, Heath LS, Birdsey R, Lal R (eds) (2003) The potential of U.S. forest soils to sequester carbon and mitigate the greenhouse effect. CRC Press, Boca Raton, FL
Kindermann GE, McCallum I, Fritz S, Obersteiner M (2008) A global forest growing stock, biomass and carbon map based on FAO statistics. Silva Fennica 42:387–396
Kintisch E (2009) Projections of climate change go from bad to worse, scientists report. Science 323:1546–1547
Kozlowski TT, Kramer PJ, Pallardy SG (1991) The physiological ecology of woody plants. Academic, San Diego, CA
Kriegler E, Hall JW, Held H, Dawson R, Schellnhuber HJ (2009) Imprecise probability assessment of tipping points in the climate system. Proc Natl Acad Sci USA 106:5041–5046
Kuzyakov Y, Subbotina I, Chen H, Bogomolova I, Xu X (2009) Black carbon decomposition and incorporation into soil microbial biomass estimated by ^{14}C labeling. Soil Biol Biochem 41:210–219
Laganière J, Angers DA, Paré D (2009) Carbon accumulation in agricultural soils after afforestation: a meta-analysis. Glob Change Biol doi: 10.1111/j.1365-2486.2009.01930.x
Lal R (2004) Soil carbon sequestration to mitigate climate change. Geoderma 123:1–22
Lal R (2008) Carbon sequestration. Phil Trans R Soc B 363:815–830
Lean JL, Rind DH (2008) How natural and anthropogenic influences alter global and regional surface temperatures: 1889 to 2006. Geophys Res Lett 35:L18701. doi:10.1029/2008GL034864
Leigh Mascarelli A (2009) What we've learned in 2008. Nat Rep Clim Change 3:4–6
Lenton TM, Held H, Kriegler E, Hall JW, Lucht W, Rahmstorf S, Schellnhuber HJ (2008) Tipping element's in the Earth's climate system. Proc Natl Acad Sci USA 105:1786–1793
Lewis SL, Lopez-Gonzalez G, Sonké B, Affum-Baffoe K, Baker TR, Ojo LO, Phillips OL, Reitsma JM, White L, Comiskey JA, M-N DK, Ewango CEN, Feldpausch TR, Hamilton AC, Gloor M, Hart T, Hladik A, Lloyd J, Lovett JC, Makana J-R, Malhi Y, Mbago FM, Ndangalasi HJ, Peacock J, S-H PK, Sheil D, Sunderland T, Swaine MD, Taplin J, Taylor D, Thomas SC, Votere R, Wöll H (2009) Increasing carbon storage in intact African tropical forests. Nature 457:1003–1007
Lindeman RL (1942) The trophic-dynamic aspects of ecology. Ecology 23:399–418
Lorenz K, Lal R, Preston CM, Nierop KGJ (2007) Strengthening the soil organic carbon pool by increasing contributions from recalcitrant aliphatic bio(macro)molecules. Geoderma 142:1–10
Lowe JA, Huntingford C, Raper SCB, Jones CD, Liddicoat SK, Gohar LK (2009) How difficult is it to recover from dangerous levels of global warming? Environ Res Lett 4, doi:10.1088/1748-9326/4/1/014012

Lukac M, Lagomarsino A, Moscatelli MC, De Angelis P, Cotrufo MF, Godbold DL (2009) Forest soil carbon cycle under elevated CO_2 – a case of increased throughput? Forestry 82:75–86

Lund HG (1999) A 'forest' by any other name. Environ Sci Policy 2:125–133

Luyssaert S, Schulze E-D, Börner A, Knohl A, Hessenmöller D, Law BE, Ciais P, Grace J (2008) Old-growth forests as global carbon sinks. Nature 455:213–215

Mann ME (2009) Defining dangerous anthropogenic interference. Proc Natl Acad Sci USA 106:4065–4066

Mann ME, Zhang Z, Hughes MK, Bradley RS, Miller SK, Rutherford S, Ni F (2008) Proxy-based reconstructions of hemispheric and global surface temperature variations over the past two millennia. Proc Natl Acad Sci USA 105:13252–13257

Marschner B, Brodowski S, Dreves A, Gleixner G, Gude A, Grootes PM, Hamer U, Heim A, Jandl G, Ji R, Kaiser K, Kalbitz K, Kramer C, Leinweber P, Rethemeyer J, Schäffer A, Schmidt MWI, Schwark L, Wiesenberg GLB (2008) How relevant is recalcitrance for the stabilization of organic matter in soils? J Plant Nutr Soil Sci 171:91–110

Martin RM (2008) Deforestation, land-use change and REDD. Unasylva 230:3–11

Mayle FE, Burbridge R, Killeen TJ (2000) Millenial-scale dynamics of southern Amazonian rainforests. Science 290:2291–2294

Mollicone D, Freibauer A, Schulze E-D, Braatz S, Grassi G (2007) Elements for the expected mechanisms on 'reduced emissions from deforestation and degradation, REDD' under UNFCCC. Environ Res Lett 2, doi:10.1088/1748-9326/2/4/045024

Nabuurs GJ, Masera O, Andrasko K, Benitez-Ponce P, Boer R, Dutschke M, Elsiddig E, Ford-Robertson J, Frumhoff P, Karjalainen T, Krankina O, Kurz WA, Matsumoto M, Oyhantcabal W, Ravindranath NH, Sanz Sanchez MJ, Zhang X (2007) Forestry. In: Metz B, Davidson OR, Bosch PR, Dave R, Meyer LA (eds) Climate Change 2007: mitigation. Contribution of working group III to the fourth assessment report of the intergovernmental panel on climate change. Cambridge University Press, Cambridge/New York, pp 541–584

Nieder R, Benbi DK, Isermann K (2003) Soil organic matter dynamics. In: Benbi DK, Nieder R (eds) Handbook of processes and modeling in the soil-plant system. Haworth, New York, pp 345–408

Odum EP (1967) The strategy of ecosystem development. Science 164:262–270

Ovington JD (1962) Quantitative ecology and the woodland ecosystem concept. Adv Ecol Res 1:103–192

Pacala S, Socolow R (2004) Stabilization wedges: solving the climate problem for the next 50 years with current technologies. Science 305:968–972

Parr JF, Sullivan LA (2005) Soil carbon sequestration in phytoliths. Soil Biol Biochem 37:117–124

Perry DA, Oren R, Hart SC (2008) Forest ecosystems. The John Hopkins University Press, Baltimore, MD

Prescott C (2008) Heat-proof carbon compound. Nat Geosci 1:815–816

Purves D, Pacala S (2008) Predictive models of forest dynamics. Science 320:1452–1453

Ramanathan V, Feng Y (2008) On avoiding dangerous anthropogenic interference with the climate system: formidable challenges ahead. Proc Natl Acad Sci USA 105:14245–14250

Raupach MR, Canadell JG, Le Quéré C (2008) Anthropogenic and biophysical contributions to increasing atmospheric CO_2 growth rate and airborne fraction. Biogeosciences 5:1601–1613

Reichstein M (2007) Impacts of climate change on forest soil carbon: principles, factors, models, uncertainties. In: Freer-Smith PH, Broadmeadow MSJ, Lynch JM (eds) Forestry and climate change. CAB International, Wallingford, UK, pp 127–135

Rigby M, Prinn RG, Fraser PJ, Simmonds PG, Langenfelds RL, Huang J, Cunnold DM, Steele LP, Krummel PB, Weiss RF, O'Doherty S, Salameh PK, Wang HJ, Harth CM, Mühle J, Porter LW (2008) Renewed growth of atmospheric methane. Geophys Res Lett 35:L22805. doi:10.1029/2008GL036037

Rosenzweig C, Karoly D, Vicarelli M, Neofotis P, Wu Q, Casassa G, Menzel A, Root TL, Estrella N, Seguin B, Tryjanowski P, Liu C, Rawlins S, Imeson A (2008) Attributing physical and biological impacts to anthropogenic climate change. Nature 453:353–358

Roston E (2008) The carbon age: how life's core element has become civilization's greatest threat. Walker & Co, New York

Ruddiman WF (2003) The anthropogenic greenhouse area began thousands of years ago. Clim Change 61:261–293
Schlesinger WH (1990) Evidence from chronosequence studies for a low carbon-storage potential of soils. Nature 348:232–234
Schlesinger WH (2006) Inorganic carbon and the global C cycle. In: Lal R (ed) Encyclopedia of soil science. Taylor & Francis, London, pp 879–881
Schröter D, Cramer W, Leemans R, Colin Prentice I, Araújo MB, Arnell NW, Bondeau A, Bugmann H, Carter TR, Gracia CA, de la Vega-Leinert AC, Erhard M, Ewert F, Glendining M, House JI, Kankaanpää S, Klein RJT, Lavorel S, Lindner M, Metzger MJ, Meyer J, Mitchell TD, Reginster I, Rounsevell M, Sabaté S, Sitch S, Smith B, Smith J, Smith P, Sykes MT, Thonicke K, Thuiller W, Tuck G, Zaehle S, Zierl B (2005) Ecosystem service supply and vulnerability to global change in Europe. Science 310:1333–1337
Schulze E-D, Beck E, Müller-Hohenstein K (2005) Plant ecology. Springer, Berlin
Seth MK (2004) Trees and their economic importance. Bot Rev 69:321–376
Shine KP, Sturges WT (2007) CO_2 is not the only gas. Science 315:1804–1805
Smernik R, Skjemstad J (2009) Mechanisms of organic matter stabilization and destabilization in soils and sediments: conference introduction. Biogeochemistry 92:3–8
Smith JB, Schneider SH, Oppenheimer M, Yohe GW, Hare W, Mastrandrea MD, Patwardhan A, Burton I, Corfee-Morlot J, Magadza CHD, Füssel H-M, Barrie Pittock A, Rahman A, Suarez A, van Ypserle J-P (2009) Assessing dangerous climate change through an update of the Intergovernmental Panel on Climate Change (IPCC) "reasons for concern". Proc Natl Acad Sci USA 106:4133–4137
Solomon S, Plattner G-K, Knutti R, Friedlingstein P (2009) Irreversible climate change due to carbon dioxide emissions. Proc Natl Acad Sci USA 105:14239–14240
Steffen W, Crutzen PJ, McNeill JR (2007) The Anthropocene: are humans now overwhelming the great forces of nature? Ambio 36:614–621
Tans P (2009) Trends in atmospheric carbon dioxide – global. http://www.esrl.noaa.gov/gmd/ccgg/trends/Accessed August 25, 2009
Tansley AG (1935) The use and abuse of vegetational concepts and terms. Ecology 16:284–307
Trumbore SE, Czimczik CI (2008) An uncertain future for soil carbon. Science 321:1455–1456
Valentini R, Matteucci G, Dolman AJ, Schulze E-D, Rebmann C, Moors EJ, Granier A, Gross P, Jensen NO, Pilegaard K, Lindroth A, Grelle A, Bernhofer C, Grünwald T, Aubinet M, Ceulemans R, Kowalski AS, Vesala T, Rannik Ü, Berbigier P, Loustau D, Guðmundsson J, Thorgeirsson H, Ibrom A, Morgenstern K, Clement R, Moncrieff J, Montagnani L, Minerbi S, Jarvis PG (2000) Respiration as the main determinant of carbon balance in European forests. Nature 404:861–865
Valladares F (2008) A mechanistic view of the capacity of forests to cope with climate change. In: Bravo F, LeMay V, Jandl G, von Gadow K (eds) Managing forest ecosystems: the challenge of climate change. Springer, New York, pp 15–40
Vaughan NE, Lenton TM, Sheperd JG (2009) Climate change mitigation: trade-offs between delay and strength of action required. Clim Change 96:29–43
Von Lützow M, Kögel-Knabner I, Ekschmitt K, Flessa H, Guggenberger G, Matzner E, Marschner B (2007) SOM fractionation methods: relevance to functional pools and to stabilization mechanisms. Soil Biol Biochem 39:2183–2207
Waring RW, Running SW (2007) Forest ecosystems – analysis at multiple scales. Elsevier Academic, Burlington, MA
Willis AJ (1997) The ecosystem: an evolving concept viewed historically. Funct Ecol 11:268–271
Zhou T, Luo Y (2008) Spatial patterns of ecosystem carbon residence time and NPP-driven carbon uptake in the conterminous United States. Global Biogeochem Cy 22, GB3032, doi:10.1029/2007GB002939

Chapter 2
The Natural Dynamic of Carbon in Forest Ecosystems

Forest Ecosystems exchange energy, water, and nutrients and, in particular, carbon (C) with surrounding ecosystems, and play a major role in the global C cycle. Forests are major terrestrial C sinks, have large C densities and sequester large amounts of atmospheric carbon dioxide (CO_2). By various natural processes, C is entering forest ecosystems in dissolved, gaseous and particulate form. The C is temporarily stored, and sequestered in above- and belowground pools in vegetation, detritus and soil. Efflux processes result in C losses to adjacent ecosystems. This chapter describes C in- and efflux processes, the C turnover within forest ecosystems and how C is sequestered in the different forest ecosystem pools with a focus on processes occurring in trees and soil.

Without human interference, a natural forest ecosystem C cycle is in effect. The following chapter begins with a description of major processes responsible for C input into forest ecosystems. Furthermore, possible effects of increasing atmospheric CO_2 concentrations, increasing temperatures and alterations in precipitation intensity and distribution are discussed. The major C influx occurs by assimilation of atmospheric CO_2 during tree and understory plant photosynthesis, either by the C_3, C_4 or the crassulacean acid metabolism (CAM) pathway. Minor C inputs originate from microbial processes, and deposition of dissolved, gaseous and particulate C. After entering the forest ecosystem, C is transferred and distributed among different above- and belowground pools in plant, animal and microbial biomass, and in soil organic matter (SOM). Increases in atmospheric CO_2 concentrations and temperature, and changes in rainfall patterns by ACC may significantly alter the C input processes. Within trees, stem wood contains the largest C pool. Litter fragmentation, root exudates and microbial metabolism are the major pathways for C entering the soil organic C (SOC) pool. In particular, the forest floor and mineral soil horizons contain often large C pools within forest ecosystems. Respiration by plants and microbial respiration during decomposition of litter and SOM are the major pathways for C losses from forest ecosystems. Minor C losses occur also during efflux of dissolved, gaseous and particulate C, for example, by emissions of isoprene and other biogenic volatile organic compounds (BVOCs) and methane (CH_4) from tree canopies, soot

K. Lorenz and R. Lal, *Carbon Sequestration in Forest Ecosystems*,
DOI 10.1007/978-90-481-3266-9_2,

emission during fires, and leaching of dissolved organic C (DOC) and dissolved inorganic C (DIC) from the soil. The C efflux processes may also be altered by increases in atmospheric CO_2 concentrations, temperature increases and changes in rainfall patterns.

2.1 Carbon Input into Forest Ecosystems

The long-term monitoring of atmospheric CO_2 concentrations began about 45 years ago at Mauna Loa in the Pacific Ocean (Keeling 1960). At this site, a massive withdrawal of CO_2 was observed in the summer, and this seasonal change in CO_2 concentrations was later observed in all parts of the world (Houghton et al. 2001). Thus, ecosystems 'breathe', that is, they take up CO_2 during the summer and release it in winter, pointing towards the requirement for sufficient sunlight for the uptake process (Baldocchi 2008). This process of CO_2 fixation in sunlight is called photosynthesis, and is mainly performed by plants beside the other photoautotrophs algae, cyanobacteria, purple and green bacteria. Autotrophs can grow using CO_2 as their sole C source, and generate the biomass on which all other organisms thrive (Thauer 2007). The photoautotrophs use exclusively C assimilated in their photosynthetic active tissues for biomass production. The gross primary production (GPP) is defined as the gross uptake of CO_2 used for photosynthesis by autotrophic C-fixing tissues (Chapin et al. 2006). Approximately 120 Pg C (annual terrestrial GPP) is taken up by vegetation through photosynthesis and an almost identical amount released (Denman et al. 2007). By the various processes discussed in Section 2.3, and, in particular, autotrophic respiration (R_a), photosynthetically fixed C is lost and the remaining C is used for other processes, that is, build-up of plant biomass. Thus, the net primary production (NPP) represents the imbalance between GPP and R_a (Eq. (2.1)).

$$GPP = NPP + R_a \tag{2.1}$$

The amount of C cycling between forests and the atmosphere (GPP) has been estimated to be 75 Pg C $year^{-1}$ (WBCSD 2007). Global terrestrial NPP has been estimated to be 63 Pg C $year^{-1}$, and about half of this occurs in forests (Grace 2005). The annual C sink in forest biomes has been estimated to be 1.58 Pg C or 59% of the total global C sink of terrestrial biomes (Robinson 2007). Only 0.7% of terrestrial NPP are sequestered in soils (Schlesinger 1990). About 85% of the total terrestrial aboveground C, and about 74% the total belowground terrestrial C are, however, stored in forest ecosystems (Robinson 2007).

NPP includes also the C transfer to herbivores and root symbionts, the excretion of organic C from algae, and the production of root exudates and biogenic volatile organic compounds (BVOCs) (Fig. 2.1; Chapin et al. 2006). Thus, the net ecosystem C balance (NECB) describes the net rate of C accumulation (or loss) from

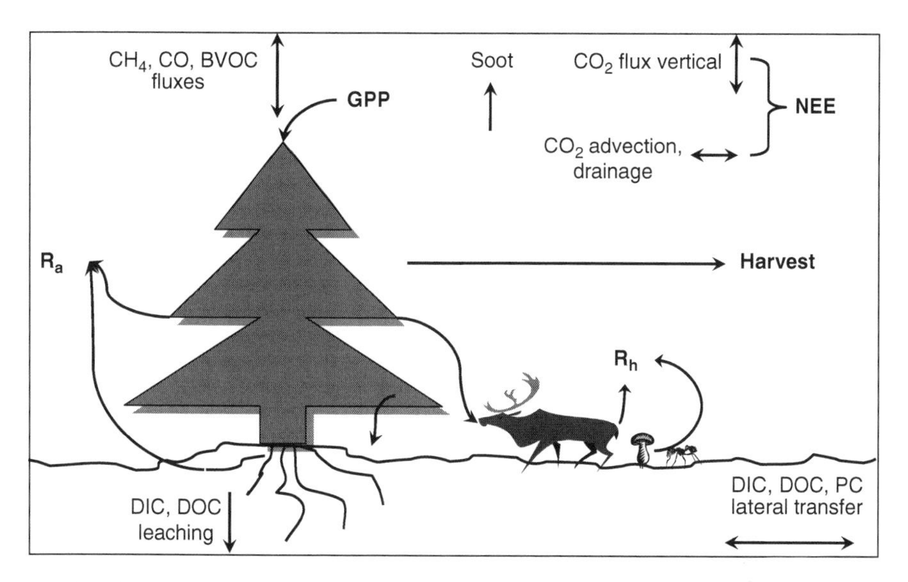

Fig. 2.1 Carbon fluxes associated with the net ecosystem C balance (GPP = gross primary production, NEE = net CO_2 exchange, R_a = autotrophic respiration, R_h = heterotrophic respiration; CH_4 = methane, CO = carbon monoxide, BVOC = biogenic volatile organic compounds, CO_2 = carbon dioxide, DIC = dissolved inorganic carbon, DOC = dissolved organic carbon, PC = particulate carbon) (Modified from Chapin et al. 2006)

ecosystems (Eq. (2.2)). NECB represents the overall ecosystem C balance from all sources and sinks – physical, biological, and anthropogenic (Chapin et al. 2006).

$$NECB = dC/dt = -NEE + F_{CO} + F_{CH4} + F_{BVOC} + F_{DIC} + F_{DOC} + F_{PC} \tag{2.2}$$

NEE is the net CO_2 exchange, F_{CO} the net carbon monoxide (CO) exchange, F_{CH4} the net CH_4 exchange, F_{BVOC} the net BVOC exchange, F_{DIC} the net DIC exchange, F_{DOC} the net DOC exchange, and F_{PC} the net particulate (non-dissolved, non-gaseous) carbon (PC) exchange.

In the following chapter, the net exchange of C by forest ecosystems will be discussed with a focus on the tree and plot scale based on the gaseous, dissolved and particulate influx and efflux of C (Fig. 2.2). Furthermore, the chemical composition of organic compounds in trees and soil will be discussed in more detail as C sequestration implies increasing the proportion of organic compounds with long residence time (Chapter 1). Finally, the possible effects of increasing atmospheric CO_2 concentrations, increasing temperatures and alterations in precipitation intensity and distribution on the C dynamics in forest ecosystems are discussed (Field et al. 2007). The effects of disturbances, forest management and land use changes on the NECB on the landscape scale are discussed in Chapter 3.

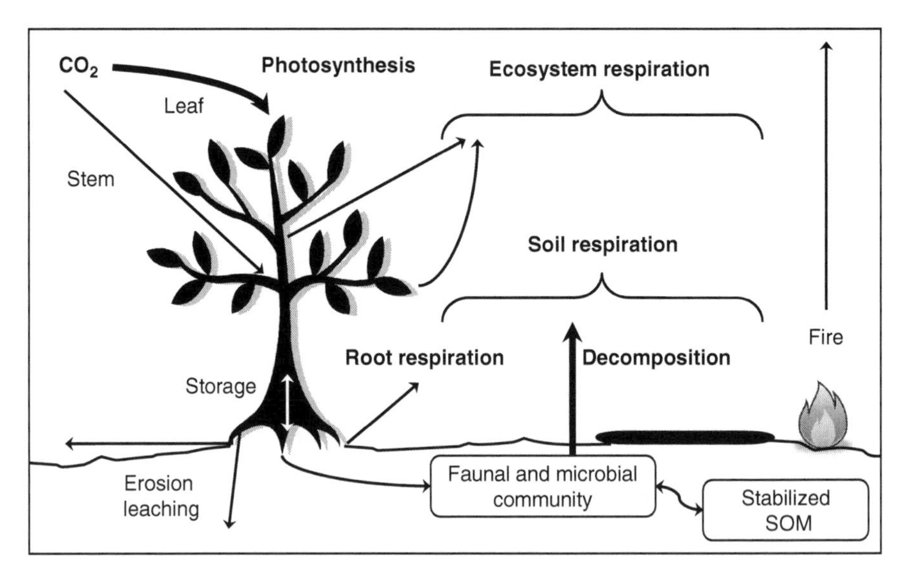

Fig. 2.2 Carbon flow through forest ecosystems (SOM = soil organic matter) (Modified from Trumbore 2006)

2.1.1 Carbon Assimilation

2.1.1.1 Eukaryotic Photosynthesis

The majority of photosynthetic forest organisms (i.e., trees, understory plants, algae, green algae associated with fungi in lichen) are eukaryotes (Schulze 2006). The major features of eukaryotic cells are membrane-enclosed structures called organelles (Madigan and Martinko 2006). Specifically, photosynthetic cells contain the organelles chloroplasts. In contrast, organelles are absent from the prokaryotes which consist of the two phylogenetic groups Bacteria and Archaea.

Plant photosynthesis creates the forest biomass and is therefore the major C input into forest ecosystems (Waring and Running 2007). By this process, atmospheric CO_2 is assimilated and C is fixed in plant biomass and, thus, entering the forest ecosystem C pool. The principal photosynthetic pigment chlorophyll and about 100 proteins are embeded in two large protein clusters, photosystem I and photosystem II (Leslie 2009). Light energy (i.e., the photosynthetically active portion of sunlight (PAR)) is absorbed and starts an electrical circuit in which electrons flow from the photosystems through protein chains that make the energy-rich molecules adenosine triphosphate (ATP) and nicotinamide adenine dinucleotide (NADPH). These molecules power the synthesis of carbohydrates that plants depend on to grow and multiply (Raven et al. 2005). Photosystem II regains its lost electrons by swiping them from water. This process is accompanied by the release of O_2 as a waste product. The total energy captured from solar energy by plant photosynthesis is about 2×10^{23} J year^{-1} (Ito and Oikawa 2004). The basic equation of photosynthesis can be written as Eq. (2.3):

$$[6\,CO_2 + 12\,H_2O + \text{Light energy} \rightarrow C_6H_{12}O_6 + 6\,H_2O + 6\,O_2] \quad (2.3)$$

The overall equation for the light-dependent reactions in green plants can be written as Eq. (2.4):

$$2\,H_2O + 2\text{ NADP}^+ + 2\text{ ADP} + 2P_i + \text{light} \rightarrow 2\text{ NADPH} + 2H^+ + 2\text{ ATP} + O_2 \quad (2.4)$$

Plants use the chemical energy to fix CO_2 into carbohydrates and other organic compounds through light-independent reactions. Thus, the overall equation for C fixation in green plants can be written as Eq. (2.5):

$$3\,CO_2 + 9\,ATP + 6\,NADPH + 6\,H^+ \rightarrow C_3H_6O_3 - \text{phosphate} + 9\,ADP + 8\,P_i + 6\,NADP^+ + 3\,H_2O \quad (2.5)$$

Within the plant, the main photosynthetic activity occurs inside of leaves. Thus, CO_2 has to enter the leaf through small pores called stomata (Heldt 2005). This uptake of CO_2 through stomata is accompanied, however, by an escape of water vapor. In particular, in well-adapted plants stomata play a relatively small part in determining the rate of photosynthesis whereas avoidance of damaging plant water deficits appears to be the primary role of stomata (Jones 1998). To reduce water loss through stomata during CO_2 fixation, plants have therefore also evolved the C_4 and the crassulacean acid metabolism (CAM) photosynthesis aside from the predominant C_3 photosynthetic pathway. As shown later, nitrogen (N) availability, light, temperature, air humidity and water potential are primary factors controlling GPP at the leaf level, and, in addition, phenology, leaf area density and canopy structure control GPP at the stand level (Farquhar 1989; Wright et al. 2004; Lindroth et al. 2008). In the following sections, the principal differences among the photosynthetic pathways are discussed with special emphasis on the effects of ACC on photosynthesis. The effects of N, phosphorus (P) and water on photosynthesis are discussed in Chapter 5.

Photosynthesis by C_3 Plants

The C_3 pathway of photosynthesis occurs in 95% of all plant species (Schulze et al. 2005). All trees and nearly all plants of cold climates are C_3 plants (Houghton et al. 2001). Photosynthesis in plant leaves takes place in the plant organelles called chloroplasts (Berg et al. 2007). The photosynthetic reaction in chloroplasts can be divided into two partial reactions. First, during the light reactions light is absorbed by chlorophyll. The trapped solar energy is used to yield reductive power in the form of NADPH, and to yield chemical energy in the form of ATP (Eq. (2.4)).

Secondly, NADPH and ATP are consumed during the following dark reactions to reduce CO_2 and fix C by carboxylation of the C_5 sugar ribulose 1,5 bisphosphate (Eq. (2.5)). The product of the carboxylation reaction 3-phosphoglycerate contains 3 C atoms, hence the name C_3 pathway photosynthesis (Schlesinger 1997). Finally, 3-phosphoglycerate is reduced to triose phosphate, and a fraction converted into fructose 6-phosphate and exported from the chloroplast (Berg et al. 2007). Atmospheric CO_2 is, thus, fixed in a C_6 sugar, a hexose. The larger fraction of triose phosphate, however, is used for the regeneration of the CO_2 acceptor ribulose 1,5 bisphosphate. Specifically, based on three molecules CO_2 fixed, six molecules of triose phosphate are synthesized but only one is ultimatively provided for various biosynthetic pathways in the plant (Heldt 2005). This cyclic process of carboxylation, reduction and regeneration reactions is called the Calvin cycle. This cycle operates in plants, algae, cyanobacteria, and in autotrophic proteobacteria (Thauer 2007). The enzyme D-ribulose-1,5- bisphosphate carboxylase/oxygenase (RUBISCO) plays the central role by catalyzing the binding of atmospheric CO_2 to the acceptor ribulose 1,5 bisphosphate. Thus, RUBISCO is the rate limiting enzyme for the photosynthesis reaction (Spreitzer and Salvucci 2002). Among woody plants, the photosynthetic capacity (i.e., CO_2 uptake m^{-2} h^{-1}) under conditions of light saturation, optimal temperature, water, and ambient CO_2 concentration is decreasing from deciduous broad-leaved plants to evergreen broad-leaved trees and to ericaceae shrubs, semi-arid sclerophyllous shrubs and evergreen conifers (Mooney 1972).

Important for the total net C assimilation in forest ecosystems are three processes which potentially limit CO_2-C capture during photosynthesis. First, the absorption of light by chlorophyll is limited by the irradiance or amount of radiant flux on a unit leaf surface area and the amount of chlorophyll (Waring and Running 2007). The following electron excitation that finally traps the energy of the excited electron in the reaction center is sensitive to temperature but independent of CO_2 concentrations (Berg et al. 2007). The second limiting process is the rate at which CO_2 in ambient air is supplied to the site of reduction in the chloroplast (Waring and Running 2007). CO_2 diffuses into plant leaves through stomata, and the rate of photosynthesis depends on the stomatal conductance which is controlled primarily by the availability of water to the plant and the CO_2 concentration inside the leaf (Schlesinger 1997). The pore size of the stomata, however, is actively regulated by plants to minimize the loss of water by transpiration. Under optimal water supply, internal CO_2 inside the leaf is relatively low and stomata show maximum conductance. Third, under optimal water supply the photosynthetic rate of C assimilation may be limited by the amount and activity of the CO_2-fixing enzyme RUBISCO. RUBISCO is the most abundant protein in nature but exhibits a sluggish catalytic performance (Tcherkez et al. 2006). In particular, RUBISCO confuses the substrate of photosynthesis, CO_2, with one of the photosynthesis products, O_2. Thus, in addition to C assimilation (carboxylase reaction) the oxygenase reaction occurs during which phosphoglycolate and 3-phosphoglycerate are synthesized by the addition of O_2 to ribulose 1,5 bisphosphate. More important for forest C sequestration, three of the four C atoms of two molecules phosphoglycolate are recycled by an energy-wasting salvage pathway.

Thus, CO_2 is released without the production of energy-rich metabolites (Berg et al. 2007). This photorespiration pathway consumes O_2 and releases CO_2.

Because of the interactive effects of the various limiting factors, net C assimilation by photosynthesis follows an asymptotic curve with increase in irradiance (Waring and Running 2007). The maximum rate of assimilation decreases from the top to the bottom of a tree canopy, and is higher for deciduous than for evergreen species. The most accurate measure of photosynthetic capacity is the actual concentration of chlorophyll pigments per unit of leaf surface area. Optimum photosynthesis temperatures for temperate species ranges between 15°C and 25°C. Drought limits photosynthesis mainly through reduction in stomatal conductance and, to a lesser extent, by stomata closure, drought-related RUBISCO degradation, and the reduction in pigments that are important to photochemical reactions.

Climate Change and Photosynthesis by C_3 Plants

Increases in atmospheric CO_2 concentrations, in temperature, in solar radiation and changes in precipitation have the potential to affect C_3 photosynthesis by forest plants (Table 2.1). The global atmospheric concentration of CO_2 has increased from a pre-industrial value of about 280 to 385 ppm by the end of 2008, and continues to increase at a rate of about 2 ppm year^{-1} (Tans 2009). The C_3 photosynthesis is, however, not saturated at current atmospheric CO_2 concentrations (Körner 2006). The cornerstone enzyme RUBISCO emerged early in evolution when the atmosphere was much richer in CO_2 than it currently is (Berg et al. 2007). Thus, when light is abundant the CO_2 concentration generally limits the photosynthetic rate and C_3 leaf photosynthesis saturates when the CO_2 concentration approaches 1,000 ppm (Körner 2006; Berg et al. 2007). Therefore, exposure of plants to elevated CO_2 concentrations results in short-term increases in the carboxylation reaction and competitive inhibition of the oxygenation reaction of RUBISCO, and decreases in stomatal opening (Long et al. 2004). Most angiosperms show a progressive decrease in stomatal conductance with increase in CO_2 but conifers and *Fagus* species appear insensitive to variations in CO_2 concentrations (Long et al. 2004).

Table 2.1 Climate change effects on plant photosynthesis

Effect	C_3 photosynthesis	C_4 photosynthesis	CAM photosynthesis
Elevated CO_2 <1,000 ppm	++ Young trees ± Mature trees	+ Indirectly due to increased water-use-efficiency	++
Increasing temperature	+ Below optimum T – Above maximum T	+ Below optimum T – Above maximum T but higher tolerance than C_3	+ Below optimum T – Above maximum T but higher tolerance than C_4
Drought	–	– Higher tolerance than C_3	– Higher tolerance than C_3 and C_4
Increasing solar radiation	± Due to photoinhibition	+ But photoinhibition	+ But photoinhibition

Furthermore, in many tree species no changes in stomatal density and/or stomatal index were observed at elevated CO_2 concentrations (Urban 2003). Thus, the increase in intercellular CO_2 concentration was mainly responsible for the reduction in stomatal conductance.

Important effects of increasing CO_2 concentrations on the forest ecosystem C pool originate from the observation that, in particular, young trees experimentally exposed to elevated levels of CO_2 show almost univocally stimulation of photosynthesis (Hyvönen et al. 2007). Thus, the anthropogenic enrichment of atmospheric CO_2 by fossil fuel combustion and land-use change may stimulate net C assimilation by CO_2-'fertilization' below 800–1,000 ppm CO_2 (Denman et al. 2007). Experimental evidence derived from free-air CO_2 enrichment (FACE) experiments with young trees indicate that photosynthesis is in fact RUBISCO limited, and light-saturated photosynthesis may increase by as much as 50% at elevated levels of CO_2 (Ainsworth and Rogers 2007). Leaf photosynthesis is, however, more sensitive to CO_2 concentration at lower stomatal conductance and is less sensitive at lower leaf N concentration (McMurtrie et al. 2008). The stimulation depends also on the tree species (Nowak et al. 2004). Furthermore, biochemical adjustments in photosynthetic capacity ('down regulation') are important in governing tree production responses to elevated CO_2 (Nowak et al. 2004). Photosynthetic acclimation to elevated CO_2 is driven by the reduced expression of photosynthetic genes (Moore et al. 1999). Specifically, greater rates of carboxylation at elevated CO_2 and reduced rates of oxygenation catalyzed by RUBISCO cause the accumulation of carbohydrates, and the resultant sugar signaling triggers the reduced expression of photosynthetic genes.

In young *Populus* trees elevated CO_2 enhanced leaf level photosynthetic activity and C uptake during the senescence period (Taylor et al. 2008). Only three FACE experiments, however, fumigate trees higher than 5 m (Leuzinger and Körner 2007). Mature forest trees under long-term exposure to elevated CO_2 show some reduction in photosynthetic capacity (Körner 2006). Thus, in contrast to preliminary conclusions based on studies with young trees, a down regulation of photosynthesis may occur in trees by long-term exposure to elevated CO_2 (Karnosky 2003). For example, in lodgepole pine (*Pinus contorta* Douglas) stands exposed for centuries to elevated CO_2 near geologic vents, the photosynthetic capacity of pine saplings was down-regulated (Sharma and Williams 2009). Furthermore, it is not clear if the observed gradual increases in atmospheric CO_2 concentrations over the past centuries have the same effect on photosynthesis as do abrupt experimental exposures to much higher CO_2 concentrations. In general, the extrapolation of CO_2 effects observed after very sudden experimental changes in CO_2 concentrations to the real world needs to be critically assessed (Drigo et al. 2008).

Aside C assimilation, the water flux through plants is also altered by elevated CO_2 concentrations. As C_3 photosynthesis is RUBISCO limited, stomata don't have to be wide open at increased CO_2 concentrations for maximal CO_2 diffusion but may be partially closed (Denman et al. 2007). Thus, water loss by transpiration is reduced and the ratio of C gain to water loss, the water-use-efficiency, is increased (Mooney 1972). Theoretically, trees growing in semi-arid or arid conditions could benefit from increased water-use-efficiency at increased atmospheric CO_2 (Huang et al. 2007). This may lengthen the photosynthesis duration in seasonally dry ecosystems, and increase net C assimilation by C_3 plants. Otherwise, recent increases in surface specific humidity in response to rising temperatures may also contribute to increases in net photosynthesis (Willett et al. 2007). Specifically, stomata respond to changes in atmospheric humidity (Schulze et al. 2005). Thus, stomata may open with decreasing water vapor pressure deficit between leaf and air as a result of increasing specific humidity. This may promote C assimilation by ensuring high CO_2 concentrations inside the leaf. For the whole leaf, however, the flux of CO_2 and water vapor may be partially decoupled (Schulze et al. 2005).

The projected increase in temperature, as a consequence of ACC, may also substantially increase the C_3 net photosynthesis (Hyvönen et al. 2007). In particular, warmer periods may lengthen the photosynthetically active period (Delpierre et al. 2009). The increase in photosynthesis with increase in temperature continues until the optimum temperature for a plant is approached (Luo 2007). Then, the photosynthetic rate decreases rapidly as the maximum temperature is exceeded (Waring and Running 2007). Specifically, photorespiration is stimulated and net photosynthesis reduced at high temperatures (Rennenberg et al. 2006; Hickler et al. 2008). Trees are, however, often adapted to wide seasonal temperature ranges, and may rapidly adapt to high temperatures (Rennenberg et al. 2006). Above 40°C, however, changes in chloroplast and enzyme activity cause abrupt decreases in net C assimilation, and the decrease in CO_2 fixation is further intensified by protein denaturation at 55°C (Waring and Running 2007). In total, frequency and duration of temperature extremes are more critical in limiting photosynthesis than increases in mean annual temperature (Waring and Running 2007). Temperature and CO_2 interact frequently and, most important, the stimulating effect of increasing CO_2 concentrations on C_3 leaf photosynthesis is more pronounced at high temperatures (Hyvönen et al. 2007). Heat waves, on the other hand, may cause negative temperature effects with increasing CO_2 concentrations (Rennenberg et al. 2006). Thus, ACC-related moderate atmospheric CO_2 and temperature changes are likely to stimulate the C_3 photosynthetic rate at the leaf scale whereas acclimation processes may be important at the forest stand scale (Xu et al. 2007).

Changes in rainfall patterns by ACC have also the potential to affect C_3 photosynthesis. The stomata openings of C_3 plant leaves are actively regulated to reduce plant water loss by transpiration. Therefore, changes in rainfall pattern affect the duration of the photosynthesis season, particularly in arid and semi-arid regions (Houghton et al. 2001). Cloud cover may be beneficial to photosynthesis in dry areas with higher solar radiation, but detrimental in areas with lower solar radiation. In contrast, moderate drought down-regulates C_3 photosynthesis, and reduces or halts CO_2 diffusion into the leaf (Rennenberg et al. 2006). Further decreases in photosynthesis may occur during prolonged droughts when RUBISCO, and photochemical pigments including chlorophyll are degraded (Waring and Running 2007). Different tree species, however, respond differently to warming and heat stress (Saxe et al. 2001).

In humid regions, the photosynthetic rate can also be limited by light and, in particular, diffuse radiation beside CO_2 and RUBISCO (Schulze 2006; Mercado et al. 2009). The surface solar radiation can be partitioned in diffuse and direct components but diffuse radiation is relevant for photosynthesis as it can penetrate deeper into the vegetation canopy (Wild 2009). The long term trends in the amount of solar radiation reaching photosynthetic active pigments are, however, not fully known (Romanou et al. 2007). During the 1950–1980 period, a decrease in surface solar radiation was observed consistent with the impacts of anthropogenic aerosols on cloud properties, water vapour and cloud feedbacks due to global warming (Mercado et al. 2009). Variations in the diffuse fraction of PAR associated with this 'global dimming' enhanced the land C sink. However, between 1986 and 2000 surface net radiation over land has rapidly increased ('brightening'), and this was caused by increases in both downward solar radiation and downward thermal radiation (Wild et al. 2008). A recent analysis of both surface and satellite measurements of aerosol optical depth indicated that since the mid-1980s visibility has increased over Europe but a net dimming over land was observed as visibility substantially decreased over south and east Asia, South America, Australia, and Africa (Wang et al. 2009). How these decadal variations in solar radiation impact the land C sink is not fully known but global dimming and brightening contributed to an increase and decrease in the land C sink over the 20th century, respectively (Mercado et al. 2009). Otherwise, when leaves are exposed to more light than they can utilize, photoinhibition retards the CO_2 fixation, independent of the photosynthetic pathway (Taiz and Zeiger 2006). For example, willow populations may produce 10% less biomass during the growing season as result of depressed photosynthetic rates caused by photoinhibition.

Photosynthesis by C_4 Plants

Many warm-climate plants and 2% of all plant species conduct net C assimilation by the C_4 photosynthetic pathway (Schlesinger 1997; Schulze et al. 2005). Some woody shrubs, small trees, and some grasses in forest shade perform C_4 photosynthesis (Sage 2001; Millard et al. 2007). This pathway evolved because at higher temperatures the problem of water loss through stomata during CO_2 assimilation becomes very serious for plants (Heldt 2005). In addition, the energy wasting photorespiration by the oxygenase reaction of RUBISCO increases more rapidly with increase in temperature than does its carboxylase activity (Berg et al. 2007). These challenges have been met in C_4 plants by ensuring high local CO_2 concentrations at the site of the Calvin cycle that are not directly dependent on stomata opening.

The essence of C_4 photosynthesis is the spatial separation of the initial CO_2 assimilation in contact with air in mesophyll cells from the Calvin cycle located in bundle-sheath cells. To achieve high local concentrations of CO_2 at the site of the Calvin cycle, CO_2 is transported from mesophyll to bundle-sheath cells by carrier compounds at the expenditure of ATP (Berg et al. 2007). The intial step in the mesophyll cells is the condensation of CO_2 with phosphoenolpyruvate (PEP) to form the carrier oxaloacetate, a reaction catalyzed by the enzyme PEP carboxylase.

In some plant species, the carrier compound oxaloacetate is converted into the carrier malat. Both carrier compounds contain four C atoms, hence the name C_4 photosynthesis. At the expenditure of energy stored in ATP, the carriers are transported into bundle-sheath cells. In bundle-sheath cells, oxaloacetate or malat are decarboxylated and the released CO_2 enters the Calvin cycle operating in the same manner as discussed for C_3 photosynthesis. The remaining pyruvate then returns to the mesophyll cells. The net reaction can be written as Eq. (2.6):

$$CO_2\,(\text{in mesophyll cell}) + ATP + H_2O \rightarrow CO_2\,(\text{in bundle-sheath cell}) + AMP + 2\,P_i + H^+ \quad (2.6)$$

For the production of one hexose compound from 6 CO_2, 30 ATP are consumed in total during the C_4 photosynthetic pathway (Berg et al. 2007). In contrast, only 18 ATP are required in the C_3 photosynthetic pathway. The C_4 pathway is, however, energy saving as the high CO_2 concentration at the site of the Calvin cycle in C_4 plants accelerates the carboxylase reaction relative to the oxygenase reaction of RUBISCO. Thus, the energy-wasting photorespiration is minimized (Mooney 1972). Similar to C_3 plants, however, RUBISCO is the rate-limiting factor in photosynthesis in C_4 plants (Spreitzer and Salvucci 2002).

Climate Change and Photosynthesis by C_4 Plants

The increases in atmospheric CO_2 concentrations are not directly causing increases in CO_2 uptake by C_4 plants (Table 2.1). Specifically, the central enzyme in the C_4 photosynthetic pathway, PEP carboxylase, is almost saturated at the current ambient CO_2 concentrations (Long et al. 2004, 2006). Therefore, either no direct photosynthetic response to increased CO_2 concentration, or less response than for C_3 plants is observed for C_4 plants (Denman et al. 2007; Malhi et al. 2002). Comparable to C_3 photosynthesis, however, C_4 photosynthesis is enhanced indirectly as a result of increased water-use-efficiency due to conservation of soil moisture by decreased stomatal conductance (Long et al. 2004; Ainsworth and Rogers 2007). Furthermore, C_4 plants are adapted to higher temperatures than C_3 plants (Schulze et al. 2005). One reason is that C_4 plants have higher quantum yields at higher temperatures, that is, they assimilate more CO_2 molecules per number of absorbed photons than C_3 plants (Taiz and Zeiger 2006). Also, the enzyme systems of C_4 plants are more heat-stable than those of C_3 plants (Mooney 1972). The ACC-induced temperature increases may, thus, have less severe effects on C_4 compared to C_3 photosynthesis. At elevated CO_2 concentrations, however, C_3 plants can also dominate even at high temperatures (Schulze et al. 2005). C_4 plants are less efficient in the utilization of light compared to C_3 plants (Taiz and Zeiger 2006). Thus, enhanced light energy from solar radiation may promote the C_4 pathway relative to the C_3 pathway. In addition, at high temperatures, C_4 plants have higher light-use efficiency than C_3 plants, and, thus, C_4 photosynthesis may be enhanced (Malhi et al. 1999).

Photosynthesis by Crassulacean Acid Metabolism

The crassulacean acid metabolism (CAM) pathway evolved because in very dry and often hot habitats, increases in temperature may result in very severe plant water losses (Heldt 2005). The essence of CAM is that the plants open their stomata only at night when temperatures are lower, and temporarily store CO_2 in an acid. During the following day, CO_2 is released from the acid and light energy is used to assimilate CO_2 by the Calvin cycle, but the stomata are closed. The differences to the C_4 photosynthetic pathway are that carboxylation and decarboxylation are separated in time rather than spatially, and that CAM is not light dependent (Mooney 1972). Another advantage of this pathway is that the water requirement for CO_2 assimilation by CAM amounts to only 5–10% of the water needed for C_3 photosynthesis (Heldt 2005). In neotropical trees of the genus *Clusia*, photosynthetic processes follow the CAM pathway (Lüttge 2006). Furthermore, epiphytes in tropical rain forests may also perform CAM (Lewis 2006). In total, 3% of all plant species are CAM plants (Schulze et al. 2005).

Climate Change and Photosynthesis by CAM Plants

The photosynthetic responses of CAM plants to elevated CO_2 concentrations are probably comparable to those of C_3 plants but much greater than those of C_4 plants (Table 2.1; Drennan and Nobel 2000; Fernandez Monteiro et al. 2009). In particular, increasing atmospheric CO_2 concentrations increase the net CO_2 uptake by leaves of CAM plants through enhancing the carboxylase activity of RUBISCO, and reducing its oxygenase activity (Nobel 2005). Thus, a significant stimulation of growth under higher CO_2 concentrations has been observed for horticultural or agricultural CAM plants (Fernandez Monteiro et al. 2009). In contrast, the response of ephiphytic CAM plants in tropical forests is relatively small and taxa specific. Similar to C_3 and C_4 photosynthesis, the water-use efficiency is predicted to increase substantially at elevated CO_2 levels in CAM plant leaves. Drought may, however, less severely affect CAM compared to C_3 or C_4 plants as CAM plants have a greater resistance to water loss (Mooney 1972). In addition, ACC-induced increases in temperature may enhance the net CO_2 uptake by CAM plants (Drennan and Nobel 2000). Whether CAM represents a competitive advantage in tropical forests where higher temperatures and longer droughts are predicted in a future climate is a matter of debate (Reyes-García and Andrade 2009).

Photosynthesis by Woody Plant Stems

Aside green leaves, photosynthesis can be measured in petioles, green flowers, calyces, green fruits, cones, stem tissues and roots (Pfanz et al. 2002). Specifically, most stems of woody plants possess greenish, chlorophyll-containing tissues which use the stem internal CO_2 and light to produce sugars and starch. Net photosynthetic CO_2 uptake, however, is rarely found. Thus, stem photosynthesis is thought to be an effective mechanism to recapture respiratory CO_2 before it diffuses out of the stem. This process may increase the ratio of NPP to production plus respiration (i.e., the C-use efficiency) (Ryan et al. 1997).

2.1.1.2 Carbon Monoxide Uptake by Plants

The major uptake of gaseous forms of C in forest ecosystems occurs through photosynthetic assimilation of CO_2. However, forest vegetation may also take up carbon monoxide (CO) (Schlesinger 1997). For the NECB, CO absorption must therefore also be considered (Chapin et al. 2006). However, atmospheric CO is rapidly oxidized to CO_2. Thus, in most accounts of the C cycle CO is included as a component of the CO_2 flux (Schlesinger, 1997). As no credible picture of the CO budget exists, the importance of CO uptake by forest vegetation and how ACC may interact with it remains to be studied (Denman et al. 2007).

2.1.1.3 Prokaryotic Carbon Assimilation

Prokaryotes (Bacteria and Archaea) lack a membrane-enclosed nucleus and organelles (Madigan and Martinko 2006). Their great metabolic capacity allows them to be present everywhere on Earth where life is supported. Thus, soil, water, animals and plants are their common habitats. Due to their resistance against cultivation on growth media, however, our knowledge about their entire metabolic capabilities is still poor.

Carbon Dioxide Fixation by Microorganisms

In addition to photosynthetic active eukaryotes, phototrophic bacteria are capable of performing photosynthesis (Madigan and Martinko 2006). Widely distributed in terrestrial environments are the phototrophic cyanobacteria. Cyanobacteria contain pigments that allow them to use light as an energy source. They are commonly found in neutral to alkaline soils but rarely under acidic conditions (Frey 2007). In addition, cyanobacteria are the phototrophic component of some lichens (Madigan and Martinko 2006). Bacteria and lichens together with fungi, algae, and bryophytes form microbiotic crusts on forest plants and soil, and are able to fix C from the atmosphere (Elbert et al. 2009). Similar to plant photosynthesis, phototrophic bacteria use energy from light to reduce atmospheric CO_2 and fix C in organic compounds. Cyanobacterial photosynthesis resembles C_3 plant photosynthesis and, thus, CO_2 is fixed by the activity of RUBISCO in the Calvin cycle (Killops and Killops 2005; Madigan and Martinko 2006). Cyanobacteria contain chlorophyll but also light-absorbing phycobilisomes that enables them to harvest also green and yellow light passing through plant chlorophyll molecules. This is particularly important for cyanobacterial photosynthesis under tree canopies (Berg et al. 2007).

Nitrifying bacteria are widely distributed in soils but contain no photosynthetic pigments (Madigan and Martinko 2006). Their non-phototrophic CO_2 fixing activity may, however, contribute to the C uptake in forest ecosystems. Aerobic nitrifying bacteria, and sulfur and iron bacteria use also the Calvin cycle to fix CO_2. In addition, other bacterial groups and some Archaea use different pathways to fix CO_2.

Non-vegetated soil is commonly regarded as a CO_2 source (Miltner et al. 2005; Kuzyakov 2006). However, recent studies indicate that bare soil may also serve as a CO_2 sink as CO_2 may be assimilated non-phototropically by soil microorganisms (Miltner et al. 2005). This process may, however, be of minor importance for the soil C budget. The processes involved in non-phototrophic CO_2 fixation have not yet been principally understood.

Bacterial Methane Oxidation

Methane-oxidizing bacteria play an important role in the forest C cycle by converting methane (CH_4) produced during anoxic decomposition, that is, in the absence of oxygen, into bacterial biomass C and CO_2 (Schlesinger 1997). Aerobic methanotrophic bacteria readily oxidize CH_4 as a sole source for C and are particularly widespread in aerobic soils (Madigan and Martinko 2006). Aerobic soils are, thus, an important sink for atmospheric CH_4 (Kaye et al. 2004). The largest CH_4 uptake rates are observed in temperate forests with coarse soil texture (Dutaur and Verchot 2007). Globally, soils are, however, only a minor sink for CH_4 (Denman et al. 2007).

Bacterial Carbon Monoxide Oxidation

Green plants, soils and, in particular, the (photo)degradation of cellulose, lignin and polyphenols from plant litter are natural carbon monoxide (CO) sources in forests (Schade et al. 1999; Austin and Vivanco 2006). Carboxydotrophic bacteria use the CO as energy source and oxidize it to CO_2 which then enters the Calvin cycle (Madigan and Martinko 2006). Carboxydotrophic bacteria in the upper layers of soil are probably the most significant sink for CO in nature but detailed studies on their properties are scanty (Houghton et al. 2001).

Climate Change and Prokaryotic Carbon Assimilation

The physiological characteristics of the majority of Archaea and Bacteria are unknown as cultivated representatives are missing (Killham and Prosser 2007). Less than 5% of all soil bacteria and Archaea have been brought into culture (van der Heijden et al. 2008). Therefore, any effects of ACC on prokaryotic C assimilation are speculative. In particular, the non-phototrophic CO_2 fixation by soil microorganisms and the characteristics of carboxydotrophic bacteria are not well known. On the other hand, cyanobacteria are very tolerant to environmental extremes and, for example, occur in hot springs (Madigan and Martinko 2006). Thus, anthropogenic increases in CO_2 concentrations, and the attendant increases in temperature and alterations in water availability have probably negligible effects on C assimilation by cyanobacteria although detailed studies are lacking. Otherwise, temperature increases associated with global warming may increase the prokaryotic CH_4 consumption in forest soils (Hart 2006). Soil water contents may, however, have stronger influences on this process than temperature. In particular, excessively high or low soil water contents (extreme events) following ACC-induced variations in precipitation may inhibit CH_4 oxidation (Kaye et al. 2004). A mechanistic understanding of CH_4 uptake processes in soils is, however, missing (Dutaur and Verchot 2007). For example, the sources of the large degree of variability of CH_4 uptake in temperate forest soils are unknown.

2.1.2 *Influx of Gaseous Carbon Compounds*

The predominant natural C uptake in forest ecosystems occurs in gaseous form as CO, CO_2 and CH_4 (Fig. 2.1; Kimmins 2004). The C-containing gases may be produced by various sources in adjacent ecosystems. Lateral transport powered by thermal gradients causes their influx into forest ecosystems (Schulze 2006). For example, cities are often covered by a plume characterized by elevated CO_2 concentrations (Kaye et al. 2006; Pataki et al. 2007). Wind may transport air with high CO_2 concentrations to adjacent forest ecosystems. As discussed in the Sections 2.1.1.1 and 2.1.1.3, elevated CO_2 concentrations may accentuate plant and bacterial photosynthesis, and non-phototrophic bacterial CO_2 fixation. The forest ecosystem C pool may therefore be enhanced by CO_2 influx. Anthropogenic activities and, in particular, agriculture and fossil fuel use are very likely responsible for recent increases in atmospheric CH_4 concentrations (Denman et al. 2007). Similar to CO_2, lateral fluxes of air masses with higher CH_4 concentrations into forest ecosystems may result in higher forest ecosystem C inputs after microbial CH_4 oxidation. Agricultural and heavily populated and industrialized areas are responsible for CO production (Denman et al. 2007). However, the importance of lateral transported CO for increased C inputs into forest ecosystem by bacterial CO oxidation can only be discussed after a consistent picture of the CO budget is available.

2.1.3 *Deposition of Dissolved and Particulate Carbon*

Gaseous C forms are the major input pathway for C entering forest ecosystems. Therefore the minor C fluxes associated with transport of dissolved and particulate C have received less attention (Chapin et al. 2006; Ciais et al. 2008). Forest surfaces receive small amounts of dissolved C (<0.45 μm particle size) by wet deposition of dissolved inorganic C (DIC), and by wet deposition of dissolved organic C (DOC) with the bulk precipitation by rain and snow (Fahey et al. 2005). Atmospheric CO_2 hydrolyzes in contact with precipitation water, and dissociate to carbonate and bicarbonate

(Seinfeld and Pandis 1998). In the absence of strong acids, the atmospheric CO_2 in bulk precipitation reaching the forest surface is in equilibrium with DIC (Schlesinger 1997). Thus, DIC enters forest ecosystems with the precipitation. In contrast to DIC, the exact nature of the complex organic molecules as sources for DOC in bulk precipitation is less well known (Dai et al. 2001). The chemical complexity impairs the elucidation of processes responsible for DOC production through dissolution of organic C. However, there is some indication that DOC in rainfall is mainly derived from airborne soil and plant particulate matter (Likens et al. 1983).

In the absence of precipitation, dry deposition transports gaseous and particulate C (>0.45 μm particle size) species from the atmosphere onto forest surfaces (Seinfeld and Pandis 1998). The quantitative contribution of dry deposited organic C to C sequestration in forest ecosystems may be, however, of minor importance (Fahey et al. 2005). Aerosols are the particulate component of atmospheric particles and affect precipitation (Seinfeld and Pandis 1998; Rosenfeld et al. 2008). Aerosols have a (predominantly) cooling effect on the global mean temperature by reducing solar radiation at Earth's surface by upward reflection and absorption, and by modifying cloud cover and other cloud properties (Ramanathan and Feng 2008; Wang et al. 2009). Organic C in aerosols originates from natural sources such as windborne soil dust and volcanoes, and from anthropogenic sources such as fossil fuel and fuelwood combustion (Schlesinger 1997). Naturally produced aerosols consist of organic compounds that are also found in biomass and soil.

Aerosols may contain graphitic or black carbon (BC) from fossil fuel and biomass combustion, and organic material that agglomerates in soot (Seinfeld and Pandis 1998; Gustafsson et al. 2009). Other combustion derived particles are charred necromass or char which have received a lot of attention as important global C sinks (Kuhlbusch 1998). Char consist presumably of a mixture of heat-altered biopolymers with domains of relatively small polyaromatic clusters, and is therefore less biochemically recalcitrant than BC (Knicker 2007). Otherwise, BC is composed mainly of highly condensed polyaromatic clusters with high chemical resistance (Schmidt and Noack 2000; Hammes et al. 2007). In contrast to non-BC aerosols, BC has a large positive radiative forcing (Ramanathan and Feng 2008). Specifically, BC is the dominant absorber of visible solar radiation and forms widespread atmospheric brown clouds in a mixture with other aerosols (Ramanathan and Carmichael 2008). Global emissions of BC increased almost linearly between 1850 and 2000, and global biofuel use (woodfuel, agricultural residues, charcoal, and dung) is a major contributor (Bond et al. 2007; Fernandes et al. 2007). How much forest fires contributed to the historical trend in BC emissions is, however, unclear. BC emissions are, in particular, the second strongest contributor to current global warming after CO_2 emissions, and BC aerosols can substantially contribute to regional rapid warming (Shindell and Faluvegi 2009).

2.1.3.1 Climate Change and Carbon Deposition

The partial pressure of CO_2 in the atmosphere determines the DIC concentration in the precipitation (Fahey et al. 2005). Thus, increases in atmospheric CO_2 concentrations by anthropogenic activities may result in higher DIC inputs onto forests.

In contrast, increasing temperatures may decrease the concentration of dissolved CO_2. Thus, a new equilibrium among the gas and the dissolved phase will determine DIC concentrations in bulk precipitation. The growth of plant biomass may be promoted by ACC and this may cause higher concentrations of plant particulate matter derived DOC. The C input onto forests with DOC in precipitation may, thus, also be enhanced. Drought events associated with ACC favor the detachment of soil particles. After atmospheric conversion to DOC, an increased C input with precipitation may be observed. Drought events also increase the frequency of biomass burning and enhance aerosol concentrations (Denman et al. 2007). Forest ecosystems may, thus, receive higher loads of BC (char and soot), and increasing atmospheric concentrations of BC may warm both regional and global climate (Shindell and Faluvegi 2009).

2.2 Carbon Dynamics in Forest Ecosystems

2.2.1 Carbon Dynamics in Trees

Carbon fixed recently by photosynthesis is subsequently translocated and partitioned within the various compartments in trees (Fig. 2.3; Körner 2006; Taiz and Zeiger 2006). Partitioning can be defined as the flux of C to a particular component as a fraction of total photosynthesis or GPP (Litton et al. 2007). In forest trees C is used for growth, defense, reserves, accumulation, and storage but the genetic control of C partitioning in trees is poorly understood (Stitt and Schulze 1994; Novaes et al. 2009). Carbon sinks within the tree are located, in particular, in immature foliage, stems and branches, reproductive organs, and roots (Table 2.2; Aber and Melillo 2001). The major tissues synthesized in plant organs are dermal tissues (i.e., epidermis), ground tissues (i.e., parenchyma, collenchyma, sclerenchyma), and vascular tissues (i.e., xylem, phloem) (Taiz and Zeiger 2006). Tree stems contain the largest pool of C in form of wood, and lignin is richest in C (Novaes et al. 2009). A fraction of the C assimilated by photosynthesis is exported from the tree to mycorrhiza and soil microorganisms (Körner 2006).

2.2.1.1 Carbon Storage in Chloroplasts

As discussed in Section 2.1.1.1, the photosynthetic reaction in plants occurs in chloroplasts. In these organelles the initial storage of fixed CO_2 occurs. Specifically, the Calvin cycle intermediate fructose 6-phosphate is used to synthesize starch, a polymer of glucose residues (Table 2.2; Berg et al. 2007). This C store is transitory as starch is usually extensively degraded during the night, in particular for the regeneration of ribulose 1,5-bisphosphate, the substrate for fixing atmospheric CO_2

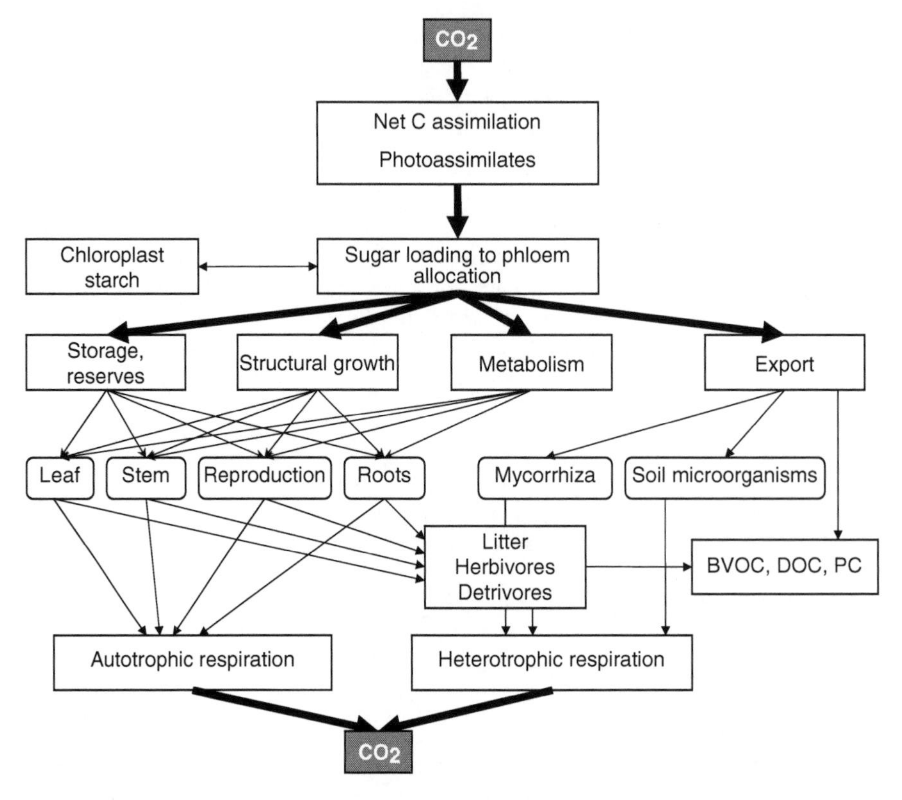

Fig. 2.3 The fate of C in plants – overview of uptake, allocation and export of C (CO_2 = carbon dioxide, BVOC = biogenic volatile organic compounds, DOC = dissolved organic carbon, PC = particulate carbon) (Modified from Körner 2006)

Table 2.2 Main organic compounds as carbon sinks within trees

Tree compartment	Carbon sink
Chloroplast	Starch
Foliage	Proteins, cellulose, hemicellulose, pectin, lignin, wax, cutin, tannin
Stem	Starch, lignin, cellulose, hemicellulose, pectin, tannin, suberin
Reproductive materials (infloresences, fruits/seeds, nectar)	Sucrose, fructose, glucose, proteins, starch, fat, sporopollenin, pectin, tannin
Roots	Lignin, suberin, carbohydrates

(Heldt 2005). Besides starch, chlorophyll, accessory pigments, lipids and proteins are synthesized in chloroplasts from recently fixed C. For example, 16% of the chloroplast protein is RUBISCO (Berg et al. 2007).

2.2.1.2 Transport of Assimilated Carbon

Beside transient storage in chloroplasts, photosynthetically fixed C is transported into adjacent mesophyll cells (Heldt 2005). For the transfer of C to mesophyll cells, the Calvin cycle intermediate 3-phosphoglycerate serves as carrier compound. In the mesophyll cells, the disaccharide sucrose is then synthesized (Berg et al. 2007). Sucrose is a readily transportable and mobilizable sugar. Sucrose is the main shuttle for the translocation of photoassimilates from leaves to other parts in the tree. Other shuttle compounds are, for example, the sugar alcohol sorbitol which is used for the translocation of assimilated CO_2 in temperate orchard trees (Heldt 2005). Mannitol is another translocated sugar alcohol (Taiz and Zeiger 2006). Furthermore, oligosaccharides of the raffinose family are translocated in several deciduous tree species (e.g., lime (*Tilia* spp.), hazelnut (*Corylus avellana* L.), elm (*Ulmus* spp.), and olive trees (*Olea europaea* L.). Sugars and sugar alcohols are exported from the mesophyll cells and loaded to the phloem (Taiz and Zeiger 2006). In the phloem the photoassimilates synthesized in the leaves (C source) are transported to the sites of consumption or storage (C sink; Heldt 2005). The overall pattern of transport in the phloem is, thus, a source-to-sink movement (Taiz and Zeiger 2006).

2.2.1.3 Carbon Storage in Foliage

Growing and developing leaves are C sinks because they carry out photosynthesis but also import C from older leaves (Mooney 1972). When the leaf grows to approximately 25% of its full size, a transition from C sink to C source status begins (Nobel 2005). Thus, older leaves apparently cannot import C from younger leaves. Sucrose and other photoassimilates may be exported upwards or downwards to C sinks within the tree, either directly or after transitory storage of starch.

Carbon is stored in developing leaves primarily through the synthesis of proteins and polysaccharides beside secondary metabolites (Table 2.2; Mooney 1972). Proteins, in particular, are the most abundant group of substances in the plant cell (Kögel-Knabner 2002). Mature leaves contain C in parenchymatic tissues in cellulose walls, the protein-rich protoplast, and the vacuola. Collenchyma tissues in leaves contain also pectin whereas sclerenchyma tissues contain cellulose, pectin, and polyoses (hemicelluloses). Furthermore, woody leaf epidermis and leaf ribs contain lignin (Taiz and Zeiger 2006). The waxy cuticle covers the leaf epidermis to prevent water loss. The leaf cuticle consists of a cutin matrix with waxes embedded in and deposited on the surface of the matrix (Heredia 2003). Cutin serves predominantly as the matrix holding the wax together that provides the major diffusion barrier (Kolattukudy 2001). Plant leaves contain also secondary metabolites for protection against herbivores and pathogenic microbes. For example, terpenoids are a class of lipids that are abundant in higher plant leaves with toxic effects on herbivores (Killops and Killops 2005). Other secondary metabolites are phenolic compounds (Taiz and Zeiger 2006). Anthocyanins and flavonoids are examples of phenolic compounds that serve as leaf pigments. Another class of phenols are tannins that occur in both condensed and hydrolyzable forms in leaf tissues (Hernes and Hedges 2004). Gymnosperms contain only condensed tannins whereas angiosperms may also contain hydrolyzable tannins in addition (Nierop and Filley 2008). Leaves of certain tree species may contain up to 39% tannin but plant-specific concentrations of condensed tannins may vary condiserably (Seth 2004; Schweitzer et al. 2008). Furthermore, leaves contain various N-containing secondary metabolites including alkaloids, cyanogenic glycosides, glucosinolates, and non-protein amino acids for antiherbivore defense (Taiz and Zeiger 2006).

2.2.1.4 Carbon Storage in Stems

Besides developing leaves, the tree stem is a C sink (Mooney 1972). In plant stems, structural materials are synthesized from imported polysaccharides but also from polysaccharides like starch stored in the stem (Table 2.2; Aber and Melillo 2001). Stem wood, in particular, represents the dominant C pool in trees at sapling stage or larger (Lamlom and Savidge 2003). The primary function of the stem is structural support to suspend photosynthetic tissues above the ground (Taiz and Zeiger 2006). Perennial plants like trees always hold a significant amount of carbohydrates in reserve in stems to buffer catastrophic losses and as source of assimilates for new spring growth (Wardlaw 1990; Aber and Melillo 2001).

Plant stems contain generally similar major tissue types and compounds as leaves (Taiz and Zeiger 2006). Woody tissues form the woody part and the supporting tissue of stems. In developing stems, the surface of plant cells consists mainly of primary cell walls. When most cell enlargement has ended, secondary woody cell walls are synthesized. Wood is manufactured by a sucession of cell division, cell expansion (elongation and radial enlargement), cell wall thickening (involving cellulose, hemicelluose, cell wall proteins, and lignin biosynthesis and deposition), programmed cell death, and heartwood formation (Plomion et al. 2001). Major compounds in wood are polysaccharides (cellulose, hemicellulose), lignins, cell wall proteins and other minor compounds (pectins, stilbenes, flavonoids, tannins, terpenoids). The wood produced by different tree species varies substantially in chemical, micro- and macromorphological traits (Cornwell et al. 2009).

Median values for cellulose and glucose concentrations of angiosperm and gymnosperm wood are 46.2% and 44.1%, respectively (Cornwell et al. 2009). Between 38% and 53% of wood consists of cellulose (Chave et al. 2009). The fundamental cellulose units are microfibrils which are the result of strong association between different chains of β-linked glucose residues (Plomion et al. 2001). The water-insoluble cellulose microfibrils are associated with mixtures of soluble hemicelluloses which account for about 25% of the dry weight of wood. Hemicelluloses contain many different sugar monomers. Median values for concentrations of non-cellulose and non-glucose sugars of angiosperm and gymnosperm wood are 20.8% and 23.3%, respectively (Cornwell et al. 2009). Median values for lignin concentrations of angiosperm and gymnosperm wood are 26.2% and 29.3%, respectively. Between 16% and 35% of wood consists of lignin (Plomion et al. 2001; Chave et al.

2009). Lignins are large, complex polymer derived mainly from the polymerization of the hydroxycinnamyl alcohols *p*-coumaryl alcohol, coniferyl alcohol and sinapyl alcohol (Amthor 2003). Lignin gives rigidity and cohesiveness to the wood tissue, and provides the hydrophobic surface for water transport. Gymnosperm lignins are based mainly on coniferyl alcohol, and dicot lignins on coniferyl and sinapyl alcohols whereas monocot lignins are a mixture of all three alcohols.

As wood ages, the inward part of sapwood is converted into heartwood through polymerization (Chave et al. 2009). Thus, wood density of heartwood is often higher than those of sapwood. Furthermore, wood density varies within individuals and with height within the plant. Roots tend to have lighter wood. Angiosperms have higher wood densities than gymnosperm wood (Weedon et al. 2009). A highly significant genetic correlation between tree growth and lignin/cellulose composition of wood was observed in *Populus* but studies with other tree species are scanty (Novaes et al. 2009). Minor components in the woody stem are resins as protective agents, for sealing wounds and discouraging insect and animal attack (Killops and Killops 2005). Resins of conifers contain diterpenoids. Wood contains also tannins but systematic studies on wood tannin concentrations are missing (Seth 2004). In the wood bark, however, tannin concentrations may be higher than those of lignin. The phellem or cork representing the outer tissue of bark contains suberin (Bernards 2002). Suberized tissues consist of poly(aliphatic) and poly(phenolic) domains, and are a barrier to diffusion (Kolattukudy 2001). However, very little is known about the cellular, molecular, and developmental processes that underlie wood formation (Plomion et al. 2001).

2.2.1.5 Carbon Partitioning for Reproduction

Beside partitioning to sinks in foliage and stems, C fixed during photosynthesis is also partitioned to C sinks in inflorescences, fruits/seeds and nectar for reproduction (Clark et al. 2001; Körner 2006). Woody perennials, in particular, partition lesser resources to reproduction than herbaceous perennials whereas annuals partition the greatest proportion of C to reproductive organs (Mooney 1972). Numerous compounds and structures such as attractants and pollen are C sinks. For example, flowers contain colored flavonoids such as anthocyanins and anthocyanidins but also flavones and flavonoids to attract pollinators (Taiz and Zeiger 2006). Carotenoids may serve as attractants, and volatile chemicals, particularly monoterpenes, may provide attractive scents and act as C sinks. Nectar is a readily available C sink and contains a solution of sucrose, fructose, and glucose (Table 2.2; Mooney 1972). Nectar may be resorbed by the tree and directed to the developing ovules.

Pollen is another C sink in trees and contains proteins beside sugars, starch and fat. The protective coating of pollen grains is made of sporopollenin and may be a long-term C sink in the terrestrial environment (Killops and Killops 2005; Lorenz et al. 2007). Specific compounds identified in sporopollenin are unbranched aliphatic chains, polyhydroxybenzene units similar to lignin, and carotenoids. Furthermore, a substantial contribution of aliphatic moieties to sporopollenin composition is indicated by several studies (Poirier et al. 2006). Sporopollenin consists of oxygenated aromatic building blocks, in particular *p*-coumaric acid and ferrulic acid (De Leeuw et al. 2006).

Strong C sinks for photosynthates in woody plants are developing fruits and seeds (Kozlowski et al. 1991). Fruits may, however, also be built from C reserves stored in trees (Chapin et al. 1990). The binding substance between fruit cells contains pectins, complex strongly branched polysaccharides (Kögel-Knabner 2002). Fruits may also contain tannins (Seth 2004). Tree seeds, otherwise, contain variable proportions of carbohydrates, fats and proteins but also phenolic compounds (Kozlowski 1971a). For example, seeds of *Acer* and *Quercus* species are high in carbohydrates whereas *Pinus* seeds are high in fat contents. Furthermore, some seeds may also contain special proteins that protect them from digestion by fungi and animals (Heldt 2005).

2.2.1.6 Carbon Partitioning to Roots

About half of the 120 Pg C fixed annually by terrestrial plants is transported to the belowground environment (Giardina et al. 2005). In particular, trees relocate large quantities of C belowground for the construction and maintenance of roots. Coarse root (>10 mm diameter) biomass in mature forest stands can range from 10% to 90% of aboveground biomass but typical values range between 20% and 30% (Burton and Pregitzer 2008). The fine root (<10 mm diameter) biomass C stock is often much lower than that in the aboveground biomass or in the soil.

Root growth is always decreased when trees become C limited (Ericsson et al. 1996). Most studies on C partitioning in plants, however, have focused on shoots rather than roots (Friend et al. 1994; Chapin et al. 2006). Scientific knowledge on C sinks in roots is, thus, still rudimentary (Robinson 2007). For example, new-root growth is thought to be fueled primarily by recent photosynthate but new roots of temperate trees not under acute stress contain a lot of stored C which is young in age (0.4 years) (Gaudinski et al. 2009). Also, previous estimates that fine roots contribute up to 40% of the total ecosystem production and that fine-root turnover represents 33% of the annual global NPP are probably unrealistically high (Vogt et al. 1996; Strand et al. 2008). The factors shaping the vertical root distribution are also poorly understood. The root profiles tend to be as shallow as possible and as deep as needed to fulfill evapotranspirational demands (Schenk 2008). Roots contain the major structural components also found in foliage and stems (Table 2.2; Horwath 2007). However, the proportions of lignin and suberin in roots may be higher than in aboveground plant organs (Lorenz and Lal 2005). On the other hand, tannin contents in roots are low (Seth 2004). Tree roots can also serve as C source. For example, C partitioned and stored in tree roots may have important role as the source of assimilates for spring growth (Wardlaw 1990). The mobilization of amino acids and carbohydrates from the woody parenchyma in roots, in particular, contributes to leaf emergence and early tree ring growth (Schulze et al. 2005).

Carbon partitioned to tree roots can be relocated by vascular connections to roots of adjacent trees by root grafting (Kozlowski 1971b). This process has been observed in many woody species, and organic substances such as carbohydrates are exchanged among trees by this process.

2.2.1.7 Climate Change and Carbon Partitioning in Trees

Atmospheric CO_2 concentrations are increasing (Tans 2009). As CO_2 fixation by photosynthesis is the major C input into forests, increasing CO_2 concentrations may alter C sequestration in forest ecosystems. Thus, effects of elevated CO_2 concentrations on C partitioning in trees will be discussed in more detail in the following section.

The response of plants to CO_2 enrichment in ecosystems may differ considerably from that in controlled settings (Neilson and Drapek 1998). As discussed in Section 2.1.1.1, elevated atmospheric CO_2 concentrations enhance the C assimilation by young trees while experimental evidence for mature trees shows mixed results (Körner 2006; Huang et al. 2007). Furthermore, the responses of whole-plant properties are also variable (Table 2.3; Franklin 2007). In particular, the relation between observed increased NPP at elevated CO_2 and C partitioning and storage in trees is not understood (Norby et al. 2005). Most of the CO_2-enrichment responses observed to-date have been studied in young trees that are most likely C limited (Körner et al. 2007). For example, by doubling CO_2 concentrations a 23% increase in NPP was observed in young tree stands in the temperate zone (Norby et al. 2005).

The response of boreal and tropical trees is unknown as current long-running forest FACE experiments are all located in the temperate zone (Hickler et al. 2008). Yet, future FACE sites are projected for tropical and high-latitude forests (Ledford 2008). Also, if growth responses will be maintained through the life cycle of a tree is not known because down regulation of photosynthetic capacity may occur (Karnosky 2003; Nowak et al. 2004). Thus, enhanced juvenile growth under

Table 2.3 Climate change effects on carbon partitioning in trees

Effect	Aboveground partitioning	Belowground partitioning
Elevated CO_2	Wood: ± stem growth, ± lignin, hemicellulose, and cellulose	Fine-roots: ± growth, lignin, condensed tannins, phenolics
	Leaf: ± growth, ± lignin, + starch, sugar, condensed tannins, phenolics	Coarse roots: unknown
		Reproduction: ±
Increased temperature	Wood: ±	Fine-roots: + turnover
		Coarse roots: ± turnover
Drought	Wood: –	(Fine-) roots: +
	Leaf: –	
Increased solar radiation	+ (?)	+ (?)

elevated CO_2 does not necessarily result in more biomass in adult trees (Körner 2006). For example, tree ring growth of Holm Oak (*Quercus ilex* L.) growing for decades in the vicinity of natural CO_2 springs was not enhanced, and the trees were only marginally taller than control trees growing in ambient CO_2 (Hättenschwiler et al. 1997). On the other hand, increases in branch number and stem diameter at elevated CO_2 indicated greater C partitioning to wood and structure in young and rapidly growing tree species in FACE experiments (Ainsworth and Long 2005; Norby et al. 2005). Long-term observations of the growth response of mature trees to elevated CO_2 are, however, rare but are needed to evaluate the response of forest biomass production and C sequestration in forest ecosystems to elevated CO_2 (Körner et al. 2005). However, several large-scale FACE experiments to study mature trees including the oldest of the forest FACE experiments will be closed by 2011 (Ledford 2008).

The structural growth of trees is controlled by a morphogenetic plan, by developmental stage and by the availability of resources other than C (Körner 2006). For example, plant N may constrain tree growth responses to elevated CO_2 but long-term experiments manipulating both CO_2 and N in tree stands are rare (Reich et al. 2006; Franklin 2007). Soil N availability reduced below a critical level, in particular, may diminish all tree responses to elevated CO_2 (Franklin et al. 2009). Otherwise, tree-growth response to elevated CO_2 is amplified when water is limiting (McMurtrie et al. 2008). Thus, future FACE sites are projected to vary not only CO_2 concentration but also temperature, precipitation and nutrients (Ledford 2008). However, the mechanisms responsible for partitioning of extra C among aboveground pools assimilated at elevated CO_2 are not yet elucidated for different tree species (McCarthy et al. 2006). Thus, elevated CO_2 concentrations may not necessarily result in higher tree biomass and enhanced C partitioning to wood (Aber and Melillo 2001; Hill et al. 2006). In general, tree species may differ in their response of C partitioning to elevated CO_2 concentrations (Körner et al. 2005). Any switches in tree species assemblage triggered by elevated CO_2 and mediated by environmental context have, thus, the potential to alter C partitioning and forest C cycling (Bradley and Pregitzer 2007).

Carbon partitioning into foliage, stems, roots and reproductive organs is driven by the availability of light, water and soil nutrients (Körner 2006). The enhanced rate of photosynthesis by elevated CO_2 does not necessarily translate into the extra amount of C accumulating in the different plant organs (Luo et al. 1997). In experiments with temperate zone forest trees, elevated CO_2 concentrations initially stimulated C partitioning to stems and stem growth (Asshoff et al. 2006). This stimulation was, however, transitory and predominantly restricted to young trees. Furthermore, no overall growth response was observed for young tropical trees, but responses of mature tropical trees to elevated CO_2 have not yet been studied (Körner 2004; Lewis 2006). The stem growth of young trees in deep forest shade may, however, be enhanced under elevated CO_2 (Körner et al. 2007). Thus, it remains uncertain if mature trees in natural boreal, temperate and tropical forests accrete more biomass C in stems during their lifetime in response to elevated CO_2 (Körner 2004; Ainsworth and Long 2005; Körner et al. 2005; Lewis 2006; Feeley et al. 2007;

Hyvönen et al. 2007). Studies are also needed to elucidate if elevated CO_2 causes changes in the chemical composition of stem wood. The few published data indicate that lignin, hemicellulose and cellulose contents are not subject to rapid change (Karnosky 2003).

In some cases there can be a sustained growth response under elevated CO_2 concentrations (Davey et al. 2006). Contrary to the general understanding as discussed before, some fast-growing poplar trees under elevated CO_2 could sequester the additional C in wood without any long-term photosynthetic acclimation or down-regulation. The underlying genomic regions determining above-and below-ground biomass responses to future high CO_2 atmospheres have been recently identified in *Populus* (Rae et al. 2007). *Populus tremuloides* genotypes vary in expression patterns in response to long-term exposure to elevated CO_2 concentrations (Cseke et al. 2009). Thus, the ability of poplar plantations to sequester the additional C needs additional research.

The extra C assimilated by trees at elevated CO_2 may be partitioned into C sinks in young leaves and roots. This response depends probably on the effects of elevated CO_2 on the canopy leaf area index (LAI) (McCarthy et al. 2006). LAI is defined as the projected one-sided leaf area per unit ground area in deciduous canopies, or as the projected needle area per unit ground area in coniferous canopies (Teske and Thistle 2004). Any growth response must probably derive from the enhancement of growth per unit leaf area as total foliage area of fully developed tree canopies does not increase at elevated levels of CO_2 (Körner et al. 2007). It has been observed that partitioning of C into leaf production in young trees increased at elevated CO_2, which may also result in larger leaf area (McCarthy et al. 2006; Hyvönen et al. 2007). Mature temperate forest trees, however, may show no consistent leaf growth enhancement as CO_2 response (Körner et al. 2005). Species-specific effects in the leaves of the mature trees are, however, observed, which may result in a shift in C fractions from more recalcitrant (i.e., lignin) to more labile compounds (i.e., starch and sucrose; Urban 2003). In contrast, in leaves of European Beech (*Fagus sylvatica* L.) seedlings and fresh leaf litter of woody species, higher lignin concentrations are observed at elevated CO_2 levels (Norby et al. 2001; Blaschke et al. 2002). Furthermore, wood lignin increased in *Populus nigra* at elevated CO_2 (Luo et al. 2008), but changes in wood lignin were not significant in conifers (Runion et al. 1999; Atwell et al. 2003; Kilpelainen et al. 2003). On the other hand, the production of phenylpropanoid-derived compounds in foliage such as condensed tannins and total phenolics may be enhanced at elevated CO_2 levels (Kasurinen et al. 2007). In some tree species, increased epicuticular wax deposition at elevated CO_2 levels is observed (Urban 2003). Changes in biochemistry resulting from elevated CO_2 are, however, often small (Bradley and Pregitzer 2007). More likely, elevated CO_2 may alter biochemistry by ecosystem-level shifts in genotype and species composition. Thus, the current understanding of the accumulation of foliar compounds in response to elevated CO_2 needs to be improved (Rogers and Ainsworth 2006).

Part of the additional assimilated C at elevated CO_2 may result in enhanced C partitioning to (fine-) roots but whole tree root-to-shoot allometries may not change (Giardina et al. 2005; Norby et al. 2005). However, there is a lack of mechanistic understanding of shoot:root partitioning in closed forest stands (Hill et al. 2006). For many tree species, increases in fine-root production at elevated CO_2 levels have been observed (Zak et al. 2000; Pendall et al. 2004). Only if the additional C is partitioned to woody root biomass, it may persist for many years (Hyvönen et al. 2007). Variations in the coarse-root biomass pool in relation to elevated CO_2 are, however, poorly quantified (Karnosky 2003). Furthermore, increased root production does not always result in larger standing root biomass as root turnover may also increase (Norby and Jackson 2000). For example, in CO_2-enriched plots deciduous trees with highly dynamic root systems partitioned C primarily to fine-root production with the greatest increase in root production in deeper soil (Norby et al. 2004). In contrast, trees with less dynamic root systems may respond differently. Thus, species-specific belowground fine-root responses to elevated CO_2 need to be addressed (Zak et al. 2000). Similar to wood and foliage, the chemical composition of roots may change at elevated CO_2. Fine-root biochemical responses are, however, also species-specific (King et al. 2005). For example, concentrations of total phenolics, soluble phenolics, condensed tannins, and lignin in fine-roots may change depending on the tree species (Zak et al. 2000).

Elevated CO_2 concentrations have the potential to alter the C partitioning in trees for reproduction. Experimental data on effects on reproduction and, in particular, fruit and seed production have not been widely published as FACE experiments fumigate mostly young trees (Ainsworth and Long 2005). The age of forest trees when reproduction occurs for the first time, however, varies greatly in length among species (Kozlowski 1971a). For example, the first fruiting of temperate forest trees occurs after 15 years of growth while others may take 50 years (Körner 2005). Mature trees partition some C resources to reproduction and, in particular, fruits and seeds are thought to be strong C sinks (Kozlowski et al. 1991; Lewis et al. 2004). The sink strength of fruit and seeds is, however, probably lower than previously thought. Specifically, the non-structural carbohydrate (NSC) pool on a whole tree basis reflects the sink/source relationship including the C capital for reproduction (Würth et al. 2005). In years with heavy fruiting, NSC pools in wood of temperate deciduous trees are rather stable, indicating that flowering and fruiting do not necessarily deplete NSC pools (Körner 2003). This was also shown for tropical trees (Würth et al. 2005). At elevated CO_2, increased NSC pools in leaves are observed. Whether this results also in larger whole tree NSC pools and C partitioning for reproduction needs to be tested (Körner et al. 2005). For example, Loblolly Pine (*Pinus taeda* L.) trees grown in elevated CO_2 commence producing pollen while they were younger and smaller than trees grown at ambient CO_2 levels (Ladeau and Clark 2006). This may enhance production of viable seeds but pollen dispersal by wind may be restricted because of the smaller tree sizes at elevated CO_2 levels. However, it is not clear if individual trees produce more pollen at elevated CO_2.

As was discussed before, young trees may respond to long-term CO_2 enrichment. This may have also implications for forest regeneration and succession in forest ecosystems. In a FACE experiment with seedlings of temperate tree species, only the growth of less-productive and shade-tolerant species such as southern Sugar Maple (*Acer barbatum* Michx.) and Winged Elm (*Ulmus alata* Michx.) was favored (Mohan et al. 2007). Thus, the future dynamics among temperate forest trees may be shifted towards late successional species.

Global air temperatures are rising and may exacerbate droughts in some forest ecosystems (Trenberth et al. 2007). The responses of mature forest trees to heat and drought are, however, less understood (Rennenberg et al. 2006). Temperature controls the amount of photosynthesis that can take place in trees, but heat also redirects metabolic processes and leads to changes in metabolite pools (Schulze et al. 2005; Boisvenue and Running 2006). Extension of the growing season duration due to warming may, thus, result in stimulation of tree productivity, increased C partitioning and storage in wood (Rustad et al. 2001; Norby et al. 2007). Growing season in mid to upper latitudes may also get longer through delayed autumnal leaf senescence as was observed in FACE experiments with *Populus* species (Taylor et al. 2008). Otherwise, the effect of recent increases in air temperatures on the growth of tropical trees is unknown (Lewis 2006).

Different tree species respond differently to warming and heat stress (Saxe et al. 2001). For example, experimental air and soil warming results in species-dependent positive growth response up to an optimum. Positive responses are indicated by higher relative growth rates, height increments, increased stem diameters and higher total dry mass. As temperatures warm, a general shift of dominance from belowground to aboveground pools in boreal, temperate, and tropical forests is projected (Apps et al. 2006). Air-warming experiments, on the other hand, point towards the importance of acclimation of photosynthetic processes for the growth responses to warming (Hyvönen et al. 2007). Furthermore, ACC is likely to increase NPP more in cold northern regions but the direct CO_2 response is likely to be stronger in warm regions (Hickler et al. 2008). The fine-root production and mortality increases associated with higher mean annual air temperatures (Gill and Jackson 2000). However, no relationship apparently exists between air temperatures and coarse root turnover. Also, root acclimation may be one reason why belowground partitioning of C in trees is less responsive to thermal perturbations than previously thought (Giardina et al. 2005).

Over wide areas, and particularly in the tropics and subtropics more intense and longer droughts have been recently observed (Trenberth et al. 2007). Drought influences canopy gas exchange and productivity in forests (Waring and Running 2007). Trees may respond with anatomical and morphological changes to ACC-induced drought stress (Schulze et al. 2005; Bréda et al. 2006). For example, root systems may be enlarged to enhance water uptake and leaves may be shed prematurely to reduce transpiring surfaces (Kozlowski et al. 1991). Otherwise, enhanced root turnover but constant leaf biomass in drier environments has been observed (Meier and Leuschner 2008). The physiological processes in trees related to drought are, however, poorly understood.

To summarize, across different climate projections the enhancement of NPP in forest ecosystems in future climate is projected (Alo and Wang 2008). This finding depends, however, on the temperature and CO_2 fertilization response of the vegetation and is, thus, disputable (Hickler et al. 2008). In particular, elevated CO_2 effects on forest productivity vary largely depending on tree species, age and co-occuring stresses (Karnosky et al. 2007).

2.2.2 *Carbon Dynamics Outside of Trees*

2.2.2.1 Fluxes of Dissolved Carbon

Dissolved C enters the forest ecosystem with bulk precipitation. A fraction of bulk precipitation is intercepted by the tree canopy, flows along the stem (stemflow), drips from the foliage and branches or passes through canopy openings (throughfall), or is further intercepted by the forest floor (Section 5.3; Chang 2006). By leaching compounds from plant surfaces, the precipitation reaching the soil becomes enriched mainly in DOC whereas DIC may be similar to that in bulk precipitation (Fahey et al. 2005). Thus, dissolution of compounds in surface tissues of plants, and soil and plant particulate matter deposited on plant surfaces contribute to DOC in throughfall. Numerous organic compounds such as carbohydrates, amino and organic acids, alkaloids and phenolics are leached from plant leaves (Mooney 1972). The quantities of materials lost by leaching can be considerable but data for different tree species and different organic compounds are lacking.

Throughfall quantity is highly variable among and within forests (Chang 2006). For example, the C flux with throughfall in a northern hardwood forest was very small, accounting for less than 2% of the C flux with litterfall (Fahey et al. 2005). Younger leaves are less susceptible to leaching than older leaves, and conifers are less susceptible than deciduous trees whereas tropical rain-forest plants are not easily leached (Mooney 1972). Furthermore, the stemflow is another major fraction of water that infiltrates into the soil (Crockford and Richardson 2000). Incident precipitation flows down the stem and, similar to leaves, may be enriched with organic compounds leached from the tree bark. The importance of stemflow, however, varies greatly among tree species and stands, but is probably a minor pathway of C flux into the forest soil (Fahey et al. 2005). Studies on stemflow chemistry are, however, scanty (Levia and Frost 2003).

Climate Change and Fluxes of Dissolved Carbon

When elevated atmospheric CO_2 concentrations enhance plant growth and, in particular, increase the plant surface area and alter the canopy structure, the quantity and quality of leached compounds may also change. Furthermore, the composition of DOC in stemflow and throughfall for individual trees may be altered by changes in C partitioning patterns among foliage, stems and roots. Over many large global regions significant increase in precipitation has been observed as is also increase in drought in other regions (Trenberth et al. 2007). Thus, DIC and DOC fluxes into forest soils may also change with the change in moisture regime.

2.2.2.2 Litter Input

Above- and belowground plant litter inputs deliver a large fraction of NPP to the soil (Schlesinger 1997). Senescence is a normal process in the life cycle of plants and results in the death of plant parts or the entire plant and, after abscission, in C deposition in the soil (Taiz and Zeiger 2006). Woody plants, in particular, periodically produce litter by shedding a variety of above- and belowground tissues and organs (Kozlowski et al. 1991). Vegetative parts such as branches, leaves, bark and root, and reproductive parts such as fruits, cones and seeds are shed. Another litter input is pollen unsuccessful in terms of reproduction (Greenfield 1999). Beside mature trees, the contribution of understory vegetation may be significant, and amount up to 25% of aboveground litterfall (Kimmins 2004). Leaves are the major component of aboveground litterfall but bark can also be important litter in, for example, *Eucalyptus* forests. Furthermore, the deposition of woody litter tends to increase with forest age (Schlesinger 1997). Globally, litterfall declines with increase in latitude from tropical to boreal forests, and the patterns are similar to global NPP patterns although a recent reevaluation suggests that annual NPP in tropical forests is not different from annual NPP in temperate forests (Schlesinger 1997; Huston and Wolverton 2009). Beside the normal process of senescence, strong mechanical forces by wind, hail, ice and snow, and plant infestation with herbivores deliver a fraction of NPP as litter to the soil (Waring and Running 2007). For example, large-scale insect outbreaks result in widespread tree mortality but the impact of insects on C dynamics in forests is not well documented (Kurz et al. 2008). Another natural disturbance in forests affecting litter input is fire (Kimmins 2004). Especially boreal forests are affected by biomass burning and one-third of global boreal forest NPP may be consumed (Wirth et al. 2002). Burning of aboveground plant parts reduces the litter input, and very severe fires may even burn the roots and, thus, reduce also the root litter input. However, fire increases the input of BC (char and soot) which is a potential C sink (Section 2.2.2.5; Kuzyakov et al. 2009).

The deposition of NPP belowground by root turnover can be larger than that from the aboveground turnover (Vogt et al. 1996). The root turnover in forest ecosystems is, however, not completely understood (Trumbore and Gaudinski 2003). In particular, fine root turnover may significantly contribute to the forest ecosystem C budget but technical difficulties restrict direct quantification (Eissenstat and Yanai 2002). The contribution of fine-root turnover to annual global NPP is probably lower than previously assumed (Strand et al. 2008). Otherwise, the C flux via mortality of coarse roots has not been measured in forest ecosystems (Fahey et al. 2005). Studies on root turnover are challenging as roots can reach maximum depths of up to 68 m with the global average estimated to be 4.6 m (Canadell et al. 1996). Globally only 56 direct measurements of complete root profiles down to the maximum rooting depth have been reported (Schenk and Jackson 2005). Thus, our knowledge on root turnover is rudimentary and the conclusions on belowground C input from root litter regarded as preliminary. Similar to deposition of litterfall, however, root turnover decreases from tropical to high-latitude forests (Gill and

Jackson 2000). In summary, a large fraction of NPP may be delivered via above- and belowground litter input to the soil, but the estimates of annual addition of C to the soil are probably low due to largely unmeasured belowground root C input (Schlesinger 1997). A fraction of the litter C deposited by the various processes is accumulating in the forest floor, in coarse woody debris (CWD) and the mineral soil (Pregitzer and Euskirchen 2004).

Climate Change and Litter Input

Enhanced C fixation due to increases in atmospheric CO_2 concentrations may enhance plant growth and, thus, the fraction of C delivered to the soil in aboveground litter (Norby et al. 2005). The amount but also the dynamics of aboveground litterfall may, therefore, change (Hyvönen et al. 2007). Beside surface litter quantity, the relative proportion of the various plant biomacromolecules in litter may change (Section 2.2.1.6). In addition to affecting aboveground litter, elevated CO_2 concentrations may cause an increase in fine-root production (Giardina et al. 2005; Hyvönen et al. 2007). The presumably higher input of root litter may be accompanied by altered chemical root litter composition at elevated CO_2. When individual trees respond to increase in atmospheric CO_2 concentrations with changes in root-to-shoot allometries, the relative C flux from above- and belowground litter will also change. Within forest ecosystems, important indirect changes may arise from the varying effects of elevated CO_2 on C_3, C_4 and CAM photosynthesis (Table 2.1). This may shift the fraction of plant species belonging to different photosynthetic pathways among the forest plant community. The fraction of C delivered to the soil may, thus, also be altered. However, projections of tree species responses to elevated CO_2 to forest responses regarding plant litter may be confounded by interactions among different tree taxa (Körner 2005). Only in early 1999, manipulative biodiversity experiments with multiple tree species at ambient CO_2 concentrations were initiated (Scherer-Lorenzen et al. 2005). Similar studies at elevated CO_2 are, however, required to test CO_2-induced diversity effects on litter input.

Increase in temperature results in species-dependent positive growth responses up to an optimum (Section 2.2.1.6). This may be accommodated by increase in the biomass fraction delivered to the soil in litter (Rustad et al. 2001). In particular, fine root mortality and, thus, litter input increases with increase in temperature but responses of coarse root litter production remain to be determined as those of temperature effects on tropical tree litter production. Similar, the physiological processes in trees as response to drought are poorly understood. Increased fine root mortality in drier stands may cause higher belowground C inputs (Meier and Leuschner 2008). Enlarged root systems in drier stands may also result in higher root litter input to the soil. Drought stress may generally alter quality and quantity of plant litter input. For example, increases in drought intensity may affect litter input by premature mortality of roots or twigs, and ultimately tree death (Bréda et al. 2006). Regional warming and consequent increases in water deficits are likely contributors to increases in tree mortality rates in temperate forests (van Mantgem et al. 2009).

The ACC may bring changes in the disturbance regimes affecting forest ecosystems (Apps et al. 2006). Long-term changes in wind patterns have been observed at continental, regional, and ocean basin scales (Trenberth et al. 2007). In particular, the strength of mid-latitude westerly winds has increased as has the intensity of tropical cyclone activity. Enhanced mechanical forces by stronger winds can result in larger litter inputs from severed tree parts, by stembreak or windthrow (Waring and Running 2007). If ACC intensifies the effects of hail, ice, and snow on the litter production by stripping foliage and branches, by stembreak and uprooting of whole trees is, however, less clear (Kimmins 2004). Otherwise, climate extremes may enhance the probability that trees become more susceptible to damages from defoliating insects, bark beetles and pathogens (Waring and Running 2007; Kurz et al. 2008). Thus, the litter input to soil may also increase. When the disturbance is accompanied by higher mortality of entire trees a higher root litter input is observed. Furthermore, the increase of some weather events and extremes due to ACC are projected to increase the severity and frequency of wildfires (Parry et al. 2007). In the future, alterations of above- and belowground litter input into forests may, thus, occur.

2.2.2.3 Non-gaseous Carbon Efflux from Roots

Dead plant roots are a substrate or C source for litter decomposition. Living plant roots, on the other hand, release organic compounds by rhizodeposition (Nguyen 2003). Rhizodeposition includes a wide range of processes by which C enters the soil (Jones et al. 2009). The processes are: (i) root cap and border cell loss, (ii) death and lysis of root cells, (iii) flow of C to root-associated symbionts living in the soils (e.g., mycorrhiza), (iv) gaseous losses, (v) leakage of solutes from living cells (root exudates), and (vi) insoluble polymer secretion from living cells (mucilage). Fungal partners in the rhizosphere receive carbonic compounds from roots (Jones et al. 2009). Specifically, ectomycorrhizal (ECM) roots are formed by interactions between soil fungi and the roots of woody plants whereas arbuscular mycorrhizal (AM) associations are symbioses between soil fungi and the roots of most terrestrial plants (Hodge et al. 2009). The tree benefits from ECM and AM by increased uptake of nutrients, particularly phosphate.

The C flow at the soil-root interface is bidirectional with C being lost from roots and taken up from the soil simultaneously. Most C lost during root growth is in the form of complex polymers. The amount of C lost by rhizodeposition of trees in forest ecosystems is, however, virtually unknown. In major temperate forest trees, for example, fixed C is transferred within a few days from the crown to the root zone (Steinmann et al. 2004). Most plant roots are linked in symbiosis with ECM or AM fungi which receive a significant fraction of fixed C from the host plant in exchange for critical services in resource foraging in the soil (Gilbert and Strong 2007). The strongest C sink for photosynthates in a boreal *Pinus sylvestris* forest, for example, were ECM roots (Högberg et al. 2008). However, the poor knowledge about rhizodeposition is problematic with regard to quantifying C sequestration in forests (Jones et al. 2009).

Jones et al. (2004) hypothesized that plant exerts little direct control over large components of their C efflux by down-regulation of exudation. In contrast, exudation can be directly up-regulated to alleviate biotic and abiotic stresses. The indirect impacts of root-derived fluxes on the biogeochemical cycle of C through the 'priming effect' on microbial activity and nutrient cycling are probably enormous (Section 2.2.2.5; Lerdau 2003). The 'priming effect' can be defined as a short-term change in the turnover of soil organic matter (SOM) caused by addition of organic C to the soil (Kuzyakov et al. 2000). Thus, C fluxes from roots can influence ecosystem C exchange. Another impact of root-derived organic acids is desorption and chelation of P, in particular in tropical soils, which may promote NPP as P is often limiting growth.

Climate Change and Carbon Efflux from Roots

The rhizosphere may play a significant role in the major challenge of GHG mitigation (Jones et al. 2009). Enhanced photosynthesis at elevated CO_2, in particular, may result in enhanced C input to the rhizosphere (Cheng 1999). However, the fate and function of this extra C input into the soil is not clear. For example, increased C losses from root at elevated CO_2 may increase the decomposition of labile C but decrease the decomposition of more recalcitrant SOC fractions (Lerdau 2003). Elevated CO_2 concentrations, otherwise, can also alter the partitioning of assimilates to the rhizosphere by altering plant structure, modification of root to shoot ratio and alteration in root morphology (Nguyen 2003). An increase in root biomass at elevated CO_2 levels is expected to result in higher amounts of C deposited by rhizodeposition (de Graaff et al. 2006). Furthermore, pathogen attacks enhanced by ACC may cause the loss of the entire root cap (Nguyen 2003). Other indirect effects of ACC on root C efflux may originate from temperature and drought effects on root morphology and root to shoot ratios. Depending on effects of higher temperatures on root growth, the C discharge from roots may increase or decrease. The database on temperature effects on C loss from roots is, however, very small (Lerdau 2003). Changes in species composition associated with ACC may have dramatic effects on C losses from roots as many synergistic relationships occur between trees including mycorrhizal interactions. Thus, ACC effects on rhizosphere processes in mixed forests need to be studied in detail (Jones et al. 2009).

2.2.2.4 Decomposition

In terrestrial ecosystems, up to 90% of the photosynthetically fixed C ultimately enters the decomposer food web (Bardgett et al. 2005). Dead organic matter serves as the substrate for decomposition (Table 2.4). By this process, detritus such as plant litter, rhizodeposits, faunal and microbial residues are broken down.

Table 2.4 Resources for decomposition and their chemical composition

Resource	Composition
Plants	Aboveground litter: celluose, hemicelluoses, pectin, lignin, tannins, cutin, suberin, lipids, waxes
	Belowground litter: similar to aboveground but relatively more lignin, suberin
	Rhizodeposits: sugars, proteins, amino and organic acids
Microorganisms	Polysaccharide and protein-type compounds; various potentially recalcitrant biomacromolecules, for example, melanins, waxes, terpenoids, chitin, sporopollenin
Soil animals	Lipids, waxes, chitin, sclerotin, keratin, fibrinogen, collagen

Decomposition of plant litter includes the processes of litter leaching, litter fragmentation and chemical alteration (Chapin et al. 2002).

The physicochemical environment, the litter properties, and the composition of the decomposer community mainly control litter decomposition (Berg and McClaugherty 2003). Globally, the sum of litter N, P, K, Ca and Mg concentrations together with the C:N ratios account for 70% of the variation in litter decomposition rates (Zhang et al. 2008). Plant species traits are predominant control on litter decomposition rates, and litter quality (i.e., N:C ratios) alters decomposers' respiration rates and the decomposition of single-species litters (Cornwell et al. 2008; Manzoni et al. 2008). However, the composition and diversity of chemical compounds in litter are potentially important functional traits and affect decomposition (Epps et al. 2007). For example, condensed tannins in plants consistently slow rates of litter decomposition (Schweitzer et al. 2008).

During decomposition, the above- and belowground detritus is chemically altered by the activity of heterotrophic organisms, i.e., fauna and microorganisms (Swift et al. 1979). Thus, the biota is also important driver of litter decomposition. A small proportion of the detrital C is reassembled in new molecules during SOM formation (Section 2.2.2.5). Mineralization, on the other hand, leads to the complete oxidation of detrital C to CO_2 and CH_4 (Section 2.3.2; Schulze et al. 2005). Much less attention has been given in the past to belowground processes such as root decomposition in terrestrial ecosystems compared to foliage despite its importance for C sequestration (Silver and Miya 2001; Rasse et al. 2005; Bird et al. 2008). For example, the great majority of soil animal taxa in forest soils acquire C from belowground and less from aboveground C input (Pollierer et al. 2007).

The substrates for decomposition include all above- and belowground plant parts (i.e., litter such as foliage, branches, stems, roots), faunal and microbial residues, and secreted organic compounds. Pollen may also be substrate for decomposition (Webster et al. 2008). Plant residues are primary substrates for decomposition as plant-derived C represents also the major part of SOC (Kögel-Knabner 2002). The microbial biomass, on the other hand, represents only between 2% and 5% of SOC although these estimates are probably too low (Frey 2007; Simpson et al. 2007). Thus, microbial residues together with rhizodeposits are regarded as secondary substrates for decomposition. The contribution of C from decomposed faunal residues to

SOC is probably minor, at least quantitatively, because soil animals immobilize less than 1% organic C (Wolters 2000). However, cadaver decomposition in soils and associated C transfer into SOC is less well studied (Carter et al. 2007).

The major plant compounds subject to decomposition are polysaccharides (cellulose, hemicelluloses, pectin; 50–60% of plant biomass), lignin (15–20%), tannins, cutin, suberin, lipids and waxes (together 10–20%) (von Lützow et al. 2006). Woody plant leaves and twigs, in particular, may contain between 2% and 27% lignins, and woody stems may contain between 15% and 36% lignin (Amthor 2003). However, the variability of organic compounds among plant species and, in particular, those of lipids and tannins is less well known. For C sequestration in forest ecosystems, the amount of less biodegradable or biochemically recalcitrant compounds in litter is important. The intrinsic structural composition of lignin, lipids, waxes, cutin and suberin contributes to their resistance against decomposition, at least during initial decomposition stages (Lorenz et al. 2007). Furthermore, root residues may be more resistant to decomposition as roots contain relatively more lignin, cutin and suberin than aboveground residues (Rasse et al. 2005). Major compounds in rhizodeposits are sugars, proteins, amino acids and organic acids which are, however, regarded as less recalcitrant (Section 2.2.2.3).

Most of the microbial and faunal resources are polysaccharide- or protein-type macromolecules (von Lützow et al. 2006). Potentially recalcitrant compounds in microorganisms that belong to the domain Archaea are pseudomurein and (tetra) ether lipids (Lorenz et al. 2009). Examples of recalcitrant compounds in bacterial biomass are murein, lipopolysaccharides, glycolipids, teichoic acids, phospholipids, melanins, waxes, terpenoids and tetrapyrrole pigments. Chitin, melanins, glomalin, hydrophobins, ergosterol and sporopollenin are potentially recalcitrant compounds found in fungal biomass. Animal residues contain potentially recalcitrant lipids, waxes, chitin, sclerotin, keratin, fibrinogen and collagen (Lorenz et al. 2009). For detailed descriptions of the chemical structure of detrital C inputs into soil see Kögel-Knabner (2002) and Killops and Killops (2005).

The initial process of decomposition is the leaching of soluble organic material by water (Chapin et al. 2002). Mainly labile and not recalcitrant compounds such as low-molecular weight sugars, polyphenols and amino acids are leached from plant foliage, and a significant litter mass loss occurs (Waring and Running 2007). Leaching is particularly important during tissue senescence and when fresh litter falls to the forest floor. The leached material may be absorbed by soil organisms, adsorbed to SOM and the soil mineral phase, or lost as DOC with the percolating water.

Soil animals carry out the litter fragmentation (Chapin et al. 2002). Specifically, the macrofauna (Ø>2 mm) and the mesofauna (0.2<Ø<2 mm) feed on detritus (Wolters 2000). By chewing activities and physical transformations in the animal gut, the litter is fragmented and this causes increases in the litter surface area. Many invertebrates preferentially feed litter that is high in carbohydrates but low in polyphenols and tannins. Aside fragmenting the litter, species such as earthworms, ants and termites transport litter deeper into the soil. However, before the litter is consumed by soil animals considerable chemical alteration of the litter must occur (Wolters 2000). In particular, the levels of aromatic compounds decrease by microbial activity. During the passage through the animal gut, the litter is further chemically altered, and gut microorganisms are added to the detritus. After excretion, the casts are further digested mainly by the activity of the microfauna (Ø<0.2 mm).

Litter decomposition can entirely depend on bacteria and fungi (van der Heijden et al. 2008). Fungi, in particular, are the main initial decomposers of fresh plant litter, and account for 60–90% of the microbial biomass in forest soils (Chapin et al. 2002; Amelung et al. 2008). Quantitatively, litter layers and forest topsoil horizons are dominated by fungi relative to bacteria biomass (Joergensen and Wichern 2008). Greater C storage in fungi-dominated soils is, however, not caused by the greater growth yield efficiency (biomass and metabolite C production per unit substrate C utilized) of fungi compared to those of bacteria (Thiet et al. 2006). However, beside detrital C from the immediate environment where fungal growth occurs, extended hyphal networks enables fungi also to transport a substantial proportion of C (25%) for growth from elsewhere in the hyphal network. Fungi can, thus, actively grow to detritus and their enzymes can break down virtually all classes of plant compounds. White-rot fungi, for example, completely decompose lignin whereas soft and brown rot fungi only partially decompose lignin (Kögel-Knabner 2002). Fungi have a competitive advantage at low pH commonly observed in forest soils but are often absent from or dormant in anaerobic soils (Chapin et al. 2002). Furthermore, the mycorrhizae not only serve as vectors for plant C input to soils but also act as decomposer by producing extracellular lytic enzymes and metabolizing SOC (Talbot et al. 2008).

In contrast to fungi, bacteria depend on the movement of substrates to them for decomposition. Bacteria mainly decompose labile substrates. Thus, bacterial dominated food webs are often characterized by high leaf litter quality whereas fungal dominated are characterized by low leaf litter quality (van der Heijden et al. 2008). However, actinomycetes, slow-growing bacteria that have filamentous structures similar to fungal hyphae, can break-down relatively recalcitrant compounds such as lignin (Chapin et al. 2002). Important is the bacterial decomposition in the rhizosphere, the decomposition of animal carcasses and those of bacterial and fungal cells. Soil bacteria are often embedded in a matrix of bacterial polysaccharides called biofilm. Bacteria in biofilms may act as a consortium in the breakdown of complex macromolecules.

Soil animals increase decomposition rates in temperate and wet tropical climates, but have neutral effects where the biological activity is constrained by temperature or moisture (Wall et al. 2008). Beside their importance for litter fragmentation, soil animals contribute to the chemical alteration of detrital C. During passage through the animal gut, forest litter may loose sterols, short-chain fatty acids, triacyclgycerols, amino acids and polysaccharides, and may accumulate triterpenoids, wax esters, aromatic-C, methoxyl-C and lignin (Hopkins et al. 1998; Rawlins et al. 2006). This indicates that gut passage increases the recalcitrance or stability of litter OM against further decomposition (Osler and Sommerkorn 2007). However, C transfer from stable to more labile SOC pools during gut passage is also observed (Fox et al. 2006). For example, during gut passage in wood-feeding insects strinking modifications to wood lignin occur in a short period (Geib et al.

2008). Also, invasive non-native litter-feeding earthworms in North American forests contribute to a shift in the aliphatic and aromatic composition of litter and particulate organic matter (POM) which may influence SOC stabilization (Crow et al. 2009). The soil animals indirectly alter the chemical composition of litter through their grazing activity on bacteria and fungi (Chapin et al. 2002). The major effect of the soil fauna on chemical alteration of litter during decomposition occurs, however, through enhancement of microbial activity by litter fragmentation (Wolters 2000). Predators may indirectly cause the accumulation of total phenolics in forest leaf litter, and the loss of cellulose and condensed tannins (Hunter et al. 2003).

In the course of decomposition, surface litter mass approximately decreases exponentially but may be better described by a curve with at least three phases (Chapin et al. 2002). First, cell solubles are predominantly lost by leaching followed by a second slower phase of decomposition through biotic activity accompanied by leaching. During the third very slow final phase, OM is further chemically altered and mixed with soil minerals resulting in the formation of SOM. Thus, cellulose and hemicelluose are lost whereas microbial products and recalcitrant litter compounds accumulate (Moorhead and Sinsabaugh 2006). However, rates of litter decomposition may better be predicted by a holistic approach based on plant life attributes rather than correlations based on individual initial litter chemistry parameters (Prescott 2005).

For the less studied root decomposition, exponential and linear decay models, and two-stage root decomposition process models have been used to describe root mass loss over time (Silver and Miya 2001). During the first phase, primarily inorganic constituents and water-soluble C are lost by microbial activity and leaching. The second phase of slower mass loss is regulated largely by recalcitrant root materials. In contrast to fine roots, however, C fluxes from coarse root decomposition have not been measured (Fahey et al. 2005).

Thus, both leaf and fine root litter decomposition may undergo a relatively slow phase. Long-term decomposition experiments indicate that this slow phase is common but not universal (Harmon et al. 2009). Specifically, the decomposition rate of the slowly decomposing litter component is higher than those of mineral SOM but lower than those of the faster phase of litter decomposition. Thus, using short-term decomposition rates potentially underestimates the global pool of litter by at least one-third in comparison to estimates based on long-term integrated decomposition rates (Harmon et al. 2009).

Another important decomposition process in forest ecosystems occurs in standing dead trees and decaying logs, that is, CWD (Kimmins 2004). Up to 20% of the aboveground biomass in mature forests may be CWD (Brown 2002). Thus, the CWD pool may be equal in size to the fine litter pool (Matthews 1997). Globally, coarse dead wood contains between 36 and 72 Pg C (Cornwell et al. 2009). Microbial decomposition, combustion, consumption by insects, and physical degradation determine the fate of CWD. CWD may contribute more to recalcitrant SOM fractions than fine litter (Kimmins 2004). Angiosperm wood is denser, has larger conduits, lower lignin concentrations, higher N and P concentrations, and lower lignin:N ratios than gymnosperm wood (Weedon et al. 2009). Thus, gymnosperm wood generally decomposes slower than angiosperm wood. Decomposition of angiosperm wood is faster at higher N and P concentrations, and slower at higher

C:N ratios. In contrast, no relationship between these wood traits and decomposition rate is observed for gymnosperm wood. Therefore, differences in lignin alcohols, the quality of hemicelluoses, the microscopic distribution of lignin and other structural compounds in tracheids and the presence of non-structural phenolics and other aromatics may be more important controls of wood decomposition rates than macronutrient or lignin concentrations.

Tree roots generally decompose more slowly than leaves (Gholz et al. 2000). The primary controller of root decomposition appears to be root chemistry whereas temperature, moisture, soil properties and substrate quality mainly control leaf litter decomposition. Thus, leaf litter decomposes faster in the humid tropical than in cold and wet boreal forests (Parton et al. 2007). Furthermore, ambient solar radiaton may contribute to photochemically derived CO_2 fluxes during litter decomposition (Brandt et al. 2009). However, little is known about the contribution of photodegradation to CO_2 production and C turnover in terrestrial ecosystems (Austin and Vivanco 2006). Photochemical mineralization to CO_2 appears the primary mechanism as opposed to leaching, other direct gaseous losses, or biological facilitation (Brandt et al. 2009). In ecosystems with high levels of solar radiation, low litter inputs, and low levels of microbial activity, the direct abiotic mineralization of litter to CO_2 may be a major mechanism for litter turnover. For example, in arid ecosystems, solar radiation can be an important contributor to leaf litter degradation (Gallo et al. 2006).

Climate Change and Decomposition

In Section 2.2.1.6 the various effects of elevated CO_2 concentrations on both quantity and quality of substrates for decomposition were discussed. Despite some changes in leaf litter quality or decomposability (e.g., C:N and lignin:N ratios), decomposition of leaves grown in elevated CO_2 is often not directly altered (Norby et al. 2001). Parsons et al. (2008), however, observed that the poorer quality of *Populus tremuloides* and *Betula papyferia* leaves grown at elevated CO_2 concentrations resulted in reduced decomposition rates. On the other hand, decomposition may be altered indirectly by effects of elevated CO_2 on the amount and dynamics of litter fall, on litter quality through changes in plant community composition and on soil properties (Reich et al. 2005; Hobbie et al. 2006; Hyvönen et al. 2007). Thus, potential shifts in the identity and traits of the dominant plant species may affect the C cyle through alterations in decomposition rates (Cornwell et al. 2008). Increased dominance of angiosperms, for example, may accelerate the decomposition of woody material and cause a 44% reduction in the woody debris pool (Weedon et al. 2009). Together with the generally faster decomposition of angiosperm tree leaves this may result in reduced forest C pools. However, long-term decomposition experiments to study these effects are lacking.

Indirect effects of elevated CO_2 may also occur through alterations in the microbial community composition (Billings and Ziegler 2008; van der Heijden et al. 2008). If elevated CO_2 concentrations favor fungi relative to bacteria, initial decomposition rates may be higher, and decomposed litter C may be more persistent as fungal cell walls are more recalcitrant and decompose more slowly than bacterial cell walls (Chapin et al. 2002; Six et al. 2006). In a *Populus* forest, however, a fungal decomposition pathway was not favored at elevated CO_2 (Van Groenigen et al. 2007). Furthermore, an in-depth comparative analysis of soil microbial diversity in this forest indicated that heterotrophic decomposers and ECM fungi increased at elevated CO_2 whereas Archaea of the phylum *Crenarchaea* decreased (Lesaulnier et al. 2008). On the other hand, CO_2 enrichment resulted in increased cellulose and chitin-degrading enzyme activities under trembling aspen (*Populus tremuloides* Michx.), paper birch (*Betula papyrifera* Marsh.) and sugar maple (*Acer saccharum* Marshall), and this may alter litter C cycling although fungal community composition was not altered (Chung et al. 2006). New decomposer communities might emerge after years of CO_2 enrichment, but long-term decomposition rates in such soils have not yet been studied (Norby et al. 2001). Furthermore, litter-mediated CO_2 effects on the decomposer fauna may also exist (Kasurinen et al. 2007; Drigo et al. 2008). Herbivores are less abundant in enriched CO_2-environments of long-term experiments as plant quality changes, in particular, plant C:N ratios and concentrations of tannins and other phenolics increase at elevated CO_2 whereas N concentrations decrease (Stiling and Cornelissen 2007). The decline in herbivore abundance can be explained by reduced herbivore survival and reproduction due to decreased nutritive value of plant tissue at elevated CO_2, and longer exposition of herbivores to natural enemies as herbivore development is delayed at elevated CO_2 (Stiling et al. 2009). However, the effects of elevated CO_2 on herbivores are plant species-specific.

Much less known than effects on aboveground processes are the CO_2 effects on belowground decomposition. Important for C sequestration in forest ecosystems is the C flux from the often observed increased fine root production, mortality and standing crop at elevated CO_2 (Hyvönen et al. 2007). This may also affect the microbial community structure and activities in the rhizosphere but less in the bulk soil (Drigo et al. 2008). The fine root decomposition, however, is probably not affected at elevated CO_2 (King et al. 2005). Also, slightly increased substrate quality of roots grown at elevated CO_2 may result in faster decomposition in the short-term but effects may diminish over a longer-term decomposition period (Chen et al. 2008). Otherwise, an indirect effect of elevated CO_2 on decomposition may arise from the export of excess carbohydrates to rhizosphere microbes (Giardina et al. 2005; Körner et al. 2007). This may result in enhanced decomposition of lignin (Ekberg et al. 2007). Indirect effects on belowground decomposition, however, may also occur from sometimes observed higher fine root biomass deeper in the soil profile as decomposition is slower in deeper soil horizons (Rasse et al. 2005; Johnson et al. 2006). Increased C inputs from increased fine-root mortality at elevated CO_2 may alter SOC storage in deeper soil horizons (Iversen et al. 2008).

Increasing temperatures may increase the total plant biomass in forest ecosystems until an optimum and, in particular, increase the above- vs. the belowground biomass of trees (Section 2.2.1.6). Thus, both the quantity and quality of substrates for decomposition are likely to be altered at higher temperatures, and this may cause an increased C flux into soils from aboveground inputs (Rustad et al. 2001).

Reduced litter C:N ratios observed in warming experiments, however, have shown only little effects on decomposition (Norby et al. 2001). In the long term decomposers are likely to receive more substrate at higher temperatures when other resources are not limiting growth (Pendall et al. 2004). Temperature is a primary controller of aboveground litter decomposition, and decomposition rates are much greater at higher than at lower temperatures (Kirschbaum 2006). Forest woody detritus C pools may, thus, be CO_2 sources if temperatures increase (Woodall and Liknes 2008). In both alpine and boreal forests, the short growing season due to cool temperatures limits the annual decomposition rate (Jandl et al. 2007a). Thus, warming is likely to increase decomposition rates in these forests (Pendall et al. 2004). Litter decomposition, however, may respond variably to warming, and no general response pattern can be identified (Rustad et al. 2001). In particular, the temperature response at later stages of decomposition is a matter of discussion (Davidson and Janssens 2006; Hyvönen et al. 2007). Also, the majority of warming experiments indicate short-term responses of litter decomposition. In the long-term, however, acclimation processes may result in negligible effects of increased temperatures on decomposition (Norby et al. 2007). Otherwise, Conant et al. (2008) hypothesized that future loss of litter C at increased temperatures may be greater than previously thought as more resistant litter OM shows greater temperature sensitivity than labile litter OM. Furthermore, shifts in plant growth form composition may also affect leaf litter decomposition rates. Specifically, in cold biomes the C loss by direct warming enhancement of leaf litter decomposition may be partly compensated by the warming-induced expansion of shrubs with recalcitrant litter (Cornelissen et al. 2007).

In contrast to aboveground decomposition, temperature plays only a secondary role in controlling root decomposition (Silver and Miya 2001). Roots grown at elevated temperatures, however, may have slightly increased substrate quality but this has probably negligible effects on longer-term decomposition (Chen et al. 2008). Otherwise, soil warming may result in more rapid root mortality but plants rapidly acclimatize to warmer temperatures. Thus, root decomposition at higher temperatures may not be altered in the long-term although the additional photosynthate allocated to roots may result in higher C flux into soils (Johnson et al. 2006). However, indirect effects of soil warming on decomposition may be mediated by shifts in microbial community composition. The composition potentially shifts toward fungi dominance in response to warming and this may affect decomposition pathways (Luo 2007). One reason is that the temperature response of fungi is lower than that of bacteria (Pendall et al. 2004). On the other hand, the significance of warming-induced soil faunal community changes for litter decomposition and, in particular, for cold biomes is not known (Aerts 2006). However, warmer and wetter climates may positively feedback on soil animals and result in accelerated decomposition rates and C release, in particular, in high latitude ecosystems where relatively undecomposed OM has accumulated (Wall et al. 2008).

The rates of surface litter decomposition and, in particular, mass loss are positively correlated with soil moisture (Saxe et al. 2001). As soil moisture is affected by ACC-induced increased precipitation but also by drying in other regions, mass loss from surface litter may increase or decrease depending on changes in rainfall patterns. Only if there is sufficient soil moisture, warming may lead to increased litter decomposition rates in cold biomes (Aerts 2006). In Mediterranean forests, on the other hand, more frequent summer droughts may inhibit decomposition (Jandl et al. 2007a). Independent of temperature, decomposition is more restricted by high than low soil moisture as oxygen limitations to decomposition cause litter C accumulation in wet soils such as those in boreal peatland forests (Chapin et al. 2002; Jandl et al. 2007a). In contrast, roots and their community of decomposers are well buffered from extremes in precipitation relative to surface litter (Silver and Miya 2001). Thus, effects of ACC on the rainfall patterns may probably have only small effects on root decomposition.

If surface litter decomposition in forest ecosystems is affected by photodegradation, ACC-related changes in surface solar fluxes may also affect litter mass loss and CO_2 production (Section 2.1.1.1). Also, ACC may indirectly alter litter input dynamics by its effects on disturbance regimes such as windthrow, insect outbreaks, and wildfires (Section 2.2.2.2).

2.2.2.5 Soil Organic Matter

Soils contain major C pools in forest ecosystems. Specifically, forest ecosystems store more than 70% of global SOC, and forest soils store about 43% of total forest ecosystem C to 1-m depth (Jobbágy and Jackson 2000; Robinson 2007). A stable SOC distribution establishes in forest soils after more than 6,000 years of soil development (Schlesinger 1990). Plant, microbial and animal residues are primary and secondary resources for SOC (Kögel-Knabner 2002). The SOM dynamic may be influenced by compounds in decomposing plant residues such as condensed tannins (Schweitzer et al. 2008). This may affect SOC dynamic in the entire soil profile as tannin markers have been found throughout forest soil profiles (Nierop and Filley 2008). However, tannins undergo structural changes with time but the degradation mechanisms are virtually unknown.

The boundary between residual C and C stored in SOM is smooth. After the decomposition of labile C compounds (e.g., carbohydrates), more recalcitrant compounds are left in the remaining residues (Chapin et al. 2002). Thus, the decomposition of detritus and SOM declines with time. In models that describe SOM dynamics typically two to five conceptual C pools are defined by their specific turnover rate (Shibu et al. 2006). The active or labile SOM pool has a turnover time of 1–2 years in temperate climate and loses a quarter to two-thirds of initial C (von Lützow et al. 2006). The intermediate pool turns over in 10–100 years and causes a total loss of 90% OM. The long-term stabilization of C in SOM occurs in the passive or slow or refractory SOM pool with very slow decomposition rates and turnover times of 100 to >1,000 years (Falloon and Smith 2000). Stabilization of SOM is the result of several mechanisms (von Lützow et al. 2006). The most important mechanisms are the selective degradation (i.e., the accumulation of biochemically recalcitrant compounds relative to labile ones), the incomplete combustion of plants (i.e., the accumulation of

BC or char or soot), the encapsulation into stable microaggregates, and the association of OM with phenols, amids, metal ions and mineral surfaces (Amelung et al. 2008; Krull et al. 2003). Within soil microaggregates, the C functional group composition is complex at the nano- and micrometer scale (Lehmann et al. 2008b). The relative importance of the various mechanisms for SOM stabilization is under discussion (e.g., Ekschmitt et al. 2008; Marschner et al. 2008). Specifically, the selective preservation of recalcitrant primary biogenic compounds (e.g., lignin, lipids, and their derivatives) is probably not a major SOM stabilization mechanism. For example, high MRT of lipids have been observed only under acid conditions and old SOM fractions are mainly found in association with soil minerals (Amelung et al. 2008). Carbon stabilized on mineral fractions has a distinct composition, i.e., relatively high portion of polysaccharides and proteins and little lignin moieties (Grandy and Neff 2008). The stabilized fraction is more related to microbial-processed OM than to plant-related compounds. Thus, whether lignin belongs to the most stable components of SOM is under discussion (Dignac et al. 2005; Hofmann et al. 2009). The long-held paradigm that lignin is inherently more stable than most chemical soil fractions is likely to be incorrect (Bol et al. 2009)

BC and fossil C are probably the only not mineral-associated SOM components that may be persistent in soils (Marschner et al. 2008). The first direct estimation of BC decomposition rates in soil indicates MRTs of BC in the range of millennia (Kuzyakov et al. 2009). Furthermore, ^{14}C ages of up to 5,040 years BP have been reported for BC (Schmidt et al. 2002). However, a fraction of the condensed aromatic structures in soils detected as BC may also be produced in situ without fire or charring, indicating the uncertainities associated with BC quantification (Hammes et al. 2007; Glaser and Knorr 2008). Furthermore, recent studies indicate that BC cannot be assumed chemically recalcitrant in all soils (Hammes et al. 2008). Specifically, the C stabilization potentials of soils are site- and horizon-specific (von Lützow et al. 2008). In particular, the processes causing substrate limitations to microbial decomposers and those causing spatial inaccessibility of SOM to decomposers in the passive pool need to be better understood for improving SOM stabilization models.

Traditionally, SOM fractions have been chemically separated by extraction and fractionation into humic acids, humin and fulvic acids (Stevenson 1994). Humic acids are predominant in forest soils, and are large, relatively insoluble compounds with extensive networks of aromatic rings and few side chains, and originate in particular from the abundant plant-derived phenolics (Chapin et al. 2002). Humin is also relatively insoluble but contains more long-chain nonpolar groups derived from cutin and waxes than humic acids. Finally, fulvic acids are more water soluble, and bind readily other organic and inorganic materials. The various humus fractions, however, remain largely uncharacterized at the molecular level (Hedges et al. 2000). In particular, the fractions have been defined operationally in terms of the methods used for their extraction and isolation (Kelleher and Simpson 2006). Thus, humic substances may not be a distinct chemical category as modern multi-dimensional nuclear magnetic resonance (NMR) studies indicate. Instead, the vast majority of operationally defined soil humic material may be a very complex mixture of faunal, microbial and plant biopolymers and their degradation products (Kelleher and Simpson 2006; Lorenz et al. 2009). Proteins, lignin, carbohydrates and aliphatic biopolymers are the major components of the mixtures. However, only analyses of biomarkers are suitable to identify the sources of SOM (Amelung et al. 2008). The contribution of the microbial biomass to SOM, for example, is probably much higher than the 1–5% reported previously (Simpson et al. 2007). Fungal dominated soil food webs may be characterized by higher SOM concentrations than bacterial dominated (van der Heijden et al. 2008). Gram-positive bacteria, in particular, consume both fresh and older SOM (Amelung et al. 2008).

Data on SOM composition in forest ecosystems are not sufficient to generalize effects of vegetation cover on SOM. As determined by pyrolysis-GC/MS, for example, the chemistry of NaOH-extractable SOM in forest topsoil horizons is primarily controlled by the biome (Vancampenhout et al. 2009). Otherwise, the organic C forms in bulk forest soils are remarkably similar as identified by synchrotron-based near-edge x-ray spectromicroscopy (Lehmann et al. 2008b). Also, bulk SOM composition detectable by NMR spectroscopy is remarkably similar among surface soils (Mahieu et al. 1999). Specifically, O-alkyls constitute 46% of total C, alkyls 26%, aromatics 19%, and carbonyls 9%. Forest soils with a low C content contain smaller proportions of C in O-alkyls but more in the other functional groups. Forest soils contain large proportions of recently added OM that is very variable in composition but forest ecosystems may possess also a wide variety of synthetic pathways for SOM formation. In contrast, NMR studies on deep SOM in forests, and molecular-level analysis of sub-soil SOM are scanty. The few published studies, however, indicate that with depth O-alkyl decreases whereas alkyl C increases, that suberin-derived C increases vs. cutin-derived C, that lignin may not be stabilized at depth, and that the proportions of char C or BC may increase with soil depth (Lorenz et al. 2007, 2009). Aside direct input, aliphatic macromolecules contributing to alkyl C signals may be formed in soils upon non-enzymatic polymerization of low-molecular-weight lipids (de Leeuw 2007).

Aside non-destructive methods like NMR and chemical fractionation methods, physical fractionation methods are used to divide SOM into pools of functional relevance, that is, uncomplexed OM, primary organo-mineral complexes and secondary organo-mineral complexes (Christensen 2001). The physically uncomplexed OM is an intermediate between animal, plant and microbial residues, and physically stabilized OM (Gregorich et al. 2006). In forest soils, up to 40% of the OM in surface horizons may be uncomplexed OM because of large litter inputs (Christensen 2001). Mineral particles stabilize SOM, and the importance of organo-mineral interactions for SOM stabilization in the passive pool increases with soil depth (von Lützow et al. 2008). Similar to humic substances, however, the chemical nature of organo-mineral associations is a matter of debate (Chenu and Plante 2006; Kleber et al. 2007).

The forest SOC pool is determined by the balance between C inputs from animals, plants and microorganisms, and the C release during decomposition (Section 2.3.2). The relative contribution of the C pool in forest soils to changes in the forest ecosystem C pool remains, however, uncertain (Jandl et al. 2007b).

Furthermore, the conclusion that forest SOC pools are in equilibrium is questionable as very old soils still accumulate C (Wutzler and Reichstein 2007). Thus, the C storage capacity of forest soils may be higher than currently assumed. Otherwise, the SOM turnover depends on the C composition, site conditions, and soil properties (i.e., texture, soil moisture, pH, nutrient status; Jandl et al. 2007a). In contrast, the chemical composition of SOM detected by NMR is mainly derived from plant inputs and not caused by differences in soil inorganic chemistry, mineralogy, physical state, microbiology, and environmental conditions (Mahieu et al. 1999). Similar to litter, SOM accumulates in a relatively un-decomposed state in forest soils with low oxygen availability or low temperatures (Chapin et al. 2002). For example, some of the largest C stores in wetlands and tundra are rich in labile OM that quickly decomposes when the environmental limitations to decomposition disappear.

In the past, studies on forest SOC storage focused mainly on the forest floor and surface horizons. Tree roots, however, may grow deep into the mineral soil (Canadell et al. 1996). A higher percentage of macromolecular compounds derived from roots than those derived from needles persist in the most stable SOM fraction (Bird et al. 2008). Roots and their associated microorganisms, but also bioturbation of OM by animals and translocation of DOC are C input pathways to the SOC pool in deeper mineral soil horizons (Lorenz and Lal 2005). Although this pool may be smaller than the surface SOC pool, the higher contribution of stabilized and recalcitrant fractions as indicated by its ^{14}C age, high alkyl C contents, and high proportions of SOC associated with soil minerals highlight its importance to C sequestration in forest ecosystems. Deep forest soils are, thus, potential C reservoirs (Jobbágy and Jackson 2000; Diochon and Kellman 2009). Tropical forests, in particular, have distinctively deeper SOC profiles compared to temperate and boreal forests. The stability of the subsoil SOC pool is, however, a matter of debate. Spatial separation of decomposers or exoenzymes and substrate may control C dynamics in the subsoil (Salomé et al. 2009). Otherwise, the stability of SOC in deep soil layers may be only maintained as long as fresh C is lacking. The subsoil decomposer activity is probably stimulated by fresh C supply ('microbial priming'), and increased microbial activity may cause the loss of ancient buried SOC (Fontaine et al. 2007). In particular, fungi may contribute to the priming effect and are, therefore, important for long-term changes of SOM decomposition (Blagodatskaya and Kuzyakov 2008). Thus, increased C inputs from roots can result in a net loss of SOC through the interaction with the soil biota, and this rhizosphere priming may have long-term implications for the SOC balance in forests (Dijkstra and Cheng 2007).

Climate Change and Soil Organic Matter

The increasing atmospheric CO_2 concentrations may not directly affect forest SOC dynamics as soil CO_2 concentrations in the pore space of active soils are between 2,000 and 38,000 ppm and, thus, much higher than the atmospheric concentration

(Drigo et al. 2008). Indirectly, however, increased above- and belowground C input, and the altered litter composition due to CO_2 effects on plant production may alter SOM accumulation and composition (Section 2.2.2.2; Pendall et al. 2004; Jastrow et al. 2005). The additional C, and, in particular, the additional labile C may also promote the decomposition of the native SOC (Hyvönen et al. 2007; Taneva and Gonzalez-Meler 2008). By this priming effect, higher inputs of labile C enhance the activity of the decomposer community and, thus, promote SOM degradation (Pendall et al. 2004; Hoosbeek et al. 2007). However, these and other indirect effects of elevated CO_2 concentrations on SOC pools in forest ecosystems have been less studied (Hyvönen et al. 2007).

The CO_2-induced priming effects on SOC depend on soil texture and climate (Langley et al. 2009). The changes in soil microbial composition and activity due to elevated CO_2 may also lead to a decline in SOC (Carney et al. 2007). For example, the effects of global change on mycorrhizal fungi may control the loss of SOC pools (Talbot et al. 2008). However, compared with the life cycle of a forest and the complexity of processes involved the FACE studies are often not long enough to fully encompass the SOC changes (Lukac et al. 2009). The fate of the extra biomass from increases in NPP and increases in C storage in forest vegetation in FACE studies in the forest floor or mineral soil is less clear (Hoosbeek and Scarascia-Mugnozza 2009). Theoretically, after 6–10 years of CO_2 fumigation SOC changes may be detectable in FACE experiments (Smith 2004). In fact, after 4–6 years of growth at elevated CO_2 forest SOC increased in several studies (Jastrow et al. 2005; Hoosbeek et al. 2007). The effect of elevated CO_2 on net SOC accumulation increased, in particular, with the addition of N fertilizers (Hungate et al. 2009). Also, after 9 years experimental CO_2 fumigation, enhanced litterfall inputs combined with forest floor decomposition rates similar to the control caused an increase in the forest floor C pool in a pine forest by sequestering 30 g C m^{-2} $year^{-1}$ more than the control (Lichter et al. 2008). In contrast, mineral soil C pools to 1.5-m depth were not different from the control. Furthermore, the signatures for lignin and fatty-acid derived compounds in the forest floor and 0–15 cm depth were also not different from the control. Otherwise, physical fractionation indicated that the amount of stabilized SOC in 0–10 cm increased under FACE treatment (Hoosbeek and Scarascia-Mugnozza 2009). Any generalizations for global forest ecosystems are, however, not possible, in particular as data from tropical FACE experiments are lacking (Lewis 2006). The sometimes observed deeper root distributions at elevated CO_2, however, may be important for SOC dynamics and sequestration (Johnson et al. 2006). In summary, elevated levels of atmospheric CO_2 result in increased C allocation belowground accompanied by greater soil CO_2 efflux (Lukac et al. 2009). Whether this will result in increased SOC sequestration in temperate forests depends on soil fertility, temperature and moisture but present FACE experiments need to be run for much longer to account for effects on SOC. Less well studied are boreal and tropical forests.

The SOC pool may increase at higher temperatures due to increases in above- and belowground growth in forest ecosystems, but acclimation processes to warming may diminish this increase (Saxe et al. 2001; Section 2.2.1.6). The knowledge to link the short-term temperature dependence of SOC decomposition to a long-term one is, however, scanty (Smith et al. 2008). For example, topsoil SOC pools in British woodlands did not significantly change over 30 years despite increase in temperature (Kirkby et al. 2005). However, forest soils respond more strongly directly to warming than soils under other forms of land use (Jandl et al. 2007a). In the long term, SOC may probably be lost at higher temperatures as decomposition is more pronounced than the stimulation of productivity (Rustad et al. 2001). Some studies, however, have shown that only the smaller labile SOC pool is exhausted by warming and SOC losses are only temporary effects (Davidson and Janssens 2006). A significant SOM fraction may therefore remain at higher temperatures, but temperature effects on SOM composition are hardly studied (Jandl et al. 2007a). However, future warming could alter SOM composition by accelerating lignin degradation and increasing leaf-cuticle-derived C sequestration in forest soils with low clay content (Feng et al. 2008).

Although SOC in coniferous forests is supposedly lower in quality, increases in specific rates of soil heterotrophic respiration upon warming did not differ from those in deciduous forest soils (Paré et al. 2006). Current constraints on SOM decomposition in wetlands, peatlands and permafrost soils are most likely to change by climatic disruptions (Davidson and Janssens 2006). Decomposition rates of OM in forest mineral soils, however, are remarkably constant across a global-scale gradient in mean annual temperature (Giardina and Ryan 2000). The temperature sensitivity of SOM decomposition may increase with decreasing SOM lability (Conant et al. 2008). The temperature dependence and resistance of SOC pools to decomposition is, however, not well known (Smith et al. 2008). In particular, predicting SOC response to long-term global warming from soil-warming experiments is probably not possible. A lack of fresh C supply may prevent the decomposition of the SOC pool in deep soil layers in response to future changes in temperature (Fontaine et al. 2007). Thus, the forest floor is expected to respond to ACC differently from the mineral soil (Smith et al. 2008). The temperature sensitivity of SOM and, thus, the magnitude of the positive feedback of decomposition due to global warming is probably smaller than currently assumed as including realistic pools of BC significantly reduces projected CO_2 emissions (Lehmann et al. 2008a). In summary, the temperature dependence of SOM decomposition remains a topic of debate (Kirschbaum 2006; Smith et al. 2008)

Increasing soil moisture from ACC-induced alterations of rainfall patterns increase SOM decomposition in dry forests. Otherwise at excess moisture levels decomposition is generally retarded due to reduced oxygen availability. This may move the SOC pool towards a new dynamic equilibrium. The SOM composition, on the other hand, seems to be less variable in relation to changes in soil moisture regime (Mahieu et al. 1999). Indirectly, above- and belowground forest plant productivity may be enhanced at more favorable moisture conditions, but anatomical and morphological changes may also occur (Sections 2.2.1.6 and 2.2.2.4), which may also affect the SOC dynamics. Alterations of the wildfire regime due to ACC have also the potential to affect SOC sequestration in the form of BC (Czimczik and

Masiello 2007). Recurring disturbances of forest ecosystems induced by ACC that alter litter dynamics in the long term have also the potential to affect SOM levels and composition. The ACC-induced shifts in tree species composition may alter SOM dynamics (Hobbie et al. 2007).

2.2.2.6 Soil Inorganic Carbon

Forest soils may contain SIC as carbonates, but the total size of this C pool is not exactly known (Schlesinger 2006). About 210 Pg of SIC is probably stored in forests, but only about 22% of the global SIC pool resides in forest soils (Eswaran et al. 2000). However, temperate forest soils may store more SIC than SOC. In arid and semiarid soils with no forest cover, up to ten times more SIC than SOC is stored. The turnover time of soil carbonates is about 85,000 years, and the SIC pool is much less dynamic and effective than SOC in the storage of atmospheric CO_2 (Schlesinger 2006). SIC comprises two components, lithogenic inorganic carbon and pedogenic inorganic carbon (West et al. 1988). The lithogenic carbonates are derived from parent material whereas pedogenic carbonates are formed during chemical weathering. Formation of pedogenic carbonates occurs, for example, when plant activity (i.e., CO_2 production by respiration) increases soil DIC or bicarbonate production (Eq. (2.7) silicate weathering, plagioclase) followed by carbonate precipitation (Eq. (2.8)) during seasonal droughts. The process of weathering is actively amplified by plants and their associated fungal mycorrhizae, by lichens, and various free living soil organisms, and consumes CO_2 (Lenton and Britton 2006). By its influence on the release of CO_2, Ca^{2+} and Mg^{2+} through microbial acitivity, increase in SOC results in increased weathering and precipitation (Bronick and Lal 2005).

$$\begin{aligned} &CaAl_2Si_2O_8 + 3H_2O + 2CO_2 \rightarrow Ca^{2+} \\ &+2HCO_3^- + Al_2Si_2O_5(OH)_4 \end{aligned} \tag{2.7}$$

$$Ca^{2+} + 2HCO_3^- \rightarrow CaCO_3 \downarrow + H_2O + CO_2 \tag{2.8}$$

The Ca content in pedogenic carbonates may derive from Ca deposition by rain and dust after weathering in upwind areas (Goddard et al. 2009). Pedogenic carbonate precipitation may also bind OM (Landi et al. 2003). The soil DIC resulting from natural chemical weathering of carbonates and silicate minerals may be leached from upper soil layers and precipitated in deeper layers. Thus, SIC contents in forests increase with increase in soil depth (Mi et al. 2008). However, some of the soil DIC (dissolved CO_2, H_2CO_3 and HCO_3^-) may be taken up by the tree root, transported within the tree, and fixed either photosynthetically or anaplerotically (dark fixation). Anaplerotic reactions 'fill up' intermediates that need to be replenished because they are consumed in some biosynthetic reactions (Berg et al. 2007). The incorporation of CO_2 into plant tissue C via phosphoenolpyruvate carboxylase is an anaplerotic reaction. For example, soil DIC contributes a small percentage (<1%) of C gain in *Pinus taeda* seedlings (Ford et al. 2007). Thus, soil

DIC is likely to contribute only a small amount of C to forest trees but may be important in C fixation in newly formed stems and fine roots, and ectomycorrhizal roots assimilationg NH_4^+. Otherwise, the large flux of CO_2 from roots through xylem has been recently reported which may be part of a recycling mechanism in trees to compensate for respiratory CO_2 losses (Aubrey and Teskey 2009).

Soome soil DIC may also be transported from forest ecosystems through fluvial systems into the oceans (Hartmann and Kempe 2008). During this transport, little re-precipitation (Eq. (2.8)) occurs (Hamilton et al. 2007). In the oceans, however, carbonate precipitation returns CO_2 to the atmosphere. In contrast to CO_2 consumed by silicate minerals, carbonate dissolution is, thus, not a geologic long-term C sink (Hartmann and Kempe 2008).

Climate Change and Soil Inorganic Carbon

Increases in atmospheric CO_2 concentrations and temperature may change the rates of carbonate mineral weathering, subsequent carbonate re-precipitation, and DIC transport (Hamilton et al. 2007). Both soil CO_2 and soil DIC are expected to increase under scenarios of rising atmospheric CO_2 concentrations (Andrews and Schlesinger 2001). Aside carbonate weathering rates, silicate weathering rates may also increase due to increasing atmospheric CO_2 and surface temperatures (Lenton and Britton 2006). As the balance of the timescale for DIC transport to streams, rivers and oceans is affected, more precipitation of carbonate minerals and, thus, C sequestration in soils may occur. Furthermore, re-precipitation may be enhanced by increasing frequencies of seasonal droughts caused by ACC. On the other hand, in forest soils that receive more rainfall the Ca deposition may be enhanced, but also the export of DIC to streams, rivers and the oceans, and, thus, the SIC balance may be altered. By changes in climate, an increased downward flux of leached pedogenic carbonates may occur (Lapenis et al. 2008). Redeposition of this DIC deeper in the soil may contribute significantly to the total net C sink although carbonate precipitation in the oceans is the main C sink. Overall, enhanced carbonate and silicate weathering due to ACC accelerates the recovery from fossil fuel CO_2 perturbations (Lenton and Britton 2006).

2.3 Carbon Efflux from Forest Ecosystems

In the previous sections, the principal pathways for C entering above- and belowground forest ecosystem C pools were discussed. The processes that are responsible for C losses from forests are the focus in the following section, and how ACC may affect them (Fig. 2.1). The processes include autotrophic respiration by the living parts of photosynthetic active organisms, that is, the primary producers. Further C losses occur during decomposition of dead organic matter by the heterotrophic activity of animals and microorganisms. Through hydrological and erosional C fluxes, and by animal movement further C efflux occurs from forest ecosystems. Forest C losses following disturbances are discussed in Chapter 3.

2.3.1 *Gaseous Carbon Efflux from Plants*

Photosynthesis by trees is the major pathway for C entering forest ecosystems and respiration the major pathway for C loss. Large amounts of CO_2 are released, in particular, from trees through autotrophic respiration in leaves, stems, roots, storage and reproductive organs. Further but minor C export from living trees occurs by emissions of biogenic volatile organic C (BVOC) compounds, oxygenated hydrocarbons, CH_4 and CO.

2.3.1.1 Carbon Dioxide Release During Photosynthesis

Already in the chloroplasts, photosynthetically fixed C is lost by photorespiration (Berg et al. 2007). This process recycles C atoms that entered the wasteful oxygenase reaction pathway of RUBISCO (Section 2.1.1.1). Thus, a fraction of the C fixed during photosynthesis is lost to the atmosphere as CO_2. Furthermore, C is recycled in the mitochondria through synthesis of serine, a potential precursor of glucose (Berg et al. 2007). In total, 20–40% of the C fixed by C_3 photosynthesis is respired by photorespiration (Chapin et al. 2002). Additional CO_2 is respired to create C_6 sugars for the Calvin cycle (Waring and Running 2007). From the C_6 sugars and C_3 sugars, a C_5 sugar is synthesized for the regeneration of the principal CO_2 acceptor ribulose 1,5 bisphosphate. In total, a third of the photons absorbed during photosynthesis are consumed to reverse the consequences of the wasteful oxygenase reaction (Heldt 2005).

Climate Change and Carbon Dioxide Release During Photosynthesis

The increased ratio of CO_2 to O_2 at elevated CO_2 concentrations suppresses the photorespiration in trees and, thus, CO_2 losses during photosynthesis (Urban 2003). This is probably a main reason for increases in CO_2 assimilation under short-term exposure to elevated CO_2. The photorespiration losses in C_4 and CAM species, however, are less than those in C_3 species as the Calvin cycle is spatially isolated in C_4 photosynthesis and temporally in CAM (Section 2.1.1.1; Urban 2003; Ainsworth and Rogers 2007). Photorespiration serves as energy sink for stress protection and by providing glycine for the synthesis of glutathione involved in stress protection (Urban 2003). Thus, heat, drought and light stress induced by ACC may promote photorespiration and CO_2 losses during photosynthesis.

2.3.1.2 Carbon Dioxide Release Due to Respiration by Autotrophs

Autotrophs are organisms that can grow using CO_2 as a sole C source (Madigan and Martinko 2006). Among them are plants, algae, cyanobacteria, purple and green bacteria, and some bacteria and archaea which do not obtain energy from light (Thauer 2007). Plants are the most important photoautotrophic organisms in forests and release large amounts of CO_2 into the atmosphere (Trumbore 2006). Between 49% and 70% of GPP is released by autotrophic respiration in forest ecosystems (Luyssaert et al. 2007). In contrast, the contribution of respiration by microbial autotrophs to CO_2 efflux from forests is of minor importance. Algae, for example, are active mainly on hydromorphous and flooded soils whereas chemolithotrophs only marginally contribute to CO_2 efflux from well-aerated soils (Kuzyakov 2006).

Respiration by photoautotrophs is defined as the CO_2 production by all living parts of photosynthetically active organisms, that is, the primary producers (Section 2.1.1.1; Chapin et al. 2006). Photosynthesis and the CO_2 release during plant leaf respiration are generally strongly linked (Hartley et al. 2006). The two major components of autotrophic respiration are: (i) maintenance respiration of living tissues, and (ii) respiration associated with synthesis and growth of new tissues (Waring and Running 2007). Maintenance respiration may account for half of total plant respiration whereas the other half may be associated with growth and ion uptake, but these proportions vary (Chapin et al. 2002). Respiration of aboveground woody tissues, in particular, may account for up to 30% of annual ecosystem respiration. During respiration by autotrophs, organic substrates are oxidized to CO_2 and water accompanied by the production of energy-rich ATP and reducing power (i.e., NADH and NADPH). The respiration by autotrophs plays an important role in the C balance of individual cells, whole-plants and ecosystems (Gonzalez-Meler et al. 2004). While in temperate forests respiration by autotrophs seems to be strongly linked to photosynthesis rates, this may not be the case in boreal and tropical forests (Trumbore 2006). In particular, half of the CO_2 assimilated annually in temperate forests is released back to the atmosphere by plant respiration but up to 75% in boreal and tropical forests. The reason for this difference is, however, not yet understood. The potential role of boreal and tropical forests for sequestering C at elevated photosynthetic rates may, thus, be overestimated (Trumbore 2006).

Already in the plant chloroplast, metabolites from the Calvin cycle are used for growth and synthesis respiration, and CO_2 is released, for example, during biosynthesis of lipids. The major proportion of autotrophic respiration for maintenance, growth and synthesis occurs, however, in mitochondria in non-photosynthetic plant organs (Chapin et al. 2002). Mitochondrial respiration is the main source of CO_2 release into the atmosphere by plants (Rennenberg et al. 2006). The mitochondrial respiration in light is lower than in the dark. Among the myriads of synthesized plant compounds, the highest C costs incur for the synthesis of proteins, tannins, lignin, and lipids. For long-lived woody plants maintenance respiration increases with stand age (Schlesinger 1997). Generally, growth respiration consumes on average 25% more C than accumulates in new tissues. Similar C costs occur also in large trees whereas for smaller trees synthesis costs may rise to 35% of the C sequestered in biomass (Waring and Running 2007).

Respiration for maintenance, growth and synthesis occurs also in plant roots. With respect to forest C sequestration, the CO_2 loss by tree root respiration needs to be quantified as root-associated C fluxes are probably the major pathway to SOM. The separation of respiration by autotrophic roots from rhizomicrobial respiration, however, is challenging (Chapin et al. 2006; Kuzyakov 2006). Furthermore, not all root-respired CO_2 may diffuse into the soil atmosphere but a substantial portion may flow from tree roots through stems (Aubrey and Teskey 2009). Thus, belowground respiration may have been grossly underestimated. Also, poorly known is the amount of C allocated to root respiration vs. exudation and root growth (Trumbore 2006). To estimate the CO_2 flux from root-derived C, tree girdling in forests is used to interrupt the flow of assimilates from the leaves to the roots (Högberg et al. 2001). Using this approach it has been shown that up to 56% of CO_2 was derived from the rhizosphere in a Scots Pine (*Pinus sylvestris* L.) forest whereas 44% originated from SOM. On average, root and rhizosphere respiration amounts to 50% of belowground respiration in forest ecosystems (Hartley et al. 2006). Rhizosphere priming effects and root decomposition, however, do not allow the separation between respiration by the autotrophic roots and the respiration by the heterotrophic soil animals and microorganisms (Kuzyakov 2006; Subke et al. 2006). Furthermore, the total C flux between tree roots and their mycorrhizal partners has not yet been quantified in situ (Millard et al. 2007).

Climate Change and Respiration by Autotrophs

The respiratoy responses to elevated CO_2 are variable but the mechanism of the respiratory response is insufficiently understood (Gonzalez-Meler et al. 2004; Leakey et al. 2009). In general, the plant-specific respiration rates are not reduced. In FACE experiments higher night-time respiration rates by upper vs. lower canopy leaves of American Sweetgum (*Liquidambar styraciflua* L.) were observed, but also little direct effects of elevated CO_2 on *Pinus taeda* L. leaf tissue respiration (Hyvönen et al. 2007). Canopy respiration does, however, not increase proportionally to increases in biomass in response to elevated CO_2 (Gonzalez-Meler et al. 2004). Not well understood is, in particular, the response of wood respiration (Moore et al. 2008). The effects of elevated CO_2 on stem respiration, for example, are variable as higher stem CO_2 efflux rates, no consistent changes and decreased stem CO_2 efflux has been reported. However, previous studies may have overestimated the effects of elevated CO_2 on autotrophic respiration in tree stems. In particular, the variance in stem CO_2 efflux results partially from root and microbial respiration as the source of CO_2 dissolved in soil water which is taken up by trees and transported through the vascular system (Moore et al. 2008). Similar to the CO_2 release from SOM decomposition (Section 2.2.2.5), a direct effect of elevated CO_2 on root respiration is unlikely (Urban 2003). A larger proportion of respiration, however, takes place in the root system at elevated CO_2 as C partitioning to roots may be enhanced (Section 2.2.1.6; Hartley et al. 2006). In summary, the role of plant respiration in augmenting the sink capacity of forest ecosystems remains uncertain. Specifically, only when C_3 forest plants respond with greater foliar respiration at elevated CO_2, the plant C balance will be reduced but photoassimilate export to sink tissues enhanced (Leakey et al. 2009). Otherwise, C_4 plants may not display enhanced respiration at elevated CO_2.

Plant respiration increases exponentially with temperature in its low range, reaching a maximum at an optimal temperature, and then declines (Luo 2007). Thus, increasing temperatures may affect plant respiration in the short-term (Hartley et al. 2006). In particular, leaf respiration is extremely affected by temperature, and exponentially increases up to 40°C (Rennenberg et al. 2006). Moderately high temperatures (20–40°C), however, do not affect mitochondrial respiration in light in C_3 trees. But at extreme temperatures mitochondrial respiration in light is rapidly increasing. In the long-term, however, respiration in leaves and roots often acclimates to temperature increases (Hartley et al. 2006). Thus, mitochondrial respiration in light may not be involved in adjustments of the CO_2 release at moderately high temperatures. Furthermore, the effects of increasing soil temperatures on tree root growth are variable (Giardina et al. 2005). Belowground respiration, however, is relatively temperature insensitive (Hartley et al. 2006).

The leaf respiration in the dark is insensitive to rapidly developing water stress and, thus, probably not affected by ACC-induced drought (Rennenberg et al. 2006). The belowground autotrophic respiration in forests, however, declines during drought (Trumbore 2006). Excess soil moisture, on the other hand, may cause morphological and physiological adaptations in tree roots to water-saturated conditions (Giardina et al. 2005).

2.3.1.3 Efflux of Other Gaseous Carbon Compounds

Although CO_2 fluxes are two orders of magnitude higher than volatile organic C fluxes, the reactivity of particular molecules emitted by plants may have direct effects on element interactions in the atmosphere and the ultimate fate of C (Lerdau 2003). Specifically, volatile compounds are lost from foliage associated with leaf expansion, stress physiology, and herbivore defense. Global estimates of volatile losses are in the range of 1 Pg C year^{-1}. On an annual time scale almost all volatile C ends up as CO_2 or in SOM but because of atmospheric transport processes the release as CO_2 can occur away from surface emissions (Lerdau 2003; Ciais et al. 2008). The emission of volatiles is either actively regulated or solely related to vapor pressure and resistance to diffusion. Most important regulated VOCs in terms of abundance and impacts on the atmosphere are isoprene and methyl butenol, whereas mono- and sesquiterpenes, and methanol are most important unregulated VOCs (Lerdau 2003).

Forest plants release small amounts of BVOC compounds (isoprene, monoterpenes, sesquiterpenes), short-chain oxygenated VOCs (formic and acetic acids, acetone, formaldehyde, acetaldehyde, methanol, ethanol), CH_4 and CO (Chapin et al. 2006; Duhl et al 2008; Seco et al. 2007). The global flux of these compounds from ecosystems is, however, small compared to photosynthesis or respiration but significant compared to the net C balance of an ecosystem (Ciais et al. 2008). Specifically, BVOC emissions constitute a substantial loss of biologically fixed C from the terrestrial biosphere, significant in relation to NEP and of the same order of magnitude as NBP (Kesselmeier et al. 2002).

BVOCs have effects on the biological, chemical and physical components of the Earth system (Laothawornkitkul et al. 2009). The main BVOC constituents are isoprenoids. Forest plants are the most important isoprenoid emitters (Rennenberg et al. 2006). While monoterpenes may account for one quarter of BVOC emissions, isoprene, the most abundant isoprenoid, may account for half of the total BVOC emission. Freshly fallen needle litter and roots of conifers, in particular, contain relatively high concentrations of monoterpenes which may affect C cyling processes in forest soils (Ludley et al. 2009). Otherwise, the sesquiterpene emission rates cover a wide range of values among tree species (Duhl et al. 2008). The landscape and global sesquiterpene fluxes are, however, highly uncertain due to difficulties associated with the analysis of sesquiterpenes.

Up to 10% of assimilated C may be emitted as isoprene, in particular from trees (Sharkey and Yeh 2001). The two broad-leafed taxa most important in current efforts at commercial forestry, *Populus* and *Eucalyptus*, are both isoprene emitters (Lerdau 2003). Isoprene is de novo synthesized in plant plastids, and may protect plants from ozone damage and from short-term high temperature episodes (Lerdau 2007; Calfapietra et al. 2008). However, the ultimate reason for isoprene production in trees remains unsolved (Heald et al. 2009). The terrestrial biosphere emits 500–700 Tg isoprene annually (Beerling et al. 2007). Isoprene is precursor of tropospheric ozone, can decrease the OH radical concentration, increase the lifetime of CH_4, and contributes to formation of secondary organic aerosols. Thus, isoprene is a major player in the oxidative chemistry of the troposphere. The physiological reason why isoprenoids are formed and emitted by forest trees is, however, a matter of debate (Rennenberg et al. 2006). Aside tree leaves and forest canopies, tree roots might be also a strong source of volatile isoprenoid emissions (Lin et al. 2007).

Trees emit short-chain oxygenated VOCs via the stomata and cuticle (Seco et al. 2007). The emission rates for these compounds increase several-fold under stress but the processes involved in synthesis, emission and uptake of short-chain oxygenated VOCs are not fully understood. Methanol production in leaves, for example, increases exponentially with increase in temperature at least in the short-term (Harley et al. 2007). Furthermore, production rates in growing leaves are an order of magnitude higher than those of mature leaves. Canopy-scale fluxes of methanol have been measured above coniferous forests, pine plantations, deciduous temperate forests and tropical forests. Methanol concentrations above a variety of different ecosystems exceed the emissions of all other VOCs except terpenoids (Harley et al. 2007).

The CH_4 emissions from terrestrial plants under aerobic conditions are reported by a hitherto unrecognized process in leaf tissues (Keppler et al. 2006). Heat and light are important factors stimulating CH_4 production, and methyl groups in plant pectin have been identified as precursors of CH_4 (Keppler et al. 2008). Tropical forests are identified as main CH_4 sources. However, subsequent studies have found only modest CH_4 emissions from terrestrial plants (Butenhoff and Khalil 2007; Dueck et al. 2007). Some woody species, however, emit CH_4 under aerobic conditions (Wang et al. 2008). On the other hand, plant-mediated transport of CH_4 dissolved in groundwater through tree stem surfaces may contribute to CH_4 emissions from terrestrial plants (Terazawa et al. 2007). Thus, experiments on plants grown in controlled conditions and re-analyses of previously published data indicated that plants do not contain a known biochemical pathway to synthesize CH_4 (Nisbet et al. 2009). However, CH_4 may be released by spontaneous breakdown of plant material under high UV stress. Furthermore, plants transpire water containing dissolved CH_4. The components of the global CH_4 cycle are, however, not comprehensively understood (Beerling et al. 2007; Dueck and Van der Werf 2008).

Forest vegetation is also a source of CO emissions (Schlesinger 1997). The formation and emission of CO on or in live plant foliage is the result of direct photochemical transformation and occurs inside of leaves (Tarr et al. 1995). The factors controlling CO emissions are, however, not well known (Guenther et al. 2000).

Climate Change and Release of Gaseous Carbon Compounds

BVOCs mediate the relationship between the biosphere and the atmosphere but ACC may alter these interactions (Laothawornkitkul et al. 2009). For example, experimentally elevated CO_2 levels have either no effects or results in lower isoprenoid emissions (Rennenberg et al. 2006). Isoprene emissions from leaves, however, are often inhibited at elevated CO_2 (Beerling et al. 2007). The isoprene synthesis, thus, appears to be CO_2 sensitive (Calfapietra et al. 2008). This inhibition completely compensates possible increases in isoprene emissions due to increases in aboveground NPP (Monson et al. 2007; Heald et al. 2009). Leaf temperature, however, also increases due to stomata closure at elevated CO_2 and due to increases in ambient temperature by global warming. For protection against damages by high temperatures monoterpene, isoprene and methyl butenol emissions in forests may, therefore, increase (Lerdau 2003; Rennenberg et al. 2006). Isoprene-producing species such as oaks, spruce, and aspen may become more abundant due to ACC than non-emitting species such as maples, birches, hickory, or pine (Lerdau 2007). Otherwise, future projections suggest a compensatory balance between the effects of temperature and CO_2 on isoprene emission (Heald et al. 2009). The stimulation of isoprene emission by heat stress may, however, be diminished by concurrent drought stress (Rennenberg et al. 2006). In particular, exposure to prolonged drought increases isoprene emission from leaves despite reductions in photosynthesis (Monson et al. 2007). However, climate feedbacks of isoprene emissions in a warmer, high CO_2 environment due to enhancements in ozone, organic aerosols and CH_4 may be smaller than previously suggested (Heald et al. 2009).

The temporal monoterpene and sesquiterpene emission variations are dominated by ambient temperatures (Lerdau 2003; Duhl et al. 2008). Thus, monoterpene and sesquiterpene emissions may increase in a warmer climate. This may also be the case for methanol emissions from leaves (Harley et al. 2007). Methanol emissions on longer time scales and their temperature dependence, however, have not been studied.

Plants are not a major source of the global CH_4 production but ACC-induced alterations in soil water content may affect CH_4 production by methanogenic archaea in micro-anaerobic soil environments (Nisbet et al. 2009). This will affect plant up-take of dissolved CH_4 and its release by plant transpiration. Furthermore, temperature increases may promote heat-induced breakdown of plant-cell material with CH_4 as a by-product.

The ACC effects on direct CO emissions by terrestrial plants can only be assessed after the factors controlling its physiological formation have been identified.

2.3.2 Carbon Efflux from Organic Matter

Beside the C release from living plants by autotrophic activity, heterotrophic activity by the decomposer community releases large amounts of C during decomposition of OM (Section 2.2.2.4). Thus, C assimilated by photosynthesis is lost from the forest C pool in gaseous, dissolved and particulate form (Chapin et al. 2006). After GPP, soil respiration is the second most important C flux in forest ecosystems (Davidson et al. 2006). Only small amounts of assimilated C ultimately remain in the forest ecosystem, and are stored in the long-term in biomass and, in particular, in the SOM pool (Schlesinger 1990). The current additional SOC sequestration potential appears to be tiny against the amount of C that can be potentially lost in the future (Reichstein 2007). Thus, the vulnerability of SOC to potential losses requires special attention.

2.3.2.1 Gaseous Carbon Efflux During Organic Matter Decomposition

The efflux of CO_2 is the dominant C output from the belowground environment (Davidson and Janssens 2006). Several respiratory processes contribute to the CO_2 produced at the soil surface, and this makes interpretation of data complicated (Ryan and Law 2005). The recalcitrant SOC pool, however, contributes only a minor portion of soil CO_2 efflux.

Heterotrophs metabolize organic compounds as their C source. Thus, decomposition is a heterotrophic process (Section 2.2.2.4; Madigan and Martinko 2006). Respiration by heterotrophic organisms is a major avenue of C loss from forest ecosystems (Ryan and Law 2005). The soil heterotrophs (i.e., fauna and microorganisms) decompose plant litter, rhizodeposits, faunal and microbial residues, and SOM. Decomposition is accompanied by the release of a large fraction of the detrital C from soil by efflux as CO_2 and CH_4. The photoassimilates fixed in the forest canopy are rapidly and directly coupled to the heterotrophic consumers in the soil.

For example, 50% of all CO_2 released from soil may come from photoassimilates synthesized a few days earlier (Högberg et al. 2008). Litter is also a source for methanol and VOC emissions (Harley et al. 2007; Leff and Fierer 2008). The microbial metabolism, in particular, is likely to be the dominant source of VOC emissions from soil and decomposing litter.

Most of the CO_2 evolved by heterotrophic activity from soils originates from soil microorganisms such as bacteria, non-mycorrhizal and mycorrhizal fungi, and actinomycetes (Kuzyakov 2006). In particular, the mycorrhizal mycelia may represent a substantial component of belowground respiration in forest ecosystems (Heinemeyer et al. 2007). In contrast, the contribution of the macrofauna to CO_2 efflux is usually only a few percent. The autotrophic and heterotrophic components of the soil CO_2 efflux, however, can hardly be separated (Section 2.3.1.2). Especially, the CO_2 release from the heterotrophic metabolism of root exudates is difficult to separate from the autotrophic metabolism (Scott-Denton et al. 2006). Thus, estimated heterotrophic contributions to total respiration in forest soils vary considerably between 40% and 77% (Cisneros-Dozal et al. 2006). Most of the CO_2 production, however, occurs in the surface litter where decomposition is rapid and a large proportion of the fine roots grow (Schlesinger 1997). The surface litter layer is also the reason for large spatial and temporal variability in CO_2 fluxes (Trumbore 2006).

The CH_4 is released from soils in much lower amounts than CO_2 (Denman et al. 2007). During decomposition of OM, methanogenic microorganisms in anoxic or oxygen-free soils release CH_4 (Horwath 2007). Thus, natural wetlands are the most important CH_4 sources but anoxic microenvironments in oxic or oxygen containing forest soils harbor also methanogens (Madigan and Martinko 2006). CH_4 production in soils seems to be highly sensitive to the development of such anaerobic microsites (von Fischer and Hedin 2007).

Climate Change and Efflux of Gaseous Carbon Compounds During Decomposition

As autotrophic respiration cannot be fully partitioned from heterotrophic soil respiration (Kuzyakov 2006), ACC effects on gaseous efflux during decomposition will be discussed for total forest soil respiration (Ryan and Law 2005). Soil respiration is the diffusive flux of CO_2 and CH_4 from the soil boundary into the atmosphere (Pendall et al. 2004). Increasing atmospheric CO_2 concentrations may stimulate C assimilation by trees (Section 2.2.1.6). Thus, larger amounts of photoassimilates transferred to the decomposers may result in higher CO_2 efflux from soils as soil respiration depends on this supply (Pendall et al. 2004; Davidson et al. 2006). In forest FACE experiments, soil respiration is stimulated by elevated CO_2 concentrations, but the responses depend on stand development and tree species composition (King et al. 2004). Also, the magnitude of the CO_2 enrichment effect on soil CO_2 efflux may decline with time since fumigation began (Bernhardt et al. 2006). This can be attributed to the initial stimulation (i.e., priming) of decomposition of native

SOC by larger litter C inputs (Hyvönen et al. 2007). Elevated CO_2 concentrations, however, may stimulate CH_4 emissions from woody wetland ecosystems (Pendall et al. 2004).

Temperature is a key factor that regulates soil respiration and litter decomposition (Norby et al. 2007). Thus, soil respiration in forests is initially increased by experimental ecosystem warming and in soil transfer studies to simulate temperature increases (Rustad et al. 2001; Hart 2006). However, after long-term (i.e., >10 years) soil warming soil respiration rates are not different from the control as was observed at Harvard Forest (Melillo et al. 2002). Acclimatization and/or depletion of substrates may be responsible for declining CO_2 flux from heated soils (Luo 2007). Furthermore, effects of diurnal warming on soil respiration may not be equal to the summed effects of day and night warming which is critical for model simulation and projection of climate-carbon feedback (Xia et al. 2009). At later stages of litter and SOM decomposition the temperature response of both processes is debatable (Davidson and Janssens 2006; Kirschbaum 2006). Labile SOM fractions are probably subject to temperature-sensitive decomposition whereas a significant SOM fraction remains after warming despite intrinsic temperature sensitivity (Davidson and Janssens 2006). For example, lignin degradation was accelerated but leaf-cuticle-derived compounds were increasingly sequestered in a heated forest soil (Feng et al. 2008). Thus, the responsiveness of most SOM decomposition to global warming is probably less than currently modeled but may be altered by shifts in tree species composition (Fissore et al. 2009). The factor by which decomposition rates increase for a 10°C increase in temperature (Q_{10}) ranges substantially across soils (Fierer et al. 2006). The SOC quality, however, explains a large percentage of variations in Q_{10} values. Thus, the temperature dependence of microbial decomposition and the biochemical recalcitrance of SOC are probably directly linked.

Alterations in precipitation by ACC may affect plant and soil water content, and soil respiration. Specifically, desiccation stress in soil and plant tissue directly affects soil respiration and may result in substantial reductions in respiration (Davidson et al. 2006). On the other hand, soil water content indirectly affects respiration by its effects on substrate diffusion and availability. In particular, decomposition and associated CO_2 efflux are reduced at high soil water contents (Chapin et al. 2002). Also, with increased soil moisture the net CH_4 flux from soils strongly increases (von Fischer and Hedin 2007).

2.3.2.2 Dissolved and Particulate Carbon Efflux

Particulate OM (>0.45 μm) in forests is leached with percolating water into groundwater but also transported by surface runoff, and finally leaves forest watersheds by entering streams. Some organic and inorganic forms of C are dissolved in water and may be transported with hydrologic fluxes (Fahey et al. 2005). The dominant export of organic C from forest watersheds, however, occurs in particulate form. Surface runoff may flush DIC and DOC directly into streams (Waring and Running 2007). In fully forested watersheds, however, root channels, animal burrows and drying cracks are more important pathways. After infiltration, water percolates also through the soil, and DIC and DOC are finally released into groundwater reservoirs and streams. Important DOC sources in soils are both litter and humus (Kalbitz et al. 2000). In addition, the soil microorganisms, in particular, soil fungi, and the soil fauna contribute to DOC production (Chantigny 2003; Osler and Sommerkorn 2007). Mycorrhizal roots and decomposing O-horizons are, however, typically the most important DOC sources (Ekberg et al. 2007). Thus, any conditions that enhance decomposition and mineralization promote also high DOC concentrations. The scientific knowledge how DOC production relates to different stages of litter decomposition is, however, limited (Kalbitz et al. 2006). For example, lignin degradation may control the DOC production in coniferous but not in deciduous litter. Beside complex aromatic and other high molecular weight compounds, DOC contains also low molecular weight compounds such as organic acids, amino acids and sugars (van Hees et al. 2005).

The infiltrating water percolates through the soil profile and net leaching of DOC from upper horizons occurs (Neff and Asner 2001; Fahey et al. 2005). It has been estimated that 115–500 kg C ha^{-1} $year^{-1}$, equivalent to up to 35% of the annual litterfall C percolates with DOC from the forest floor into the mineral soil (Kalbitz and Kaiser 2008). The DOC concentrations decrease with soil depth due to microbial degradation and sorption to mineral surfaces. The turnover of low molecular weight DOC compounds, in particular, may contribute substantially to CO_2 efflux from forest soils (van Hees et al. 2005). Degradation and sorption alter the DOC composition as carbohydrates are preferentially utilized by microorganisms whereas more recalcitrant lignin-derived DOC accumulates on mineral surfaces (Kaiser and Guggenberger 2000; Kalbitz et al. 2003). Otherwise, alkyl-C may also be preferentially adsorbed by deeper soil horizons (Dai et al. 2001). Adsorption of DOC to deeper mineral soil horizons is important for C sequestration as it contributes to the long-term C sink in forest soils (Lorenz and Lal 2005). Part of the adsorbed OM may, however, also be released into the soil solution and mineralized by microorganisms (McDowell 2003). Thus, small hydrologic efflux of organic and inorganic C to streams occurs from forest ecosystems (Fahey et al. 2005).

During soil respiration, CO_2 is produced mainly by the activity of microorganisms and roots (Oh et al. 2007). Soil CO_2 reacts with water to produce carbonic acid, and net leaching of DIC from the forest floor may occur with the percolating water (Fahey et al. 2005). The concentration of inorganic C in the soil solution depends primarily upon the partial pressure of CO_2 in the gas phase. In contrast to DOC, DIC concentrations are relatively uniform through the soil profile. However, the export of both DIC and DOC from the pedosphere has rarely been studied. Specifically, the lateral transport of CO_2 may be an important pathway for C efflux from the forest pedosphere (Fiedler et al. 2006). Substantial C amounts may be transferred as respiration-derived DIC to aquatic ecosystems (Chapin et al. 2006).

In addition to water, wind contributes to soil erosion, that is, the detachment, transport, and deposition of soil particles. By erosion processes particulate C may be displaced from the forest ecosystem. During transport, decomposition may reduce the amount of particulate C that is finally deposited. If erosion contributes to the terrestrial C sink is, however, under discussion (Berhe et al. 2007). Recently, empirical evidence was reported for quantification of eroded C and its dynamic replacement in agricultural soils (Quine and Van Oost 2007). Comparable studies in forest soils are, however, missing.

Climate Change and Efflux of Dissolved and Particulate Carbon

The source of virtually all organic C in forests is photosynthesis (McDowell 2003). Thus, stimulation of C assimilation by increasing CO_2 concentrations may increase plant biomass and change its chemical composition. Both may alter the dissolved and particulate C efflux from forests. However, little is known about effects of elevated CO_2 on DOC release from litter. Initial stimulation of mineralization and DOC release by increased CO_2 is observed (Hagedorn and Machwitz 2007). This stimulation is, however, only transitory and, thus, CO_2-induced changes in tree species composition may have larger effects on DOC release from litter and forest floor (Mohan et al. 2007). The total C flux to the soil may be increased at elevated CO_2 (van Hees et al. 2005). This may result in higher soil CO_2 and DIC concentrations (Bernhardt et al. 2006). Thus, DIC concentrations are increased in a FACE experiment (Karberg et al. 2005). Furthermore, increasing alkalinity of the soil solution suggests increasing soil mineral weathering. The quantification of this effect, however, remains to be rigorously tested in the field (Oh et al. 2007). Furthermore, suppression of tree transpiration due to CO_2-induced stomatal closure may increase the C efflux to streams due to increased runoff (Gedney et al. 2006).

Direct effects of increasing temperatures on OM decomposition are under discussion (Section 2.2.2.4). As belowground processes are probably less affected, changes in dissolved and particulate C fluxes may be small. Otherwise, increasing temperatures enhance decomposition and the enzymatic breakdown of SOM to DOC, and the release of CO_2 (Bengtson and Bengtsson 2007; Norby et al. 2007). If this process enhances also the DIC and DOC efflux is not clear. Furthermore, warming increases plant productivity (Norby et al. 2007). Thus, dissolved and particulate C efflux may change. Indirectly, altered plant and decomposer community at higher temperatures may also be responsible for changes in dissolved and particulate C efflux from forests.

Changes in precipitation patterns may alter the amount and composition of surface runoff and percolating water, and thus, dissolved and particulate C efflux. Rising temperatures on one hand but, in large parts of Europe and North America, the substantial decline in atmospherically deposited anthropogenic sulfur and sea salt are, however, responsible for the observed DOC increases in surface waters (Evans et al. 2006; Montheith et al. 2007). In the past, DOC solubility was reduced by high soil water acidity and ionic strength caused by sulfur and sea salt deposition. Furthermore, the increase in mechanical forces due to stronger winds may detach more soil particles. This may result in increasing efflux of particulate C by wind erosion.

2.3.3 Carbon Efflux from Soil Carbonates

Soil CO_2 reacts with water to form carbonic acid (Section 2.2.2.6). Carbonic acid may further react with carbonate and silicate minerals. This process causes weathering (Hamilton et al. 2007). Thus, CO_2 is consumed and DIC produced. In humid and subhumid climates, DIC produced by soil mineral weathering may be transported through groundwater flowpaths, and eventually reach the oceans after entering streams and rivers (Fiedler et al. 2006). Little re-precipitation occurs during transport. Thus, the dissolution of $CaCO_3$ during pedogenesis contributes to the CO_2 efflux from forests but only on a geological timescale (Kuzyakov 2006). The abiotic CO_2 efflux is of minor importance compared with the biotic CO_2 fluxes. Aside soil respiration, CO_2 is also produced during weathering of soil minerals by stronger acids than carbonic acid (e.g., sulfuric acid, nitric acid) (Oh et al. 2007).

2.3.3.1 Climate Change and Carbon Efflux from Carbonates

Increases in atmospheric CO_2 concentration may disrupt the timescale for DIC transport from the sites of soil mineral weathering and the ultimate C sink in the oceans (Hamilton et al. 2007). Enhanced rates of carbonate mineral weathering and subsequent re-precipitation may be the consequence. Higher GPP at elevated CO_2 together with higher decomposition rates at higher temperatures may increase CO_2 concentrations in the soil pore space. Thus, carbonate weathering processes may occur more rapidly (Schlesinger 2006). On the other hand, drought events promote the precipitation reaction for pedogenic carbonates.

2.4 Conclusions

Photosynthesis, that is, the fixation of CO_2 at the expense of energy from sun light in green tissues of forest plants is the major natural input path for C entering forest ecosystems. The majority of forest plants possess the C_3 photosynthetic pathway but C_4 and CAM photosynthesis may also occur. The CO_2 concentration inside stomata limits the light-dependent C uptake during C_3 photosynthesis. It is, however, unclear if the increasing atmospheric CO_2 concentrations increase the net C assimilation during the life-cycle of a forest as down-regulation of photosynthetic activity may occur in mature trees. Photosynthesis in forests which receive higher but not excessive precipitation will benefit from ACC whereas increased drought may suppress photosynthetic CO_2 uptake. Minor C input processes into forest ecosystems are the lateral transfer of DIC, DOC and PC. The major natural C loss from forests occurs by autotrophic respiration in plant tissues, that is, in leaf, stem, reproductive organs and roots, and by heterotrophic respiration by the decomposer community. Specifically, an almost equal amount of C fixed is respired and returned back to the atmosphere as CO_2. Elevated CO_2 concentrations have variable effects on respiration in forest ecosystems. The forest soil C cycle will probably respond with an increased throughput to elevated CO_2. Furthermore, plant respiration may acclimate to temperature increases. It is unclear if the increased NPP in response to warming cause the accumulation of SOM as respiratory losses may also increase. Between acquisition and release, CO_2 is temporarily stored in the two major forest C pools in tree stems and in SOM. The most stable fraction of SOM in the sub-soil, however, may be thousands of years old and of paramount importance for long-term C storage although C sequestration rates are lower than those in aboveground pools. Most importantly, the knowledge about the processes that stabilize and destabilize the major C pool in forest soils is still at an infant state. Indirect effects of ACC, however, are likely to be more important for C dynamics in forest ecosystems than direct effects. Specifically, increased frequency and severity of wildfires and insect outbreaks, longer growing season and ACC-induced shifts in biodiversity have the potential to strongly affect the C dynamics in forest ecosystems. In summary, there is no theoretical consensus on responses of the forest C cycle to global change.

2.5 Review Questions

1. Describe the three photosynthetic pathways occurring in forest plants and compare their vulnerability to ACC, that is, against alterations in atmospheric CO_2 concentrations and temperature increases.
2. How may the proportion of C_3, C_4 and CAM plants among the forest plant community change by ACC?
3. Current FACE experiments seem to be not suitable to study effects of elevated CO_2 on C dynamics occurring in the life-cycle of forest ecosystems – why? How would you study the effects of elevated CO_2 on the C dynamics in forest ecosystems?
4. Is there a CO_2 "fertilization" effect? Compare with nutrient fertilization!
5. What are other natural C input pathways in forest ecosystems and how important are they for the C budget?
6. Recent measurements of DOC concentrations in rivers and streams indicate that forests loose higher amounts of C by this pathway than previously thought – what might be the responsible processes?
7. Why is it so difficult to separate autotrophic from heterotrophic respiration in forest ecosystems?

8. In the past, forest management focused on maximizing productivity and, thus, C storage in aboveground pools. Is there a maximum storage capacity in belowground C pools?
9. Is the present C dynamic in forest ecosystems still “natural”?

References

Aber JD, Melillo JM (2001) Terrestrial ecosystems. Academic, San Diego, CA

Aerts R (2006) The freezer defrosting: global warming and litter decomposition rates in cold biomes. J Ecol 94:713–724

Ainsworth EA, Long SP (2005) What have we learned from 15 years of free-air CO_2 enrichment (FACE)? A meta-analytical review of responses to rising CO_2 in photosynthesis, canopy properties and plant production. New Phytol 165:351–371

Ainsworth EA, Rogers A (2007) The response of photosynthesis and stomatal conductance to rising [CO_2]: mechanisms and environmental interactions. Plant Cell Environ 30:258–270

Alo CA, Wang G (2008) Potential future changes of the terrestrial ecosystem based on climate projections bei eight general circulation models. J Geophys Res 113:G01004. doi:10.1029/2007JG000528

Amelung W, Brodowski S, Sandhage-Hofmann A, Bol R (2008) Combining biomarker with stable isotope analyses for assessing the transformation and turnover of soil organic matter. Adv Agron 100:155–250

Amthor JS (2003) Efficiency of lignin biosynthesis: a quantitative analysis. Ann Bot 91:673–695

Andrews JA, Schlesinger WH (2001) Soil CO_2 dynamics, acidification, and chemical weathering in a temperate forest with experimental CO_2 enrichment. Global Biogeochem Cy 15:149–162

Apps MJ, Bernier P, Bhatti JS (2006) Forests in the global carbon cycle: implications of climate change. In: Bhatti JS, Lal R, Apps MJ, Price MA (eds) Climate change and managed ecosystems. Taylor & Francis, Boca Raton, FL, pp 175–200

Asshoff R, Zotz G, Körner C (2006) Growth and phenology of mature temperate forest trees in elevated CO_2. Glob Change Biol 12:848–861

Atwell BJ, Henery ML, Whitehead D (2003) Sapwood development in *Pinus radiata* trees grown for three years at ambient and elevated carbon dioxide partial pressures. Tree Physiol 23:13–21

Aubrey DP, Teskey RO (2009) Root-derived CO_2 efflux via xylem stream rivals soil CO_2 efflux. New Phytol (in press) doi: 10.1111/j.1469-8137.2009.02971.x

Austin AT, Vivanco L (2006) Plant litter decomposition in a semi-arid ecosystem controlled by photodegradation. Nature 442:555–558

Baldocchi D (2008) ‘Breathing’ of the terrestrial biosphere: lessons learned from a global network of carbon dioxide flux measurement systems. Aust J Bot 56:1–26

Bardgett RD, Bowman WD, Kaufmann R, Schmidt KS (2005) A temporal approach to linking aboveground and belowground ecology. Trends Ecol Evol 20:634–641

Beerling DJ, Hewitt CN, Pyle JA, Raven JA (2007) Critical issues in trace gas biogeochemistry and global change. Philos Trans R Soc A 365:1629–1642

Bengtson P, Bengtsson G (2007) Rapid turnover of DOC in temperate forests accounts for increased CO_2 production at elevated temperatures. Ecol Lett 10:783–790

Berg B, McClaugherty C (2003) Plant litter: decomposition, humus formation, carbon sequestration. Springer, Berlin

Berg JM, Tymoczko JL, Stryer L (2007) Biochemistry. WH Freeman, New York

Berhe AA, Harte J, Harden JW, Torn MS (2007) The significance of the erosion-induced terrestrial carbon sink. BioScience 57:337–346

Bernards MA (2002) Demystifying suberin. Can J Bot 80:227–240

Bernhardt ES, Barber JJ, Pippen JS, Taneva L, Andrews JA, Schlesinger WH (2006) Long-term effects of free air CO_2 enrichment (FACE) on soil respiration. Biogeochemistry 77:91–116
Billings SA, Ziegler SE (2008) Altered patterns of soil carbon substrate usage and heterotrophic respiration in a pine forest with elevated CO_2 and N fertilization. Glob Change Biol 14:1025–1036
Bird JA, Kleber M, Torn MS (2008) ^{13}C and ^{15}N stabilization dynamics in soil organic matter fractions during needle and fine root decomposition. Org Geochem 39:465–477
Blagodatskaya E, Kuzyakov Y (2008) Mechanisms of real and apparent priming effects and their dependence on soil microbial biomass and community structure: critical review. Biol Fertil Soils 45:115–131
Blaschke L, Forstreuter M, Sheppard LJ, Keith IK, Murray MB, Polle A (2002) Lignification in beech (*Fagus sylvatica*) grown at elevated CO_2 concentrations: interaction with nutrient availability and leaf maturation. Tree Physiol 22:469–477
Boisvenue C, Running SW (2006) Impacts of climate change on natural forest productivity – evidence since the middle of the 20th century. Glob Change Biol 12:862–882
Bol R, Poirier N, Balesdent J, Gleixner G (2009) Molecular turnover time of soil organic matter in particle-size fractions of an arable soil. Rapid Commun Mass Spectrom 23:2551–2558
Bond TC, Bhardwaj E, Dong R, Jogani R, Jung S, Roden C, Streets DG, Trautmann NM (2007) Historical emissions of black and organic carbon aerosol from energy-related combustion, 1850-2000. Global Biogeochem Cy 21, GB2018, doi:10.1029/2006GB002840
Bradley KL, Pregitzer KS (2007) Ecosystem assembly and terrestrial carbon balance under elevated CO_2. Trends Ecol Evol 22:538–547
Brandt LA, Bohnet C, King JY (2009) Photochemically induced carbon dioxide production as a mechanism for carbon loss from plant litter in arid ecosystems. J Geophys Res 114:G02004. doi:10.1029/2008JG000772
Bréda N, Huc R, Granier A, Dreyer E (2006) Temperate forest trees and stands under severe drought: a review of ecophysiological responses, adaptation processes and long-term consequences. Ann For Sci 63:625–644
Bronick CJ, Lal R (2005) Soil structure and management: a review. Geoderma 124:3–22
Brown S (2002) Measuring carbon in forests: current status and future challenges. Environ Pollut 116:363–372
Burton AJ, Pregitzer KS (2008) Measuring forest floor, mineral soil, and root carbon stocks. In: Hoover CM (ed) Field measurements for forest carbon monitoring. Springer, New York, pp 129–142
Butenhoff CL, Khalil MAK (2007) Global methane emissions from terrestrial plants. Environ Sci Technol 41:4032–4037
Calfapietra C, Scarascia-Mugnozza G, Karnosky DF, Loreto F, Sharkey TD (2008) Isoprene emission rates under elevated CO_2 and O_3 in two field-grown aspen clones differing in their sensitivity to O_3. New Phytol 179:55–61
Canadell J, Jackson RB, Ehleringer JR, Mooney HA, Sala OE, Schulze E-D (1996) Maximum rooting depth of vegetation types at the global scale. Oecologia 108:583–595
Carney KM, Hungate BA, Drake BG, Megonigal JP (2007) Altered soil microbial community at elevated CO_2 leads to loss of soil carbon. Proc Natl Acad Sci USA 104:4990–4995
Carter DO, Yellowlees D, Tibbett M (2007) Cadaver decomposition in terrestrial ecosystems. Naturwissenschaften 94:12–24
Chang M (2006) Forest hydrology: an introduction to water and forests. Taylor & Francis, Boca Raton, FL
Chantigny MH (2003) Dissolved and water-extractable organic matter in soils: a review on the influence of land use and management practices. Geoderma 113:357–380
Chapin FS III, Schulze E-D, Mooney HA (1990) The ecology and economics of storage in plants. Annu Rev Ecol Syst 21:423–447
Chapin FS III, Matson PA, Mooney HA (2002) Principles of terrestrial ecosystem ecology. Springer, New York
Chapin FS III, Woodwell GM, Randerson JT, Rastetter EB, Lovett GM, Baldocchi DD, Clark DA, Harmon ME, Schimel DS, Valentini R, Wirth C, Aber JD, Cole JJ, Goulden ML, Harden JW,

Heimann M, Howarth RW, Matson PA, McGuire AD, Melillo JM, Mooney HA, Neff JC, Houghton RA, Pace ML, Ryan MG, Running SW, Sala OE, Schlesinger WH, Schulze E-D (2006) Reconciling carbon-cycle concepts, terminology, and methods. Ecosystems 9:1041–1050

Chave J, Coomes D, Jansen S, Lewis SL, Swenson NG, Zanne AE (2009) Towards a worldwide wood economics spectrum. Ecol Lett 12:351–366

Chen H, Rygiewicz PT, Johnson MG, Harmon ME, Tian H, Tang JW (2008) Chemistry and long-term decomposition of roots of douglas-fir grown under elevated atmospheric carbon dioxide and warming conditions. J Environ Q 37:1327–1336

Cheng W (1999) Rhizosphere feedbacks in elevated CO_2. Tree Physiol 19:313–320

Chenu C, Plante AF (2006) Clay-sized organo-mineral complexes in a cultivation chronosequence: revisiting the concept of the 'primary organo-mineral complex'. Eur J Soil Sci 57:596–607

Christensen BT (2001) Physical fractionation of soil and structural and functional complexity in organic matter turnover. Eur J Soil Sci 52:345–353

Chung H, Zak DR, Lilleskov EA (2006) Fungal community composition and metabolism under elevated CO_2 and O_3. Oecologia 147:143–154

Ciais P, Borges AV, Abril G, Meybeck M, Folberth G, Hauglustaine D, Janssens IA (2008) The impact of lateral carbon fluxes on the European carbon balance. Biogeosciences 5:1259–1271

Cisneros-Dozal LM, Trumbore S, Hanson PJ (2006) Partitioning sources of soil-respired CO_2 and their seasonal variation using a unique radiocarbon tracer. Glob Change Biol 12:194–204

Clark DA, Brown S, Kicklighter DW, Chambers JQ, Thomlinson JR, Ni J (2001) Measuring net primary production in forests: concepts and field methods. Ecol Appl 11:356–370

Conant RT, Drijber RA, Haddix ML, Parton WJ, Paul EA, Plante AF, Six J, Steinweg JM (2008) Sensitivity of organic matter decomposition to warming varies with its quality. Glob Change Biol 14:868–877

Cornelissen JHC, van Bodegom PM, Aerts R, Callaghan TV, Van Logtesijn RSP, Alatalo J, Chapin FS, Gerdol R, Guðmundsson J, Gwynn-Jones D, Hartley AE, Hik DS, Hofgaard A, Jónsdóttir IS, Karlsson S, Klein JA, Laundre J, Magnusson B, Michelsen A, Molau U, Onipchenko VG, Quested HM, Sandvik SM, Schmidt IK, Shaver GR, Solheim B, Soudzilovskaia NA, Stenström A, Tolvanen A, Totland Ø, Wada N, Welker JM, Zhao X, Team MOL (2007) Global negative vegetation feedback to climate warming responses of leaf litter decomposition rates in cold biomes. Ecol Lett 10:619–627

Cornwell WK, Cornelissen JHC, Allison SD, Bauhus J, Eggleton P, Preston CM, Scarff F, Weedon JT, Wirth C, Zanne AE (2009) Plant traits and wood fate across the globe-rotted, burned, or consumed? Glob Change Biol doi: 10.1111/j.1365-2486.2009.01916.x

Cornwell WK, Cornelissen JHC, Amatangelo K, Dorrepaal E, Eviner VT, Godoy O, Hobbie SE, Hoorens B, Kurokawa H, Pérez-Harguindeguy N, Quested HM, Santiago LS, Wardle DA, Wright IJ, Aerts R, Allison SD, van Bodegom P, Brovkin V, Chatain A, Callaghan TV, Díaz S, Garnier E, Gurvich DE, Kazakou E, Klein JA, Read J, Reich PB, Soudzilovskaia NA, Victoria VM, Westoby M (2008) Plant species traits are the predominant control on litter decomposition rates within biomes worldwide. Ecol Lett 11:1065–1071

Crockford RH, Richardson DP (2000) Partitioning of rainfall into throughfall, stemflow and interception: effect of forest type, ground cover and climate. Hydrol Process 14:2903–2920

Crow SE, Filley TR, McCormick M, Szlávecz K, Stott DE, Gamblin D, Conyers G (2009) Earthworms, stand age, and species composition interact to influence particulate organic matter chemistry during forest succession. Biogeochemistry 92:61–82

Cseke LJ, Tsai C-J, Rogers A, Nelsen MP, White HL, Karnosky DF, Podila GK (2009) Transcriptomic comparison in the leaves of two aspen genotypes having similar carbon assimilation rates but different partitioning patterns under elevated [CO_2]. New Phytol 182:891–911

Czimczik CI, Masiello CA (2007) Controls on black carbon storage in soils. Global Biogeochem Cy 21, GB3005, doi:10.1029/2006GB002798

Dai K'OH, Johnson CE, Driscoll CT (2001) Organic matter chemistry and dynamics in clear-cut and unmanaged hardwood forest ecosystems. Biogeochemistry 54:51–83

Davey PA, Olcer H, Zakhleniuk O, Bernacchi CJ, Calfapietra C, Long SP, Raines CA (2006) Can fast-growing plantation trees escape biochemical down-regulation of photosynthesis when grown throughout their complete production cycle in the open air under elevated carbon dioxide? Plant Cell Environ 29:1235–1244

Davidson EA, Janssens IA (2006) Temperature sensitivity of soil carbon decomposition and feedbacks to climate change. Nature 440:165–173

Davidson EA, Janssens IA, Luo Y (2006) On the variability of respiration in terrestrial ecosystems: moving beyond Q_{10}. Glob Change Biol 12:154–164

De Leeuw JW (2007) On the origin of sedimentary aliphatic macromolecules: a comment on recent publications by Gupta et al. Org Geochem 38:1585–1587

De Leeuw JW, Versteegh GJM, Van Bergen PF (2006) Biomacromolecules of algae and plants and their fossil analogues. Plant Ecol 182:209–233

Delpierre N, Soudani K, François C, Köstner B, Pontailler J-Y, Nikinmaa E, Misson L, Aubinet M, Bernhofer C, Granier A, Grünewald G, Heinesch B, Longdoz B, Ourcival JM, Rambal S, Vesala T, Dufrêne E (2009) Exceptional carbon uptake in European forests during the warm spring of 2007: a data-model analysis. Glob Change Biol 15:1455–1474

Denman KL, Brasseur G, Chidthaisong A, Ciais P, Cox PM, Dickinson RE, Hauglustaine D, Heinze C, Holland E, Jacob D, Lohmann U, Ramachandran S, da Silva Dias PL, Wofsy SC, Zhang X (2007) Couplings between changes in the climate system and biogeochemistry. In: Intergovernmental panel on climate change (ed) Climate change 2007: the physical science basis, Chapter 7. Cambridge University Press, Cambridge, UK

Dignac M-F, Bahri H, Rumpel C, Rasse DP, Bardoux G, Balesdent J, Girardin C, Chenu C, Mariotti A (2005) Carbon-13 natural abundance as a tool to study the dynamics of lignin monomers in soil: an appraisal at the Closeaux experimental field (France). Geoderma 128:3–17

Dijkstra FA, Cheng W (2007) Interactions between soil and tree roots accelerate long-term soil carbon decomposition. Ecol Lett 10:1046–1053

Diochon AC, Kellman L (2009) Physical fractionation of soil organic matter: destabilization of deep soil carbon following harvesting of a temperate coniferous forest. J Geophys Res 114:G01016. doi:10.1029/2008JG000844

Drennan PM, Nobel PS (2000) Responses of CAM species to increasing atmospheric CO_2 concentrations. Plant Cell Environ 23:767–781

Drigo B, Kowalchuk GA, van Veen JA (2008) Climate change goes underground: effects of elevated atmospheric CO_2 on microbial community structure and activities in the rhizosphere. Biol Fertil Soils 44:667–679

Dueck TA, de Visser R, Poorter H, Persijn S, Gorissen A, de Visser W, Schapendonk A, Verhagen J, Snel J, Harren FJM, Ngai AKY, Verstappen F, Bouwmeester H, Voeseneck LACJ, van der Werf A (2007) No evidence for substantial aerobic methane emission by terrestrial plants: a ^{13}C-labelling approach. New Phytol 175:29–35

Dueck T, van der Werf A (2008) Are plants precursors for methane? New Phytol 178:693–695

Duhl TR, Helmig D, Guenther A (2008) Sesquiterpene emissions from vegetation: a review. Biogeosciences 5:761–777

Dutaur L, Verchot LV (2007) A global inventory of the soil CH_4 sink. Global Biogeochem Cy 21, GB4013, doi:10.1029/2006GB002734

Eissenstat DM, Yanai RD (2002) Root life span, efficiency, and turnover. In: Waisel Y, Eshel A, Kafkafi U (eds) Plant roots – the hidden half. Marcel Dekker, New York, pp 221–238

Ekberg A, Buchmann N, Gleixner G (2007) Rhizosphere influence on soil respiration and decomposition in a temperate Norway spruce stand. Soil Biol Biochem 39:2103–2110

Ekschmitt K, Kandeler E, Poll C, Brune A, Buscot F, Friedrich M, Gleixner G, Hartmann A, Kästner M, Marhan S, Miltner A, Scheu S, Wolters V (2008) Soil-carbon preservation through habitat constraints and biological limitations on decomposer activity. J Plant Nutr Soil Sci 171:27–35

Elbert W, Weber B, Büdel B, Andreae MO, Pöschl U (2009) Microbiotic crusts on soil, rock and plants: neglected major players in the global cycles of carbon and nitrogen? Biogeosciences Discuss 6:6983–7015

Epps KY, Comerford NB, Reeves III JB, Cropper Jr.WP, Araujo QR (2007) Chemical diversity – highlighting a species richness and ecosystem function disconnect. Oikos 116:1831–1840

Ericsson T, Rytter L, Vapaavuori E (1996) Physiology of carbon allocation in trees. Biomass Bioenerg 11:115–127

Eswaran H, Reich PF, Kimble JM (2000) Global carbon stocks. In: Lal R, Kimble JM, Eswaran H, Stewart BA (eds) Global climate change and pedogenic carbonates. CRC, Boca Raton, FL, pp 15–25

Evans CD, Chapman PJ, Clark JM, Monteith DT, Cresser MS (2006) Alternative explanations for rising dissolved organic carbon export from organic soils. Glob Change Biol 12:2044–2053

Fahey TJ, Siccama TG, Driscoll CT, Likens GE, Campbell J, Johnson CE, Battles JJ, Aber JD, Cole JJ, Fisk MC, Groffman PM, Hamburg SP, Holmes RT, Schwarz PA, Yanai RD (2005) The biogeochemistry of carbon at Hubbard Brook. Biogeochemistry 75:109–176

Falloon PD, Smith P (2000) Modelling refractory soil organic matter. Biol Fertil Soils 30:388–398

Farquhar GD (1989) Models of integrated photosynthesis of cells and leaves. Philos Trans R Soc B 323:357–367

Feeley KJ, Wright SJ, Nur Surpadi MN, Kassim AR, Davies SJ (2007) Decelerating growth in tropical forest trees. Ecol Lett 10:461–469

Feng X, Simpson AJ, Wilson KP, Williams DD, Simpson MJ (2008) Increased cuticular carbon sequestration and lignin oxidation in response to soil warming. Nat Geoscience 1:836–839

Fernandes SD, Trautmann NM, Streets DG, Roden CA, Bond TC (2007) Global biofuel use, 1850-2000. Global Biogeochem Cy 21, GB2019, doi:10.1029/2006GB002836

Fernandez Monteiro JA, Zotz G, Körner C (2009) Tropical epiphytes in a CO_2-rich atmosphere. Acta Oecol 35:60–68

Fiedler S, Höll BS, Jungkunst HF (2006) Discovering the importance of lateral CO_2 transport from a temperate spruce forest. Sci Total Environ 368:909–915

Field CB, Lobell DB, Peters HA, Chiariello NR (2007) Feedbacks of terrestrial ecosystems to climate change. Annu Rev Environ Resour 32:7.1–7.29

Fierer N, Colman BP, Schimel JP, Jackson RB (2006) Predicting the temperature dependence of microbial respiration in soil: A continental-scale analysis. Global Biogeochem Cy 20, GB3026, doi:10.1029/2005GB002644

von Fischer JC, Hedin LO (2007) Controls on soil methane fluxes: tests of biophysical mechanisms using stable isotope tracers. Global Biogeochem Cy 21, GB2007, doi:10.1029/2006GB002687

Fissore C, Giardina CP, Swanston CW, King GM, Kolka RK (2009) Variable temperature sensitivity of soil carbon in North American forests. Glob Change Biol 15:2295–2310

Fontaine S, Barot S, Barré P, Bdioui N, Mary B, Rumpel C (2007) Stability of organic carbon in deep soil layers controlled by fresh carbon supply. Nature 450:277–281

Ford CR, Wurzburger N, Hendrick RL, Teskey RO (2007) Soil DIC uptake and fixation in *Pinus taeda* seedlings and its C contribution to plant tissues and ectomycorrhizal fungi. Tree Physiol 27:375–383

Fox O, Vetter S, Ekschmitt K, Wolters V (2006) Soil fauna modifies the recalcitrance–persistence relationship of soil carbon pools. Soil Biol Biochem 38:1353–1363

Franklin O (2007) Optimal nitrogen allocation controls tree responses to elevated CO_2. New Phytol 174:811–822

Franklin O, McMurtrie RE, Iversen CM, Crous KY, Finzi AC, Tissue DT, Ellsworth DS, Oren R, Norby RJ (2009) Forest fine-root production and nitrogen use under elevated CO_2: contrasting responses in evergreen and deciduous trees explained by a common principle. Glob Change Biol 15:132–144

Frey SD (2007) Spatial distribution of soil organisms. In: Paul EA (ed) Soil microbiology, ecology, and biochemistry. Elsevier, Amsterdam, pp 283–300

Friend AL, Coleman MD, Isebrands JG (1994) Carbon allocation to root and shoot systems of woody plants. In: Davis TD, Haissig BE (eds) Biology of adventitious root formation. Plenum Press, New York, pp 245–273

Gallo ME, Sinsabaugh RL, Cabaniss SE (2006) The role of ultraviolet radiation in litter decomposition in arid ecosystems. Appl Soil Ecol 34:82–91

Gaudinski JB, Torn MS, Riley WJ, Swanston C, Trumbore SE, Joslin JD, Majdi H, Dawson TE, Hanson PJ (2009) Use of stored carbon reserves in growth of temperate tree roots and leaf buds: analyses using radiocarbon measurements and modeling. Glob Change Biol 15:992–1014

Gedney N, Cox PM, Betts RA, Boucher O, Huntingford C, Stott PA (2006) Detection of a direct carbon dioxide effect in continental river runoff records. Nature 439:835–838

Geib SM, Filley TR, Hatcher PG, Hoover K, Carlson JE, MdM J-G, Nakagawa-Izumi A, Sleighter RL, Tien M (2008) Lignin degradation in wood-feeding insects. Proc Natl Acad Sci USA 105:12932–12937

Gholz HL, Wedin DA, Smitherman SM, Harmon ME, Parton WJ (2000) Long-term dynamics of pine and hardwood litter in contrasting environments: toward a global model of decomposition. Glob Change Biol 6:751–765

Giardina CP, Coleman MD, Hancock JE, King JS, Lilleskov EA, Loya WM, Pregitzer KS, Ryan MG, Trettin CC (2005) The response of belowground carbon allocation in forests to global change. In: Binkley D, Menyailo O (eds) Trees species effects on soils: implications for global change. NATO Science Series. Kluwer, Dordrecht, pp 119–154

Giardina CP, Ryan MG (2000) Evidence that decomposition rates of organic carbon in mineral soil do not vary with temperature. Nature 404:858–861

Gilbert GS, Strong DR (2007) Fungal symbionts of tropical trees. Ecology 88:539–540

Gill RA, Jackson RB (2000) Global patterns of root turnover for terrestrial ecosystems. New Phytol 147:13–31

Glaser B, Knorr K-H (2008) Isotopic evidence for condensed aromatics from non-pyrogenic sources in soils – implications for current methods for quantifying soil black carbon. Rapid Commun Mass Spectrom 22:935–942

Goddard MA, Mikhailova EA, Post CJ, Schlautman MA, Galbraith JM (2009) Continental United States atmospheric wet calcium deposition and soil inorganic carbon stocks. Soil Sci Soc Am J 73:989–994

Gonzalez-Meler MA, Taneva L, Trueman RJ (2004) Plant respiration and elevated atmospheric CO_2 concentration: cellular responses and global significance. Ann Bot 94:647–656

Grace J (2005) Role of forest biomes in the global carbon balance. In: Griffiths H, Jarvis PG (eds) The carbon balance of forest biomes. Taylor & Francis, Oxon, UK, pp 19–45

de Graaff M-A, van Groenigen K-J, Six J, Hungate B, van Kessel C (2006) Interactions between plant growth and soil nutrient cycling under elevated CO_2: a meta-analysis. Glob Change Biol 12:2077–2091

Grandy AS, Neff JC (2008) Molecular C dynamics downstream: the biochemcial decomposition sequence and its impact on soil organic matter structure and function. Sci Total Environ 404:297–307

Greenfield LG (1999) Weight loss and release of mineral nitrogen from decomposing pollen. Soil Biol Biochem 31:353–361

Gregorich EG, Beare MH, McKim UF, Skjemstad JO (2006) Chemical and biological characteristics of physically uncomplexed organic matter. Soil Sci Soc Am J 70:975–985

Van Groenigen K-J, Six J, Harris D, van Kessel C (2007) Elevated CO_2 does not favor a fungal decomposition pathway. Soil Biol Biochem 39:2168–2172

Guenther A, Geron C, Pierce T, Lamb B, Harley P, Fall R (2000) Natural emissions of non-methane volatile organic compounds, carbon monoxide, and oxides of nitrogen from North America. Atmos Environ 34:2205–2230

Gustafsson Ö, Kruså M, Zencak Z, Sheesley RJ, Granat L, Engström E, Praveen PS, Rao PSP, Leck C, Rodhe H (2009) Brown clouds over South Asia: biomass or fossil fuel combustion. Science 323:495–498

Hättenschwiler S, Miglietta F, Raschi A, Körner C (1997) Morphological adjustments of mature *Quercus ilex* trees to elevated CO_2. Acta Oecol 18:361–365

Hagedorn F, Machwitz M (2007) Controls on dissolved organic matter leaching from forest litter grown under elevated atmospheric CO_2. Soil Biol Biochem 39:1759–1769

Hamilton SK, Kurzman AL, Arango C, Jin L, Robertson GP (2007) Evidence for carbon sequestration by agricultural liming. Global Biogeochem Cy 21, GB2021, doi:10.1029/2006GB002738

Hammes K, Schmidt MWI, Smernik RJ, Currie LA, Ball WP, Nguyen TH, Louchouran P, Houel S, Gustafsson Ö, Elmquist M, Cornelissen G, Skjemstad JO, Masiello CA, Song J, Peng P'A, Mitra S, Dunn JC, Hatcher PG, Hockaday WC, Smith DM, Hartkopf-Fröder C, Böhmer A, Lüer B, Huebert BJ, Amelung W, Brodowski S, Huang L, Zhang W, Gschwend PM, Xana Flores-Cervantes D, Largeau C, Rouzaud J-N, Rumpel C, Guggenberger G, Kaiser K, Rodionov A, Gonzalez-Vila FJ, Gonzalez-Perez JA, de la Rosa JM, Manning DAC, López-Capél E, Ding L (2007) Comparison of quantification methods to measure fire-derived (black/elemental) carbon in soils and sediments using reference materials from soil, water, sediment and the atmosphere. Global Biogeochem Cy 21, GB3016, doi:10.1029/2006GB002914

Hammes K, Torn MS, Lapenas AG, Schmidt MWI (2008) Centennial black carbon turnover in a Russian steppe soil. Biogeosciences 5:1339–1350

Harley P, Greenberg J, Niinemets Ü, Guenther A (2007) Environmental controls over methanol emission from leaves. Biogeosciences 4:1083–1099

Harmon ME, Silver WL, Fasth B, Chen H, Burke IC, Parton WJ, Hart SC, Currie WS, LIDET (2009) Long-term patterns of mass loss during the decomposition of leaf and fine root litter: an intersite comparison. Glob Change Biol 15:1320–1338

Hart SC (2006) Potential impacts of climate change on nitrogen transformations and greenhouse gas fluxes in forests: a soil transfer study. Glob Change Biol 12:1032–1046

Hartley IP, Armstrong AF, Murthy R, Barron-Gafford G, Ineson P, Atkin OK (2006) The dependence of respiration on photosynthetic substrate supply and temperature: integrating leaf, soil and ecosystem measurements. Glob Change Biol 12:1954–1968

Hartmann J, Kempe S (2008) What is the maximum potential for CO_2 sequestration by "stimulated" weathering on the global scale? Naturwissenschaften 95:1159–1164

Heald CL, Wilkinson MJ, Monson RK, Alo CA, Wang G, Guenther A (2009) Response of isoprene emission to ambient CO_2 changes and implications for global budgets. Glob Change Biol 15:1127–1140

Hedges JI, Eglinton G, Hatcher PG, Kirchman DL, Arnosti C, Derenne S, Evershed RP, Kögel-Knabner I, deLeeuw JW, Littke R, Michaelis W, Rullkötter J (2000) The molecularly-uncharacterized component of nonliving organic matter in natural environments. Org Geochem 31:945–958

van Hees PAW, Jones DL, Finlay R, Godbold DL, Lundström US (2005) The carbon we do not see-the impact of low molecular weight compounds on carbon dynamics and respiration in forest soils: a review. Soil Biol Biochem 37:1–13

van der Heijden MGA, Bardgett RD, van Straalen NM (2008) The unseen majority: soil microbes as drivers of plant diversity and productivity in terrestrial ecosystems. Ecol Lett 11:296–310

Heinemeyer A, Hartley IP, Evans SP, Da La Fuentes JAC, Ineson P (2007) Forest soil CO_2 flux: uncovering the contribution and environmental responses of ectomycorrhizas. Glob Change Biol 13:1786–1797

Heldt H-W (2005) Plant biochemistry. Elsevier, San Diego, CA

Heredia, A (2003) Biophysical and biochemical characteristics of cutin, a plant barrier biopolymer. Biochim Biophys Acta 1620:1–7

Hernes PJ, Hedges JI (2004) Tannin signature of barks, needles, leaves, cones, and wood at the molecular level. Geochim Cosmochim Acta 68:1293–1307

Hickler T, Smith B, Prentice IC, Mjöfors K, Miller P, Arneth A, Sykes MT (2008) CO_2 fertilization in temperate FACE experiments not representative of boreal and tropical forests. Glob Change Biol 14:1531–1542

Hill PW, Farrar JF, Boddy EL, Gray AM, Jones DL (2006) Carbon partitioning and respiration – their control and role in plants at high CO_2. In: Nösberger J, Long SP, Norby RJ, Stitt M, Hendrey GR, Blum H (eds) Managed ecosystems and CO_2 – case studies, processes, and perspectives. Springer, Berlin, pp 271–292

Hobbie SE, Ogdahl M, Chorover J, Chadwick OA, Oleksyn J, Zytkowiak R, Reich PB (2007) Tree species effects on soil organic matter dynamics: the role of soil cation composition. Ecosystems 10:999–1018

Hobbie SE, Reich PB, Oleksyn J, Ogdahl M, Zytkowiak R, Hale C, Karolewski P (2006) Tree species effects on decomposition and forest floor dynamics in a common garden. Ecology 87:2288–2297

Hodge A, Berta G, Doussan C, Merchan F, Crespi M (2009) Plant root growth, architecture and function. Plant Soil 321:153–187

Högberg P, Högberg MN, Göttlicher SG, Betson NR, Keel SG, Metcalfe DB, Campbell C, Schindlbacher A, Hurry V, Lundmark T, Linder S, Näsholm T (2008) High temporal resolution tracing of photosynthate carbon from the tree canopy to forest soil microorganisms. New Phytol 177:220–228

Högberg P, Nordgren A, Buchmann N, Taylor AFS, Ekblad A, Högberg MN, Nyberg G, Ottosson-Löfvenius M, Read DJ (2001) Large-scale forest girdling shows that current photosynthesis drives soil respiration. Nature 411:789–792

Hofmann A, Heim A, Christensen BT, Miltner A, Gehre M, Schmidt MWI (2009) Lignin dynamics in two ^{13}C-labelled arable soils during 18 years. Eur J Soil Sci 60:205–257

Hoosbeek MR, Scarascia-Mugnozza GE (2009) Increased litter build up and soil organic matter stabilization in a poplar plantation after 6 years of atmospheric CO_2 enrichment (FACE): final results of POP-EuroFACE compared to other forest FACE experiments. Ecosystems 12:220–239

Hoosbeek MR, Vos JM, Meinders MBJ, Velthorst EJ, Scarascia-Mugnozza GE (2007) Free atmospheric CO_2 enrichment (FACE) increased respiration and humification in the mineral soil of a poplar plantation. Geoderma 138:204–212

Hopkins DW, Chudek JA, Bignell DE, Frouz J, Webster EA, Lawson T (1998) Application of ^{13}C NMR to investigate the transformations and biodegradation of organic materials by wood- and soil-feeding termites, and a coprophagus litter-dwelling dipteran larva. Biodegradation 9:423–431

Horwath W (2007) Carbon cycling and formation of soil organic matter. In: Paul EA (ed) Soil microbiology, ecology, and biochemistry. Elsevier, Amsterdam, pp 303–339

Houghton JT, Ding Y, Griggs DJ, Noguer M, van der Linden PJ, Dai X, Maskell K, Johnson CA (2001) Climate change 2001: the scientific basis. Cambridge University Press, Cambridge

Hunter MD, Adl S, Pringle CM, Coleman DC (2003) Relative effects of macroinvertebrates and habitat on the chemistry of litter during decomposition. Pedobiologia 47:101–115

Huang J-G, Bergeron Y, Denneler B, Berninger F, Tardif J (2007) Response of forest trees to increased atmospheric CO_2. Crit Rev Plant Sci 26:265–283

Hungate BA, van Groenigen K-J, Six J, Jastrow JD, Luo Y, de Graaff M-A, van Kessel C, Osenberg CW (2009) Assessing the effect of elevated CO_2 on soil C: a comparison of four meta-analyses. Glob Change Biol 15:2020–2034

Huston MA, Wolverton S (2009) The global distribution of net primary production: resolving the paradox. Ecol Monogr 79:343–377

Hyvönen R, Ågren GI, Linder S, Persson T, Cotrufo MF, Ekblad A, Freeman M, Grelle A, Janssens IA, Jarvis PG, Kellomäki S, Lindroth A, Loustau D, Lundmark T, Norby RJ, Oren R, Pilegaard K, Ryan MG, Sigurdsson BD, Strömgren M, Oijen M, Wallin G (2007) The likely impact of elevated [CO_2], nitrogen deposition, increased temperature and management on carbon sequestration in temperate and boreal forest ecosystems: a literature review. New Phytol 173:463–480

Ito A, Oikawa T (2004) Global mapping of terrestrial primary productivity and light-use efficiency with a process-based model. In: Shiyomi M, Kawahata H, Koizumi H, Tsuda A, Awaya Y (eds) Global environmental change in the ocean and on land. Terrabup, Tokyo, Japan, pp 343–358

Iversen CM, Ledford J, Norby RJ (2008) CO_2 enrichment increases carbon and nitrogen input from fine roots in a deciduous forest. New Phytol 179:837–847

Jandl R, Lindner M, Vesterdahl L, Bauwens B, Baritz R, Hagedorn F, Johnson DW, Minkkinen K, Byrne KA (2007a) How strongly can forest management influence soil carbon sequestration? Geoderma 137:253–268

Jandl R, Vesterdal L, Olsson M, Bens O, Badeck F, Rock J (2007b) Carbon sequestration and forest management. CAB Rev Perspect Agric Vet Sci Nutr Nat Res . doi:10.1079/PAVSNNR20072017
Jastrow JD, Miller RM, Matamala R, Norby RJ, Boutton TW, Rice CW, Owensby CE (2005) Elevated atmospheric carbon dioxide increases soil carbon. Glob Change Biol 11:2057–2064
Jobbágy EG, Jackson RB (2000) The vertical distribution of soil organic carbon and its relation to climate and vegetation. Ecol Appl 10:423–436
Joergensen RG, Wichern F (2008) Quantiative assessment of the fungal contribution to microbial tissue in soil. Soil Biol Biochem 40:2977–2991
Johnson MG, Rygiewicz PT, Tingey DT, Phillips DL (2006) Elevated CO_2 and elevated temperature have no effect on Douglas-fir fine-root dynamics in nitrogen-poor soil. New Phytol 170:345–356
Jones HG (1998) Stomatal control of photosynthesis and transpiration. J Exp Bot 49:387–398
Jones DL, Hodge A, Kuzyakov Y (2004) Plant and mycorrhizal regulation of rhizodeposition. New Phytol 163:459–480
Jones DL, Nguyen C, Finlay RD (2009) Carbon flow in the rhizosphere: carbon trading at the soil-root interface. Plant Soil 321:5–33
Kaiser K, Guggenberger G (2000) The role of DOM sorption to mineral surfaces in the preservation of organic matter in soils. Org Geochem 31:711–725
Kalbitz K, Kaiser K (2008) Contribution of dissolved organic matter to carbon storage in forest mineral soils. J Plant Nutr Soil Sci 171:52–60
Kalbitz K, Kaiser K, Bargholz J, Dardenne P (2006) Lignin degradation controls the production of dissolved organic matter in decomposing foliar litter. Eur J Soil Sci 57:504–516
Kalbitz K, Schmerwitz J, Schwesig D, Matzner E (2003) Biodegradation of soil-derived dissolved organic matter as related to its properties. Geoderma 113:273–291
Kalbitz K, Solinger S, Park J-H, Michalzik B, Matzner E (2000) Controls on the dynamics of dissolved organic matter in soils: A review. Soil Sci 165:277–304
Karberg NJ, Pregitzer KS, King JS, Friend AL, Wood JR (2005) Soil carbon dioxide partial pressure and dissolved inorganic carbon chemistry under elevated carbon dioxide and ozone. Oecologia 142:296–306
Karnosky DF (2003) Impacts of elevated atmospheric CO_2 on forest trees and forest ecosystems: knowledge gaps. Environ Int 29:161–169
Karnosky DF, Tallis M, Darbah J, Taylor G (2007) Direct effects of elevated carbon dioxide on forest tree productivity. In: Freer-Smith PH, Broadmeadow MSJ, Lynch JM (eds) Forestry and climate change. CAB International, Wallingford, UK, pp 136–142
Kasurinen A, Peltonen PA, Julkunen-Tiitto R, Vapaavuori E, Nuutinen V, Holopainen T, Holopainen JK (2007) Effects of elevated CO_2 and O_3 on leaf litter phenolics and subsequent performance of litter-feeding soil macrofauna. Plant Soil 292:25–43
Kaye JP, Burke IC, Mosier AR, Guerschman JP (2004) Methane and nitrous oxide fluxes from urban soils to the atmosphere. Ecol Appl 14:975–981
Kaye JP, Groffman PM, Grimm NB, Baker LA, Pouyat RV (2006) A distinct urban biogeochemistry? Trends Ecol Evol 21:192–199
Keeling CD (1960) The concentration and isotopic abundances of carbon dioxide in the atmosphere. Tellus 12:200–203
Kelleher BP, Simpson AJ (2006) Humic substances in soils: are they really chemically distinct? Environ Sci Technol 40:4605–4611
Keppler F, Hamilton JTG, Braß M, Röckmann T (2006) Methane emissions from terrestrial plants under aerobic conditions. Nature 439:187–191
Keppler F, Hamilton JTG, McRoberts WC, Vigano I, Brass M, Röckmann T (2008) Methoxyl groups of plant pectin as a precursor of atmospheric methane: evidence from deuterium labelling studies. New Phytol 178:808–814
Kesselmeier J, Ciccioli P, Kuhn U, Stefani P, Biesenthal T, Rottenberger S, Wolf A, Vitullo M, Valentini R, Nobre A, Kabat P, Andreae MO (2002) Volatile organic compound emissions in relation to plant carbon fixation and the terrestrial carbon budget. Global Biogeochem Cy 16:1126. doi:10.1029/2001GB001813

Killham K, Prosser JI (2007) The prokaryotes. In: Paul EA (ed) Soil microbiology, ecology, and biochemistry. Elsevier, Amsterdam, pp 119–144
Killops S, Killops V (2005) Introduction to organic geochemistry. Blackwell, Malden, MA
Kimmins JP (2004) Forest ecology. Prentice Hall, Upper Saddle River, NJ
King JS, Hanson PJ, Bernhardt E, Deangelis P, Norby RJ, Pregitzer KS (2004) A multiyear synthesis of soil respiration responses to elevated atmospheric CO_2 from four forest FACE experiments. Glob Change Biol 10:1027–1042
King JS, Pregitzer KS, Zak DR, Holmes WE, Schmidt K (2005) Fine root chemistry and decomposition in model communities of north-temperate tree species show little response to elevated atmospheric CO_2 and varying soil resource availability. Oecologia 146:318–328
Kilpelainen A, Peltola H, Ryyppo A, Sauvala K, Laitinen K, Kellomaki S (2003) Wood properties of Scots pines (*Pinus sylvestris*) grown at elevated temperature and carbon dioxide concentration. Tree Physiol 23:889–897
Kirkby KJ, Smart SM, Black HIJ, Brunce RGH, Corney PM, Smithers RJ (2005) Long term ecological change in British woodland (1971–2001). English Nature Research Report 653, English Nature, London, UK
Kirschbaum MUF (2006) The temperature dependence of organic-matter decomposition-still a topic of debate. Soil Biol Biochem 38:2510–2518
Kleber M, Sollins P, Sutton R (2007) A conceptual model of organo-mineral associations in soils: self-assembly of organic molecular fragments into zonal structures on mineral surfaces. Biogeochemistry 85:9–24
Knicker H (2007) How does fire affect the nature and stability of soil organic nitrogen and carbon? a review. Biogeochemistry 85:91–118
Kögel-Knabner I (2002) The macromolecular organic composition of plant and microbial residues as inputs to soil organic matter. Soil Biol Biochem 34:139–162
Körner C (2003) Carbon limitation in trees. J Ecol 91:4–17
Körner C (2004) Through enhanced tree dynamics carbon dioxide enrichment may cause tropical forests to lose carbon. Philos Trans R Soc Lond B 359:493–498
Körner C (2005) An introduction to the functional diversity of temperate forest trees. In: Scherer-Lorenzen M, Körner C, Schulze E-D (eds) Forest diversity and function. Ecological studies, Vol. 176. Springer, Berlin, pp 13–37
Körner C (2006) Plant CO_2 responses: an issue of definition, time and resource supply. New Phytol 172:393–411
Körner C, Asshoff R, Bignucolo O, Hättenschwiler S, Keel SG, Peláez-Riedl S, Pepin S, Siegwolf RTW, Zotz G (2005) Carbon flux and growth in mature deciduous forest trees exposed to elevated CO_2. Science 309:1360–1362
Körner C, Morgan J, Norby R (2007) CO_2 fertilization: when, where, how much? In: Canadell JG, Pataki DE, Pitelka LF (eds) Terrestrial ecosystems in a changing world. Springer, Berlin, pp 9–21
Kolattukudy PE (2001) Polyesters in higher plants. In: Scheper Th (ed) Advances in biochemical engineering/biotechnology, Vol 71. Springer-Verlag, Berlin, pp 1–49
Kozlowski TT (1971a) Growth and development of trees – volume 1. Academic, New York
Kozlowski TT (1971b) Growth and development of trees – volume 2. Academic, New York
Kozlowski TT, Kramer PJ, Pallardy SG (1991) The physiological ecology of woody plants. Academic, San Diego, CA
Krull ES, Baldock JA, Skjemstad JO (2003) Importance of mechanisms and processes of the stabilisation of soil organic matter for modelling carbon turnover. Funct Plant Biol 30:207–222
Kuhlbusch TAJ (1998) Black carbon and the carbon cycle. Science 280:1903–1904
Kurz WA, Dymond CC, Stinson G, Rampley GJ, Neilson ET, Carroll AL, Ebata T, Safranyik L (2008) Mountain pine beetle and forest carbon feedback to climate change. Nature 452:987–990
Kuzyakov Y (2006) Sources of CO_2 efflux from soil and review of partitioning methods. Soil Biol Biochem 38:425–448
Kuzyakov Y, Friedel JK, Stahr K (2000) Review of mechanisms and quantification of priming effects. Soil Biol Biochem 32:1485–1498

Kuzyakov Y, Subbotina I, Chen H, Bogomolova I, Xu X (2009) Black carbon decomposition and incorporation into soil microbial biomass estimated by ^{14}C labeling. Soil Biol Biochem 41:210–219

Ladeau SL, Clark JS (2006) Pollen production by *Pinus taeda* growing in elevated atmospheric CO_2. Funct Ecol 20:541–547

Lamlom SH, Savidge RA (2003) A reassessment of carbon content in wood: variation within and between 41 North American species. Biomass Bioenerg 25:381–388

Landi A, Mermut AR, Anderson DW (2003) Origin and rate of pedogenic carbonate accumulation in Saskatchewan soils, Canada. Geoderma 117:143–156

Langley JA, McKinley DC, Wolf AA, Hungate BA, Drake BG, Megonigal JP (2009) Priming depletes soil carbon and releases nitrogen in a scrub-oak ecosystem. Soil Biol Biochem 41:54–60

Laothawornkitkul J, Taylor JE, Paul ND, Hewitt CN (2009) Biogenic volatile organic compounds in the Earth system. New Phytol 183:27–51

Lapenis AG, Lawrence GB, Bailey SW, Aparin BF, Shiklomanov AI, Speranskaya NA, Torn MS, Calef M (2008) Climatically driven loss of calcium in steppe soil as a sink for atmospheric carbon. Global Biogeochem Cy 22, GB2010, doi:10.1029/2007GB003077

Leakey ADB, Xu F, Gillespie KM, McGrath JM, Ainsworth EA, Ort DR (2009) Genomic basis for stimulated respiration by plants growing under elevated carbon dioxide. Proc Natl Acad Sci USA 106:3597–3602

Ledford H (2008) Forestry carbon dioxide projects to close down. Nature 456:289

Leff JW, Fierer N (2008) Volatile organic compound (VOC) emissions from soil and litter samples. Soil Biol Biochem 40:1629–1636

Lehmann J, Skjemstad J, Sohi S, Carter J, Barson M, Falloon P, Coleman K, Woodbury P, Krull E (2008a) Australian climate-carbon cycle feedback reduced by soil black carbon. Nat Geosci 1:832–835

Lehmann J, Solomon D, Kinyangi J, Dathe L, Wirick S, Jacobsen C (2008b) Spatial complexity of soil organic matter forms at nanometre scales. Nat Geosci 1:238–242

Lenton TM, Britton C (2006) Enhanced carbonate and silicate weathering accelerates recovery from fossil fuel CO_2 perturbations. Global Biogeochem Cy 20, GB3009, doi:10.1029/2005GB002678

Lerdau MT (2003) Keystone molecules and organic chemical flux from plants. In: Melillo JM, Field CB, Moldan B (eds) Interactions of the major biogeochemical cycles: global change and human impacts. Island Press, Washington, D.C., pp 177–192

Lerdau M (2007) A positive feedback with negative consequences. Science 316:212–213

Lesaulnier C, Papamichail D, McCorkle S, Ollivier B, Skierna S, Taghavi S, Zak D, Van der Lelie D (2008) Elevated atmospheric CO_2 affects soil microbial diversity associated with trembling aspen. Environ Microbiol 10:926–941

Leslie M (2009) On the origin of photosynthesis. Science 323:1286–1287

Leuzinger S, Körner C (2007) Water savings in mature deciduous forest trees under elevated CO_2. Glob Change Biol 13:2498–2508

Levia DF, Frost EE (2003) A review and evaluation of stemflow literature in the hydrologic and biogeochemical cycles of forested and agricultural ecosystems. J Hydrol 274:1–29

Lewis SL (2006) Tropical forests and the changing earth system. Philos Trans R Soc Lond B 361:195–210

Lewis SL, Malhi Y, Phillips OL (2004) Fingerprinting the impacts of global change on tropical forests. Philos Trans R Soc Lond B 359:437–462

Lichter J, Billings SA, Ziegler SE, Gaindh D, Ryals R, Finzi AC, Jackson RB, Stemmler EA, Schlesinger WH (2008) Soil carbon sequestration in a pine forest after 9 years of atmospheric CO_2 enrichment. Glob Change Biol 14:2910–2922

Likens GE, Edgerton ES, Galloway JN (1983) The composition and deposition of organic carbon in precipitation. Tellus 35(B):16–24

Lin C, Owen SM, Peñuelas J (2007) Volatile organic compounds in the roots and rhizosphere of *Pinus* spp. Soil Biol Biochem 39:951–960

Lindroth A, Lagergren F, Aurelia M, Bjarnadottir B, Christensen T, Dellwik E, Grelle A, Ibrom A, Johansson T, Lankreijer H, Launiainen S, Laurila T, Mölder M, Nikinmaa E, Pilegaard K,

Sigurdsson BD, Vesala T (2008) Leaf area index is the principal scaling parameter for both gross photosynthesis and ecosystem respiration of Northern deciduous and coniferous forests. Tellus 60B:129–142

Litton CM, Raich JW, Ryan MG (2007) Review: carbon allocation in forest ecosystems. Glob Change Biol 13:2089–2109

Long SP, Ainsworth EA, Leakey ADB, Nösberger J, Ort DR (2006) Food for thought: lower-than-expected crop yield stimulation with rising CO_2 concentrations. Science 312:1918–1921

Long SP, Ainsworth EA, Rogers A, Ort DR (2004) Rising atmospheric carbon dioxide: plants FACE the future. Annu Rev Plant Biol 55:591–628

Lorenz K, Lal R (2005) The depth distribution of soil organic carbon in relation to land use and management and the potential of carbon sequestration in subsoil horizons. Adv Agron 88:35–66

Lorenz K, Lal R, Preston CM, Nierop KGJ (2007) Strengthening the soil organic carbon pool by increasing contributions from recalcitrant aliphatic bio(macro)molecules. Geoderma 142:1–10

Lorenz K, Lal R, Preston CM, Nierop KGJ (2009) Soil organic carbon sequestration by biochemically recalcitrant biomacromolecules. In: Lal R, Follett RF (eds) Soil carbon sequestration and the greenhouse effect, 2nd edn. Soil Science Society of America Special Publication 57, Madison, WI, pp 207–222

Lüttge U (2006) Photosynthetic flexibility and ecophysiological plasticity: questions and lessons from *Clusia*, the only CAM tree, in the neotropics. New Phytol 171:7–25

von Lützow M, Kögel-Knabner I, Ekschmitt K, Matzner E, Guggenberger G, Marschner B, Flessa H (2006) Stabilization of organic matter in temperate soils: mechanisms and their relevance under different soil conditions – a review. Eur J Soil Sci 57:426–445

von Lützow M, Kögel-Knabner I, Ludwig B, Matzner E, Flessa H, Ekschmitt K, Guggenberger G, Marschner B, Kalbitz K (2008) Stabilization mechanisms of organic matter in four temperate soils: development and application of a conceptual model. J Plant Nutr Soil Sci 171:111–124

Ludley KE, Jickells SM, Chamberlain PM, Whitaker J, Robinson CH (2009) Distribution of monoterpenes between organic resources in upper soil horizons under monocultures of *Picea abies*, *Picea sitchensis* and *Pinus sylvestris*. Soil Biol Biochem 41:1050–1059

Lukac M, Lagomarsino A, Moscatelli MC, De Angelis P, Cotrufo MF, Godbold DL (2009) Forest soil carbon cycle under elevated CO_2 - a case of increased throughput? Forestry 82:75–86

Luo Y (2007) Terrestrial carbon-cycle feedback to climate warming. Annu Rev Ecol Evol Syst 38:683–712

Luo Y, Chen JL, Reynolds JF, Field CB, Mooney HA (1997) Disproportional increases in photosynthesis and plant biomass in a Californian grassland exposed to elevated CO_2: a simulation analysis. Funct Ecol 11:696–704

Luo ZB, Calfapietra C, Scarascia-Mugnozza G, Liberloo M, Polle A (2008) Carbon-based secondary metabolites and internal N pools in *Populus nigra* under Free Air CO_2 Enrichment (FACE) and N fertilization. Plant Soil 304:45–57

Luyssaert S, Inglima I, Jung M, Richardson AD, Reichstein M, Papale D, Piao SL, Schulze E-D, Wingate L, Matteucci G, Aragao L, Aubinet M, Beer C, Bernhofer C, Black KG, Bonal D, Bonnefond J-M, Chambers J, Ciais P, Cook B, Davis KJ, Dolman AJ, Gielen B, Goulden M, Grace J, Granier A, Grelle A, Griffis T, Grünwald T, Guidolotti G, Hanson PJ, Harding R, Hollinger DY, Hutyra LR, Kolari P, Kruijt B, Kutsch W, Lagergren F, Laurila T, Law BE, Le Maire G, Lindroth A, Loustau D, Malhi Y, Mateus J, Migliavacca M, Misson L, Montagnani L, Moncrieff J, Moors E, Munger JW, Nikinmaa E, Ollinger SV, Pita G, Rebmann C, Roupsard O, Saigusa N, Sanz MJ, Seufert G, Sierra C, Smith M-L, Tang J, Valentini R, Vesala T, Janssens IA (2007) The CO_2-balance of boreal, temperate and tropical forests derived from a global database. Glob Change Biol 13:2509–2537

Madigan MT, Martinko JM (2006) Brock – biology of microorganisms. Prentice Hall, Upper Saddle River, NJ

Mahieu N, Powlson DS, Randall EW (1999) Statistical analysis of published carbon-13 CPMAS NMR spectra of soil organic matter. Soil Sci Soc Am J 63:307–319

Malhi Y, Baldocchi DD, Jarvis PG (1999) The carbon balance of tropical, temperate and boreal forests. Plant Cell Environ 22:715–740
Malhi Y, Meir P, Brown S (2002) Forests, carbon and global climate. Philos Trans R Soc Lond A 360:1567–1591
van Mantgem PJ, Stephenson NL, Byrne JC, Daniels LD, Franklin JF, Fulé PZ, Harmon ME, Larson AJ, Smith JM, Taylor AH, Veblen TT (2009) Widespread increase of tree mortality rates in the western United States. Science 323:521–524
Manzoni S, Jackson RB, Trofymow JA, Porporato A (2008) The global stoichiometry of litter nitrogen mineralization. Science 321:684–686
Marschner B, Brodowski S, Dreves A, Gleixner G, Gude A, Grootes PM, Hamer U, Heim A, Jandl G, Ji R, Kaiser K, Kalbitz K, Kramer C, Leinweber P, Rethemeyer J, Schäffer A, Schmidt MWI, Schwark L, Wiesenberg GLB (2008) How relevant is recalcitrance for the stabilization of organic matter in soils? J Plant Nutr Soil Sci 171:91–110
Matthews E (1997) Global litter production, pools, and turnover times: estimates from measurement data and regression models. J Geophys Res 102:18771–18800
McCarthy HR, Oren R, Finzi AC, Johnsen KH (2006) Canopy leaf area constrains [CO_2]-induced enhancement of productivity and partitioning among aboveground carbon pools. Proc Natl Acad Sci USA 103:19356–19361
McDowell WH (2003) Dissolved organic matter in soils-future directions and unanswered questions. Geoderma 113:179–186
McMurtrie RE, Norby RJ, Medlyn BE, Dewar RC, Pepper DA, Reich PB, Barton CVM (2008) Why is plant-growth response to elevated CO_2 amplified when water is limiting, but reduced when nitrogen is limiting? A growth-optimisation hypothesis. Funct Plant Biol 35:521–534
Meier IC, Leuschner C (2008) Belowground drought response of European beech: fine root biomass and carbon partitioning in 14 mature stands across a preciptation gradient. Glob Change Biol 14:2081–2095
Melillo JM, Steudler PA, Aber JD, Newkirk K, Lux H, Bowles FP, Catricala C, Magill A, Ahrens T, Morrisseau S (2002) Soil warming and carbon-cycle feedbacks to the climate system. Science 298:2173–2176
Mercado LM, Bellouin N, Sitch S, Boucher O, Huntingford C, Wild M, Cox PM (2009) Impact of changes in diffuse radiation on the global land carbon sink. Nature 458:1014–1018
Mi N, Wang S, Liu J, Yu G, Zhang W, Jobbágy EG (2008) Soil inorganic carbon storage pattern in China. Glob Change Biol 14:2380–2387
Millard P, Sommerkorn M, Grelet G-A (2007) Environmental change and carbon limitation in trees: a biochemical, ecophysiological and ecosystem appraisal. New Phytol 175:11–28
Miltner A, Kopinke F-D, Kindler R, Selesi D, Hartmann A, Kästner M (2005) Non-phototrophic CO_2 fixation by soil microorganisms. Plant Soil 269:193–203
Mohan JE, Clark JS, Schlesinger WH (2007) Long-term CO_2 enrichment of a forest ecosystem: implications for forest regeneration and succession. Ecol Appl 17:1198–1212
Monson RK, Trahan N, Rosenstiel TN, Veres P, Moore D, Wilkinson M, Norby RJ, Volder A, Tjoelker MG, Briske DD, Karnosky DF, Fall R (2007) Isoprene emission from terrestrial ecosystems in response to global change: minding the gap between models and observations. Philos Trans.R Soc A 365:1677–1695
Montheith DT, Stoddard JL, Evans CD, de Wit HA, Forsius M, Høgåsen T, Wilander A, Skjelkvåle BL, Jeffries DS, Vuorenmaa J, Keller B, Kopácek J, Veseley J (2007) Dissolved organic carbon trends resulting from changes in atmospheric deposition chemistry. Nature 450:537–541
Mooney HA (1972) The carbon balance of plants. Annu Rev Ecol Syst 3:315–346
Moore BD, Cheng S-H, Sims D, Seemann JR (1999) The biochemical and molecular basis for photosynthetic acclimation to elevated atmospheric CO_2. Plant Cell Environ 22:567–582
Moore DJP, Gonzalez-Meler MA, Taneva L, Pippen JS, Kim H-S, DeLucia EH (2008) The effect of carbon dioxide enrichment on apparent stem respiration from *Pinus taeda* L. is confounded by high levels of soil carbon dioxide. Oecologia 158:1–10
Moorhead DL, Sinsabaugh RL (2006) A theoretical model of litter decay and microbial interaction. Ecol Monogr 76:151–174

Neff JC, Asner GP (2001) Dissolved organic carbon in terrestrial ecosystems: synthesis and a model. Ecosystems 4:29–48

Neilson RP, Drapek RJ (1998) Potentially complex biosphere responses to transient global warming. Glob Change Biol 4:505–521

Neilson RP, Drapek RJ (1998) Potentially complex biosphere responses to transient global warming. Glob Change Biol 4:505-521

Nguyen C (2003) Rhizodeposition of organic C by plants: mechanisms and controls. Agronomie 23:375–396

Nierop KGJ, Filley TR (2008) Simultaneous analysis of tannin and lignin signatures in soils by thermally assisted hydrolysis and methylation using 13C-labeled TMAH. J Anal Appl Pyrolysis 83:227–231

Nisbet RER, Fisher R, Nimmo RH, Bendall DS, Crill PM, Gallego-Sala AV, Hornibrook ERC, López-Juez E, Lowry D, Nisbet PBR, Shuckburgh EF, Sriskantharajah S, Howe CJ, Nisbet EG (2009) Emission of methane from plants. Proc R Soc B . doi:101098/rspb.2008.1731

Nobel PS (2005) Physicochemical and environmental plant physiology. Elsevier, Amsterdam

Norby RJ, Cotrufo MF, Ineson P, O'Neill EG, Canadell JG (2001) Elevated CO_2, litter chemistry, and decomposition: a synthesis. Oecologia 127:153–165

Norby RJ, Jackson RB (2000) Root dynamics and global change: seeking an ecosystem perspective. New Phytol 147:3–12

Norby RJ, Ledford J, Reilly CD, Miller NE, O'Neill EG (2004) Fine-root production dominates response of a deciduous forest to atmospheric CO_2 enrichment. Proc Natl Acad Sci USA 101:9689–9693

Norby RJ, DeLucia EH, Gielen B, Calfapietra C, Giardina CP, King JS, Ledford J, McCarthy HR, Moore DJP, Ceulemans R, De Angelis P, Finzi AC, Karnosky DF, Kubiske ME, Lukac M, Pregitzer KS, Scarascia-Mugnozza G, Schlesinger WH, Oren R (2005) Forest response to elevated CO_2 is conserved across a broad range of productivity. Proc Natl Acad Sci USA 102:18052–18056

Norby RJ, Rustad LE, Dukes JS, Ojima DS, Parton WJ, Del Grosso SJ, McMurtie RE, Pepper DA (2007) Ecosystem responses to warming and interacting global change factors. In: Canadell JG, Pataki DE, Pitelka LF (eds) Terrestrial ecosystems in a changing world. Springer, Berlin, pp 23–36

Novaes E, Osorio L, Drost DR, Miles BL, Boaventura-Novaes CRD, Benedict C, Dervinis C, Yu Q, Sykes R, Davis M, Martin TA, Peter GF, Kirst M (2009) Quantitative genetic analysis of biomass and wood chemistry of *Populus* under different nitrogen levels. New Phytol 182:878–890

Nowak SR, Ellsworth DS, Smith SD (2004) Functional responses of plants to elevated atmospheric CO_2 - do photosynthetic and productivity data from FACE experiments support early predictions? New Phytol 162:253–280

Oh N-H, Hofmockel M, Lavine ML, Richter DD (2007) Did elevated atmospheric CO_2 alter soil mineral weathering? Glob Change Biol 13:2626–2641

Osler GHR, Sommerkorn M (2007) Toward a complete soil C and N cycle: incorporating the soil fauna. Ecology 88:1611–1621

Paré D, Boutin R, Larocque GR, Raulier F (2006) Effect of temperature on soil organic matter decomposition in three forest biomes of eastern Canada. Can J Soil Sci 86:247–256

Parry ML, Canziani OF, Polutikopf JP, van der Linden PJ, Hanson CE (eds) (2007) Climate change 2007: impacts, adaptation and vulnerability. Contribution of working group II to the fourth assessment report of the intergovernmental panel on climate change. Cambridge University Press, Cambridge

Parsons WFJ, Bockheim JG, Lindroth RL (2008) Independent, interactive, and species-specific responses of leaf litter decomposition to elevated CO_2 and O_3 in a northern hardwood forest. Ecosystems 11:505–519

Parton W, Silver WL, Burke IC, Grassens L, Harmon ME, Currie WS, King JY, Adair EC, Brandt LA, Hart SC, Fasth B (2007) Global-scale similarities in nitrogen release patterns during long-term decomposition. Science 315:361–364

Pataki DE, Xu T, Luo YQ, Ehleringer JR (2007) Inferring biogenic and anthropogenic carbon dioxide sources across an urban to rural gradient. Oecologia 152:307–322

Pendall E, Bridgham S, Hanson PJ, Hungate B, Kicklighter DW, Johnson DW, Law BE, Luo Y, Patrick Megonigal J, Olsrud M, Ryan MG, Wan S (2004) Below-ground process responses to elevated CO_2 and temperature: a discussion of observations, measurement methods, and models. New Phytol 162:311–322

Pfanz H, Aschan G, Langenfeld-Heyser R, Wittmann C, Loose M (2002) Ecology and ecophysiology of tree stems: corticular and wood photosynthesis. Naturwissenschaften 89:147–162

Plomion C, Leprovost G, Stokes A (2001) Wood formation in trees. Plant Physiol 127:1513–1523

Poirier N, Derenne S, Balesdent J, Chenu C, Bardoux G, Mariotti A, Largeau C (2006) Dynamics and origin of the non-hydrolysable organic fraction in a forest and a cultivated temperate soil, as determined by isotopic and microscopic studies. Eur J Soil Sci 57:719–730

Pollierer MM, Langel R, Körner C, Maraun M, Scheu S (2007) The underestimated importance of belowground carbon input for forest soil animal food webs. Ecol Lett 10:729–736

Pregitzer KS, Euskirchen ES (2004) Carbon cycling and storage in world forests: biome patterns related to forest age. Glob Change Biol 10:2052–2077

Pregitzer KS, Euskirchen ES (2004) Carbon cycling and storage in world forests: biome patterns related to forest age. Glob Change Biol 10:2052–2077

Prescott CE (2005) Do rates of litter decomposition tell us anything we really need to know? For Ecol Manag 220:66–74

Quine TA, Van Oost K (2007) Quantifying carbon sequestration as a result of soil erosion and deposition: retrospective assessment using caesium-137 and carbon inventories. Glob Change Biol 13:2610–2625

Rae AM, Tricker PJ, Bunn SM, Taylor G (2007) Adapation of tree growth to elevated CO_2: quantiative trait loci for biomass in *Populus*. New Phytol 175:59–69

Ramanathan V, Carmichael G (2008) Global and regional climate changes due to black carbon. Nat Geosci 1:221–227

Ramanathan V, Feng Y (2008) On avoiding dangerous anthropogenic interference with the climate system: formidable challenges ahead. Proc Natl Acad Sci USA 105:14245–14250

Rasse DP, Rumpel C, Dignac M-F (2005) Is soil carbon mostly root carbon? Mechanisms for a specific stabilization. Plant Soil 269:341–356

Raven PH, Evert RF, Eichhorn SE (2005) Biology of plants, 7th edn. WH Freeman & Co, New York

Rawlins AJ, Bull ID, Poirier N, Ineson P, Evershed RP (2006) The biochemical transformation of oak (*Quercus robur*) leaf litter consumed by the pill millipede (*Glomeris marginata*). Soil Biol Biochem 38:1063–1076

Reich PB, Hungate BA, Luo Y (2006) Carbon-nitrogen interactions in terrestrial ecosystems in response to rising atmospheric carbon dioxide. Annu Rev Evol Syst 37:611–636

Reich PB, Oleksyn J, Modrzynski J, Mrozinski P, Hobbie SE, Eissenstat DM, Chorover J, Chadwick OA, Hale CM, Tjoelker MG (2005) Linking litter calcium, earthworms and soil properties: a common garden test with 14 tree species. Ecol Lett 8:811–818

Reichstein M (2007) Impacts of climate change on forest soil carbon: principles, factors, models, uncertainties. In: Freer-Smith PH, Broadmeadow MSJ, Lynch JM (eds) Forestry and climate change. CAB International, Wallingford, UK, pp 127–135

Rennenberg H, Loreto F, Polle A, Brilli F, Fares S, Beniwal RS, Gessler A (2006) Physiological responses of forest trees to heat and drought. Plant Biol 8:556–571

Reyes-García C, Andrade JL (2009) Crassulacean acid metabolism under global climate change. New Phytol 181:754–757

Robinson D (2007) Implications of a large global root biomass for carbon sink estimates and for soil carbon dynamics. Proc R Soc B 274:2753–2759

Rogers A, Ainsworth EA (2006) The response of foliar carbohydrates to elevated [CO_2]. In: Nösberger J, Long SP, Norby RJ, Stitt M, Hendrey GR, Blum H (eds) Managed ecosystems and CO_2 – case studies, processes, and perspectives. Springer, Berlin, pp 293–308

Romanou A, Liepert B, Schmidt GA, Rossow WB, Ruedy RA, Zhang Y (2007) 20th century changes in surface solar irradiance in simulations and observations. Geophys Res Lett 34:L05713. doi:10.1029/2006GL028356

Rosenfeld D, Lohmann U, Raga GB, O'Dowd CD, Kulmala M, Fuzzi S, Reissell A, Andreae MO (2008) Flood or drought: how do aerosols affect precipitation? Science 321:1309–1313
Runion GB, Entry JA, Prior SA, Mitchell RJ, Rogers HH (1999) Tissue chemistry and carbon allocation in seedlings of *Pinus palustris* subjected to elevated atmospheric CO_2 and water strees. Tree Physiol 19:329–335
Rustad LE, Campbell JL, Marion GM, Norby RJ, Mitchell MJ, Hartley AE, Cornelissen JHC, Gurevitch J, GCTE-NEWS (2001) A meta-analysis of the response of soil respiration, net nitrogen mineralization, and aboveground plant growth to experimental ecosystem warming. Oecologia 126:543–562
Ryan MG, Lavigne MB, Gower ST (1997) Annual carbon costs of autotrophic respiration in boreal forest ecosystems in relation to species and climate. J Geophys Res 102:871–883
Ryan MG, Law BE (2005) Interpreting, measuring, and modeling soil respiration. Biogeochemistry 73:3–27
Sage RF (2001) Environmental and evolutionary preconditions for the origin and diversification of the C_4 photosynthetic syndrome. Plant Biol 3:202–213
Salomé C, Nunan N, Pouteau V, Lerch TZ, Chenu C (2009) Carbon dynamics in topsoil and in subsoil may be controlled by different regulatory mechanisms. Glob Change Biol doi: 10.1111/j.1365-2486.2009.01884.x
Saxe H, Cannell MGR, Johnsen Ø, Ryan MG, Vourlitis G (2001) Tree and forest functioning in response to global warming. New Phytol 149:369–400
Schade GW, Hofmann R-M, Crutzen PJ (1999) CO emissions from degrading plant matter. Tellus 51B:889–908
Schenk HJ (2008) The shallowest possible water extraction profile: a null model for global root distributions. Vadose Zone J 7:1119–1124
Schenk HJ, Jackson RB (2005) Mapping the global distribution of deep roots in relation to climate and soil characteristics. Geoderma 126:129–140
Scherer-Lorenzen M, Potvin C, Koricheva J, Schmid B, Hector A, Bornik Z, Reynolds G, Schulze E-D (2005) The design of experimental tree plantations for functional biodiversity research. In: Scherer-Lorenzen M, Körner C, Schulze E-D (eds) Forest diversity and function. Ecological studies, Vol. 176. Springer, Berlin, pp 347–376
Schlesinger WH (1990) Evidence from chronosequence studies for a low carbon-storage potential of soils. Nature 348:232–234
Schlesinger WH (1997) Biogeochemistry – an analysis of global change. Academic, San Diego, CA
Schlesinger WH (2006) Inorganic carbon and the global C cycle. In: Lal R (ed) Encyclopedia of soil science. Taylor & Francis, London, pp 879–881
Schmidt MWI, Noack AG (2000) Black carbon in soils and sediments: analysis, distribution, implications, and current challenges. Global Biogeochem Cy 14:777–793
Schmidt MWI, Skjemstad JO, Jäger C (2002) Carbon isotope geochemistry and nanomorphology of soil black carbon: black chernozemic soils in central Europe originate from ancient biomass burning. Global Biogeochem Cy 16:1123. doi:10.1029/2002GB001939
Schulze E-D (2006) Biological control of the terrestrial carbon sink. Biogeosciences 3:147–166
Schulze E-D, Beck E, Müller-Hohenstein K (2005) Plant ecology. Springer, Berlin
Schweitzer JA, Madritch MD, Bailey JK, LeRoy CJ, Fischer DG, Rehill BJ, Lindroth RL, Hagerman AE, Wooley SC, Hart SC, Whitham TG (2008) From genes to ecosystems: the genetic basis of condensed tannins and their role in nutrient regulation in a *Populus* model system. Ecosystems 11:1005–1020
Scott-Denton LE, Rosenstiel TN, Monson RK (2006) Differential controls by climate and substrate over the heterotrophic and rhizospheric components of soil respiration. Glob Change Biol 12:205–216
Seco R, Peñuelas J, Filella I (2007) Short-chain oxygenated VOCs: emission and uptake by plants and atmospheric sources, sinks, and concentrations. Atmos Environ 41:2477–2499
Seinfeld JH, Pandis SN (1998) Atmospheric chemistry and physics: from air pollution to climate change. Wiley, New York

Seth MK (2004) Trees and their economic importance. Bot Rev 69:321–376
Sharkey TD, Yeh S (2001) Isoprene emission from plants. Annu Rev Plant Physiol Plant Mol Biol 52:407–436
Sharma S, Williams DG (2009) Carbon and oxygen isotope analysis of leaf biomass reveals contrasting photosynthetic responses to elevated CO_2 near geologic vents in Yellowstone National Park. Biogeosciences 6:25–31
Shibu ME, Leffelaar PA, Van Keulen H, Aggarwal PK (2006) Quantitative description of soil organic matter dynamics – a review of approaches with reference to rice-based cropping systems. Geoderma 137:1–18
Shindell D, Faluvegi G (2009) Climate response to regional radiative forcing during the twentieth century. Nat Geosci 2:294–300
Silver WL, Miya RK (2001) Global patterns in root decomposition: comparisons of climate and litter quality effects. Oecologia 129:407–419
Simpson AJ, Simpson MJ, Smith E, Kelleher BP (2007) Microbially derived inputs to soil organic matter: are current estimates too low? Environ Sci Technol 41:8070–8076
Six J, Frey SD, Thiet RK, Batten KM (2006) Bacterial and fungal contributions to carbon sequestration in agroecosystems. Soil Sci Soc Am J 70:555–569
Smith P (2004) How long before a change in soil organic carbon can be detected? Glob Change Biol 10:1–6
Smith P, Fang C, Dawson JJC, Moncrieff JB (2008) Impact of global warming on soil organic carbon. Adv Agron 97:1–43
Spreitzer RJ, Salvucci ME (2002) Rubisco: structure, regulatory interactions, and possibilities for a better enzyme. Annu Rev Plant Biol 53:449–475
Steinmann K, Siegwolf RTW, Saurer M, Körner C (2004) Carbon fluxes to the soil in a mature temperate forest assessed by ^{13}C isotope tracing. Oecologia 141:489–501
Stevenson FJ (1994) Humus chemistry. Wiley, New York
Stiling P, Cornelissen T (2007) How does elevated carbon dioxide (CO_2) affect plant-herbivory interactions? A field experiment and meta-analysis of CO_2-mediated changes on plant chemistry and herbivore performance. Glob Change Biol 13:1823–1842
Stiling P, Moon D, Rossi A, Hungate BA, Drake B (2009) Seeing the forest for the trees: long term exposure to elevated CO_2 increases some herbivore densities. Glob Change Biol 15:1895–1902
Stitt M, Schulze E-D (1994) Plant growth, storage, and resource allocation: from flux control in a metabolic chain to the whole-plant level. In: Schulze E-D (ed) Flux control in biological systems: from enzymes to populations and ecosystems. Academic Press, San Diego, CA, pp 57–118
Strand AE, Pritchard SG, McCormack ML, Davis MA, Oren R (2008) Irreconcilable differences: fine-root life spans and soil carbon persistence. Science 319:456–458
Subke J-A, Inglima I, Cotrufo MF (2006) Trends and methodological impacts in soil CO_2 efflux partitioning: a metaanalytical review. Glob Change Biol 12:921–943
Swift MJ, Heal OW, Anderson JM (1979) Decomposition in terrestrial ecosystems. University of California Press, Berkeley, CA
Taiz L, Zeiger E (2006) Plant physiology. Sinauer, Sunderland
Talbot JM, Allison SD, Treseder KK (2008) Decomposers in disguise: mycorrhizal fungi as regulators of soil C dynamics in ecosystems under global change. Funct Ecol 22:955–963
Taneva L, Gonzalez-Meler MA (2008) Decomposition kinetics of soil carbon of different age from a forest exposed to 8 years of elevated atmospheric CO_2 concentration. Soil Biol Biochem 40:2670–2677
Tans P (2009) Trends in atmospheric carbon dioxide - global. http://www.esrl.noaa.gov/gmd/ccgg/trends/
Tarr MA, Miller WL, Zepp RG (1995) Direct carbon monoxide photoproduction from plant matter. J Geophys Res 100D:11403–11413
Taylor G, Tallis MJ, Giardina CP, Percy KE, Miglietta F, Gupta PS, Gioli B, Calfapietra C, Gielen B, Kubiske ME, Scarascia-Mugnozza G, Kets K, Long SP, Karnosky DF (2008) Future atmospheric CO_2 leads to delayed autumnal senescence. Glob Change Biol 14:264–275

Tcherkez GGB, Farquhar GD, Andrews TJ (2006) Despite slow catalysis and confused substrate specifity, all ribulose bisphosphate carboxylases may be nearly perfectly optimized. Proc Natl Acad Sci USA 103:7246–7251

Terazawa K, Ishizuka S, Sakata T, Yamada K, Takahashi M (2007) Methane emissions from stems of *Fraxinus mandshurica* var. *japonica* trees in a floodplain forest. Soil Biol Biochem 39:2689–2692

Teske ME, Thistle HW (2004) A library of forest canopy structure for use in interception modeling. For Ecol Manag 198:341–350

Thauer RK (2007) A fifth pathway of carbon fixation. Science 318:1732–1733

Thiet RK, Frey SD, Six J (2006) Do growth yield efficiencies differ between soil microbial communities differing in fungal:bacterial ratios? Reality check and methodological issues. Soil Biol Biochem 38:837–844

Trenberth KE, Jones PD, Ambenje P, Bojariu R, Easterling D, Klein Tank A, Parker D, Rahimzadeh F, Renwick JA, Rusticucci M, Soden B, Zhai P (2007) Observations: surface and atmospheric climate change. In: Solomon S, Qin D, Manning M, Chen Z, Marquis M, Averyt KB, Tignor M, Miller HL (eds) Climate change 2007: the physical science basis. Contribution of working group I to the fourth assessment report of the intergovernmental panel on climate change. Cambridge University Press, Cambridge, UK/New York, NY, pp 235–336

Trumbore S (2006) Carbon respired by terrestrial ecosystems – recent progress and challenges. Glob Change Biol 12:141–153

Trumbore SE, Gaudinski JB (2003) The secret lives of roots. Science 302:1344–1345

Vogt KA, Vogt DJ, Palmiotto PA, Boon P, O'Hara J, Asbjornsen H (1996) Review of root dynamics in forest ecosystems grouped by climate, climatic forest types and species. Plant Soil 187:159–219

Urban O (2003) Physiological impacts of elevated CO_2 concentration ranging from molecular to whole plant responses. Phytosynthetica 41:9–20

Vancampenhout K, Wouters K, De Vos B, Buurman P, Swennen R, Deckers J (2009) Differences in chemical composition of soil organic matter in natrual ecosystems from different climatic regions – a pyrolysis-GC/MS study. Soil Biol Biochem 41:568–579

Wall DH, Bradford MA, John MGS, Trofymow JA, Behan-Pelletier V, Bignell DE, Dangerfield JM, Parton WJ, Rusek J, Voigt W, Wolters V, Zadeh Gardel H, Ayuke FO, Bashford R, Beljakova OI, Bohlen PJ, Brauman A, Flemming S, Henschel JR, Johnson DL, Hefin Jones T, Kovarova M, Kranabetter JM, Kutny L, Lin K-C, Maryati M, Masse D, Pokarzhevski A, Rahman H, Sabará MG, Salamon J-A, Swift MJ, Varela A, Vasconcelos HL, White D, Zou Z (2008) Global decomposition experiment shows soil animal impacts on decomposition are climate-dependent. Glob Change Biol 14:2661–2677

Wang K, Dickinson RE, Liang S (2009) Clear sky visibility has decreased over land globally from 1973 to 2007. Science 323:1468–1470

Wang Z-P, Han X-G, Wang GG, Song Y, Gulledge J (2008) Aerobic methane emission from plant in the Inner Mongolia steppe. Environ Sci Technol 42:62–68

Wardlaw IF (1990) The control of carbon partitioning in plants. New Phytol 116:341–381

Waring RW, Running SW (2007) Forest ecosystems – analysis at multiple scales. Elsevier Academic, Burlington, MA

WBCSD (World business council for sustainable development) (2007) The sustainable forest products industry, carbon and climate change – key messages for policy-makers. Geneva, Switzerland. www.wbcsd.org/DocRoot/IUMhw6W4Ia0Sbp4edftv/sfpi-carbon-climate.pdf

Webster EA, Tilston EL, Chudek JA, Hopkins DW (2008) Decompostion in soil and chemical characteristics of pollen. Eur J Soil Sci 59:551–558

Weedon JT, Cornwell WK, Cornelissen JHC, Zanne AE, Wirth C, Coomes DA (2009) Global meta-analysis of wood decomposition rates: a role for trait variation among tree species? Ecol Lett 12:45–56

West LT, Drees LR, Wilding LP, Rabenhorst MC (1988) Differentiation of pedogenic and lithogenic carbonate forms in Texas. Geoderma 43:271–287

Wild M (2009) Global dimming and brightening: a review. J Geophys Res 114, D00D16, doi:10.1029/2008JD011470

Wild M, Grieser J, Schär C (2008) Combined surface solar brightening and increasing greenhouse effect support recent intensification of the global land-based hydrological cycle. Geophys Res Lett 35:L17706. doi:10.1029/2008GL034842

Willett KM, Gillett NP, Jones PD, Thorne PW (2007) Attribution of observed surface humidity changes to human influence. Nature 449:710–713

Wirth C, Schulze E-D, Lühker B, Grigoriev S, Siry M, Hardes G, Ziegler W, Backor M, Bauer G, Vygodskaya NN (2002) Fire and site type effects on the long-term carbon and nitrogen balance in pristine Siberian Scots pine forests. Plant Soil 242:41–63

Wolters V (2000) Invertebrate control of soil organic matter stability. Biol Fert Soils 31:1–19

Woodall CW, Liknes GC (2008) Climatic regions as an indicator of forest coarse and fine woody carbon stocks in the United States. Carbon Bal Manag 3:5

Wright IJ, Reich PB, Westoby M, Ackerly DD, Baruch Z, Bongers F, Cavender-Bares J, Chapin T, Cornelissen JHC, Diemer M, Flexas J, Garnier E, Groom PK, Gulias J, Hikosaka K, Lamont BB, Lee T, Lee W, Lusk C, Midgley JJ, Navas M-L, Niinemets Ü, Oleksyn J, Osada N, Poorter H, Poot P, Prior L, Pyankov VI, Roumet C, Thomas SC, Tjoelker MG, Veneklaas EJ, Villar R (2004) The worldwide leaf economics spectrum. Nature 428:821–827

Würth MKR, Peláez-Riedl S, Wright SJ, Körner C (2005) Non-structural carbohydrate pools in a tropical forest. Oecologia 143:11–24

Wutzler T, Reichstein M (2007) Soils apart from equilibrium - consequences for soil carbon balance modelling. Biogeosciences 4:125–136

Zak DR, Pregitzer KS, King JS, Holmes WE (2000) Elevated atmospheric CO_2, fine roots and the response of soil microorganisms: a review and hypothesis. New Phytol 147:201–222

Zhang D, Hui D, Luo Y, Zhou G (2008) Rates of litter decomposition in terrestrial ecosystems: global patterns and controlling factors. J Plant Ecol 1:85–93

Xia J, Han Y, Zhang Z, Wan S (2009) Effects of diurnal warming on soil respiration are not equal to the summed effects of day and night warming in a temperate steppe. Biogeosciences 6:1361–1370

Xu C, Gertner GZ, Scheller RM (2007) Potential effects of interaction between CO_2 and temperature on forest landscape responses to global warming. Glob Change Biol 13:1469–1483

Chapter 3
Effects of Disturbance, Succession and Management on Carbon Sequestration

The C dynamics in forests depends on the natural processes and perturbations by ACC (see Chapter 2). In primary forests (i.e., forests of native species without clear indications of human activity and no significant disturbance of ecological processes) natural C sequestration processes are in effect (FAO 2006a). The primary forests occupy about one-third of the global forest area (Table 3.1). To fully account for the C sequestration potential of forests, however, the temporal changes in forest structure and function at the stand and landscape level, and their effects on the net primary productivity (NPP) and the net ecosystem C balance (NECB) must be assessed. Forest dynamics are one of the greatest sources of uncertainty in predicting future climate (Purves and Pacala 2008). Specifically, models of ACC effects do not incorporate episodic disturbances such as fires and insect epidemics (Running 2008). The annual C storage in forests depends, in particular, on disturbances, forest succession, and climate variation (Gough et al. 2008). Disturbance is any factor that significantly reduces the overstory leaf area index (LAI) for more than one year or an event that makes growing space available for surviving trees (Oliver and Larson 1996; Waring and Running 2007). The long-term net C flux from forests depends on changes in the rates of disturbance (Goward et al. 2008). High frequency of disturbances, for example, results in low wood biomass accumulation (Potter et al. 2008).

Natural disturbances can be defined as events that cause unforeseen loss of living biomass or events that decrease the actual or potential value of the wood or forest stand (Schelhaas et al. 2003). Major natural disturbances such as wildfires, windthrow, hurricanes, herbivore outbreaks, landslides, floods, glacial advances, and volcanic eruptions can replace the forest stand, and primary succession occurs. Secondary succession occurs on previously vegetated sites where disturbances led SOM and plants or plant parts for reproduction in place (Chapin et al. 2002). By injuring or killing some trees within a forest stand, minor natural disturbances may affect the C cycle. Globally, disturbances affected more than 104 million hectare or 3.2% of the total forest area in 2000 (FAO 2006a). Credible information is, however, lacking for many countries, and the actual disturbed forest area is probably significantly larger. Furthermore, no distinction is made in the reported data between stand-replacing and minor disturbances.

K. Lorenz and R. Lal, *Carbon Sequestration in Forest Ecosystems*,
DOI 10.1007/978-90-481-3266-9_3, © Springer Science+Business Media B.V. 2010

Table 3.1 Global forest characteristics in 2005 (FAO 2006b)

Forest characteristic	Global forest area (%)
Primary	36.4
Modified natural	52.7
Semi-natural	7.1
Productive plantation	3.0
Protective plantation	0.8

On two thirds of the global forest area, human intervention associated with forest management and silviculture directly affect C sequestration in modified natural and semi-natural forests, and in forest plantations (Table 3.1; FAO 2006a). Silviculture can be defined as the art and science of controlling the establishment, growth, composition, health and quality of forests and woodlands to meet the diverse needs and values of landowners and society on a sustainable basis (Helms 1998). Deforestation, degradation and poor forest management have the potential to reduce the C pool in managed forests whereas sustainable management, planting and rehabilitation can increase the ecosystem C pool. Productive forest plantations are increasing in extent and occupy about 3% of the total forest area (Table 3.1; FAO 2006b). Management activities include controlling species composition, genetic improvement, fertilization, weed control and irrigation (Pregitzer and Euskirchen 2004). In this chapter the effects of natural disturbances on forest stands and their C balance are discussed by comparing the characteristics of some common disturbances. Then, C dynamics and sequestration are characterized during the natural succession cycle of forest stand development. Forest management activities and their effects on the C balance are also discussed. How disturbances in peatland forests, and by mining and urbanization affects forest C sequestration are finally compared. Nutrient constraints on forest C sequestration are discussed in Chapter 5, and water constraints in Chapters 2 and 5.

3.1 Effects of Natural Disturbances on Carbon Sequestration in Forest Ecosystems

Trees are of paramount importance to C sequestration in forest ecosystems as the largest C influx into forests occurs by their photosynthetic activity (Jandl et al. 2007b). During forest stand development, dynamic temporal changes in tree growth and species composition occur over centuries and potentially affect the C sequestration (Waring and Running 2007). In Chapter 2, C dynamics and sequestration were described with focus on the physiological processes on the tree and plot scale. However, the changes in the stand development are accompanied by canopy level competition and, thus, changes in C uptake, allocation and sequestration. Pristine forests, in particular, can be distinguished based on their stand dynamics (Shorohova et al. 2009). To assess the effects of forestry on C sequestration, the landscape scale is therefore particularly pertinent (Harmon 2001). The C storage in biomass on a landscape scale depends on tree age and size class distribution along with natural gap dynamics (Körner 2006). Environmental factors and disturbances affect the spatial and temporal patterns in the storage and flux of C in forested landscapes (Fahey et al. 2005).

Disturbances may disrupt the C cycle of forest ecosystems or export C from the ecosystem (Schulze 2006). Disturbances have the potential to restart the successional cycle for new stand development and may, therefore, be major driving forces that determine the C balance of forests (Krankina and Harmon 2006). Site disturbance contributes to the global variability of NEP among boreal, temperate and tropical forests (Luyssaert et al. 2007). Forest ecosystems probably do not attain a development stage where C assimilation and respiration are balanced ("climax concept"; Schulze et al. 2002). In particular, the NEP of forests worldwide is greater than zero, indicating that they are not in equilibrium and are net C sinks (Jarvis and Linder 2007). Specifically, old-growth forests are potential net C sinks (Carey et al. 2001). Old-growth forests contain large amounts of biomass, both above- and belowground, and are expected to maintain biomass accumulation for centuries (Suchanek et al. 2004; Luyssaert et al. 2008). Otherwise, when the disturbance regime is governed by large, rare disturbance events it is difficult to capture representatively tree mortality (Fisher et al. 2008). In mature equilibrium forests with constant mortality, spatially clustered disturbances may not be correctly characterized and mortality underestimated and, thus, tree growth overestimated.

Stand growth models combining the biogeochemistry with the dynamics of competition and tolerance among trees are important tools to assess C allocation associated with the vegetation dynamics during stand development (Waring and Running 2007). Four stages of stand development can be distinguished when a new forest replaces a previous one: (i) initiation, (ii) stem exclusion, (iii) understory reinitiation, and (iv) the old-growth phase.

3.1.1 *Natural Disturbances*

A forest stand can develop naturally after a major, stand replacing disturbance removes or kills all trees above the forest floor vegetation (Oliver and Larson 1996). Following major disturbances, the successional cycle for stand development is restarted (Krankina and Harmon 2006). Wildfires, wind-throw, hurricanes, ice storms, herbivore outbreaks, landslides, floods, glacial advances, earthquakes and volcanic eruptions have the potential to replace a forest stand (Chapin et al. 2002). Otherwise, disturbances can also be classified with respect to their effects on the C cycle (Schulze et al. 1999). Specifically, disturbances can result in continuous forcing of the forest C cycle by changes in temperature, CO_2, solar radiation and precipitation. On the other hand, pests, insects and wind-throws disrupt the C cycle but OM is maintained in forests. Grazing and fire may cause flush-type losses that export C. Disturbance regimes affect the net C balance in forest ecosystems directly but also by long-distance transports of elements between ecosystems and disturbance effects on important element interactions within forest ecosystems (Hungate et al. 2003). Wind-throws, wildfires, insect or disease outbreaks, snow and ice may also cause minor disturbances if they injure or kill some trees leaving some pre-disturbance trees alive (Oliver and Larson 1996). These minor forest disturbances may occur between or instead of major disturbances.

Disturbances alter the forest species composition but photosynthetic C uptake and decomposition are often relatively insensitive against changes in species composition (Waring and Running 2007). In the short term, disturbances have very large effects on the C cycle through reducing leaf area and increased decomposition of dead material, and may turn a forest stand into a temporary C source (Hyvönen et al. 2007). In the long term, however, the net C storage may approximate zero as the C loss by decomposition is replaced by forest re-growth. The impacts of natural disturbances on the large SOC pool are less well documented (Overby et al. 2003). Also, lack of data on key ecosystem C fluxes such as root production affects the assessment of disturbance effects on the NECB (Chapin et al. 2006).

3.1.1.1 Major (Stand-Replacing) Disturbances

By killing the forest vegetation, removing the plant cover and soil, stand-replacing disturbances reduce the live biomass and change the pool of actively cycling SOM (Chapin et al. 2002). Thus, the CO_2 uptake by plant photosynthesis is interrupted, and gaseous, liquid and particulate C losses from the forest ecosystem are enhanced. Specifically, C is transferred from the living biomass to detritus (dead) OM pools (Krankina and Harmon 2006). Large amounts of woody detritus are added to the soil in the form of boles, branches and the woody root system (Fahey et al. 2005). The forest area affected by major disturbances shifts from a C sink relative to the atmosphere towards a C source (Krankina and Harmon 2006). It remains a net C source until C uptake by the new generation of forest vegetation exceeds the C losses. A patchwork of fire and wind-throw may account for the NEP in forests where such disturbances occur regularly, i.e., the boreal forest and mixed old-growth tropical forests (Jarvis and Linder 2007).

Wildfires

Wildfires are a fundamental global feature influencing ecosystem patterns and processes, including vegetation distribution and structure, the C cycle, and climate (Bowman et al. 2009). The occurrence of wildfires, their behavior and effects on forest stands have been intensively studied (Fig. 3.1; Oliver and Larson 1996). The basic wildfire requirements are vegetative resources to burn, environmental conditions that promote combustion, and ignitions (Krawchuk et al. 2009). Less is, however, known about possible releases of GHGs, and reduction of future forest productivity and CO_2 uptake (Bormann et al. 2008). Biomass burning is one of the

Fig. 3.1 Forest wildfire (Dave Powell, USDA Forest Service, Bugwood.org, http://creativecommons.org/licenses/by/3.0/us/)

dominant emission sources of CH_4 (Rigby et al. 2008). Fire also influences local, regional and global climate by releasing atmospheric aerosols and changing surface albedo (Bowman et al. 2009). Aerosols from forest fires contribute to the reduction in solar radiation reaching the forest canopy but how this may affect forest productivity is not known.

Stand-replacing wildfires have a particularly great potential to reduce the forest C pool and turn the stand into a net C source (Table 3.2; Binkley et al. 1997). Crown fires, in particular, lead to complete mortality of the overstory trees and forest stands are replaced (Waring and Running 2007). Frequently burning forest stands have little accumulation of forest floor, decaying branches and logs but standing dead trees (Kimmins 2004). By destroying forest vegetation and detritus biomass, wildfires exacerbate soil erosion by wind and water (FAO 2006a). Intense wildfire may result in the loss of fine mineral soil by water erosion and the massive smoke plume (Bormann et al. 2008). Nutrient metals are also released during forest fires (Rauch and Pacyna 2009). Less than 5% of the global forest area was damaged by wildfires in 2005, but credible data are scanty (FAO 2007b). The major cause of fire in the boreal regions of Canada and the Russian Federation is lightning whereas burning for deforestation are often the cause of wildfires in other regions.

After the immediate pulse by fire, C may be emitted into the atmosphere for up to 50 years because non-combusted, dead biomass left on the site continues to decompose (Krankina and Harmon 2006). Thus, fire results in the redistribution of C among the various ecosystem C pools but the C budgets of forest fires are difficult

Table 3.2 Qualitative effects of major and minor natural disturbances on forest carbon pools

Disturbance	Severity	Soil C	Biomass C	Ecosystem C
Forest fires	Major	– (Short term) + (Long term)[a]	– – (Short term) + (Long term)	– –(Short term) + (Long term)
	Minor	– (Short term) + (Long term)[a]	– (Short term) + (Long term)	– (Short term) + (Long term)
Mechanical forces	Major	– –	– –	– –
	Minor	?[b]	–	–
Biotic factors	Major	?	– –	– –
	Minor	?	–	–
Drought	Minor	?	–	–

[a]Long term soil C sink depends on stability of BC/charcoal
[b]Qualitative effects on SOC pool often unknown

to establish (Andreae 2004). At low burning temperature, combustion of biomass results in the formation of char, tar and volatile compounds. At higher temperatures, biomass, forest floor, CWD, roots and soil are partially converted during flaming combustion to the major C species CO_2, CO, and CH_4, but also to C_2H_4, C_2H_2, PAH, and soot particles (Andreae 2004; Knicker 2007). At extreme temperatures, unprotected SOM is also combusted. Once the flaming combustion ceases, smoldering begins and emits large amounts of CO and incompletely oxidized pyrolysis products. This stage can continue for days or weeks. Beside directly affecting the C balance, fire may also affect nutrient limitation of NPP in forests (Hungate et al. 2003). For example, volatile losses following fire in both Mediterranean and tropical forests are relatively richer in N than P, K, Na, Ca, Mg, and S compared with the pre-burn vegetation. In contrast, ash formed and materials leached following fire are relatively richer in P, K, Na, Ca, Mg, and S than in N. Thus, fine soil, soil C and soil N losses and changes in soil structure after intense wildfires may result in major declines in forest productivity (Bormann et al. 2008). It may take centuries to restore soil fertility losses following intense fires (Kashian et al. 2006).

Forest fires occur in all global regions (FAO 2006a). In the late 1990s, 1,330 Tg biomass burned annually in tropical forests, and fire-induced emissions were estimated at 2,101 Tg CO_2, 139.0 Tg CO, 9.0 Tg CH_4, 10.8 Tg none-methane hydrocarbons, and with soot 12.0 Tg particulate matter (PM) <2.5 µm, 11.3 Tg total PM, 8.7 Tg total C, 7.0 Tg organic C and 0.88 Tg BC (Andreae 2004). In extra-tropical forests, 640 Tg biomass burned annually, leading to emissions of 1,004 Tg CO_2, 68.0 Tg CO, 3.0 Tg CH_4, 3.6 Tg none-methane hydrocarbons, and with soot 8.3 Tg $PM_{2.5}$, 11.3 Tg total PM, 5.3 Tg total C, 5.8 Tg organic C and 0.36 Tg BC. These estimates are, however, extremely uncertain to at least ±50%. Between 0.56 and 0.61 g BC are emitted per kilogram forest biomass burned (Bond et al. 2004).

For the NECB of forest ecosystems, it is important to note that only a fraction of the emitted C is lost to other ecosystems after wide-range atmospheric transport of emissions (FAO (2007a). In particular, CO_2, CO, CH_4 and soot emitted during fires are partially sequestered in forests by the various influx processes (see Chapter 2). Furthermore, the changes in the SOC pool and, in particular, the importance and stability of BC produced during wildfires are strongly debated (Czimczik and Masiello 2007; Lehmann and Sohi 2008). For example, BC from forest fires is likely to have higher ageing rates than BC from charcoal making which may influence its stability in soil (Cheng and Lehmann 2009). Otherwise, no significant effects of fire on the SOC in topsoil and mineral soils were observed (Johnson and Curtis 2001). However, the first direct evidence of intense wildfire shows sharply reduced mineral soil C and N (Bormann et al. 2008). Specifically, combined losses form the O horizon and mineral soil totaled 23 Mg C ha^{-1} and 690 kg N ha^{-1}. Fire-derived charcoal, on the other hand, may cause the loss of humus (Lehmann and Sohi 2008; Wardle et al. 2008). Most important to C sequestration, the loss of chlorophyll containing tissues reduces the capacity of the burned stand for C uptake by photosynthesis until a new generation of forest vegetation is established. Fire dependent forest ecosystems which are affected by periodic and intense wildfires (such as the boreal forest) are characterized by large contiguous expanses of even-aged stands of fire-adapted tree species (Stocks 2004). Modeling studies indicate that fire-affected forest stands may be a net C sink (positive NECB) over the history of several thousand years of development (Smithwick et al. 2007). In particular, charcoal or other long-term C sinks produced during wildfires may contribute to this C sink.

Mechanical Forces

High wind events are major natural disturbances which may completely destroy forest stands through their mechanical forces (Table 3.2; Kirilenko and Sedjo 2007). Catastrophic wind-throw has the potential to destroy tree canopies and regeneration, perturb the soil and severely affect the understory (Fig. 3.2; Ulanova 2000). Trees uprooted by winds are observed in almost every global forest region (Oliver and Larson 1996). Periodic extreme wind events cause widespread forest disturbance in high latitude coastal forests of western North America and Europe (Kimmins 2004). Whole forests are, however, rarely destroyed by high-intensity windstorms associated with hurricanes, typhoons, and tornados (Waring and Running 2007). The C losses following high winds occur through decomposition of downed wood besides the loss of photosynthetic active tissues. Furthermore, uprooting causes distinct changes in the soil profile by creating pit-and-mound topography (Ulanova 2000). Thus, C-rich surface soil horizons are lost within a pit but the adjacent profile is buried by organic-mineral or organic material. The remaining bare soil may lose C by erosion. Furthermore, in the wind-throw affected stand a more active and diverse soil biota increase and may enhance C loss by decomposition (Ulanova 2000). Otherwise, periodic windstorms have an important role in maintaining soil fertility in certain forest stands as they slow-down soil degradation processes (Waring and Running 2007). Specifically, soil mixing by uprooted trees reduces the accumulation of OM on the forest floor which otherwise would lead to the development of podzolic horizons.

Fig. 3.2 Forest disturbance by strong winds (Red and white pine, Steven Katovich, USDA Forest Service, Bugwood.org, http://creativecommons.org/licenses/by/3.0/us/)

The effect of high winds on global forests and their C balance is not very well documented, and only few of the data are available (FAO 2006a). For example, over the period 1950–2000, 18.7 million cubic metre wood were damaged annually in Europe, corresponding to about 4% of the total fellings and to 0.075% of the total volume of growing stock (Schelhaas et al. 2003). However, the degree to which forest stand or stems have been affected in Europe by wind damage varies. The storms of 1990 and 1999 alone, for example, damaged exceptionally 120 and 180 million cubic metre of wood, highlighting the very large variation among years (Schelhaas et al. 2003). The storm of 1999, in particular, reduced the European C balance by an amount that is equivalent to 30% of the NBP (Lindroth et al. 2009). Over the period 1851–2000, tropical cyclones affected 97 million trees annually in the U.S. (Zeng et al. 2009). About 53 Tg biomass and 25 Tg C were released each year. Over the period 1980–1990, the CO_2 released by impacts of tropical cyclones offset the C sink in forest trees by 9–18% over the entire U.S. Exceptionally large loss of biomass were caused by hurricanes Hugo and Katrina, although flooding contributed to the severe damage caused by Katrina (Solomon and Freer-Smith 2007; Chambers et al. 2007). Both storms damaged or destroyed 1.5 to 2 million hectare of southern pine forests. The total tree biomass loss by Katrina was estimated to be 92–112 Tg C, equivalent to 50–140% of the annual net C sink in forest trees in the U.S. Large forest blowdowns occur also outside the hurricane belt. These local blowdowns, for example, account for a relatively small proportion of the forest area in the Amazon Basin (Nelson et al. 1994). However, local C cycling or biomass estimates may be affected by blowdowns on smaller areas as large amounts of litter and nutrients are added to the soil. Specifically, forests regrowing after wind damage on these sites may have lower biomass but higher productivity than mature forests and act as a local C sinks. Especially in the United States tornadoes cause also forest blowdowns but if the recently observed increase in tornado forcing is related to global warming remains mostly unexplored (Diffenbaugh and Trapp 2008).

Beside winds, mass movements of soil, water, snow and ice impose strong mechanical forces on forest stands and possess stand-replacing potentials (Waring and Running 2007). Landslides, for example, transfer biomass from live to dead respiring C pools (Ren et al. 2009). The deforestation by landslides is unique as it involves highly biased nutrient redistribution. Thus, it takes decades for forest stands to fully recover and even longer for the recovery of soil nutrients. About 235 Tg C were added from the live biomass pool to the dead respiring pool by landslides following the Wenchuan earthquake 2008 in China. The cumulative CO_2 release in the coming decades is estimated to be comparable to that caused by hurricane Kathrina (about 105 Tg C; Ren et al. 2009). Further mass movements affecting forest ecosystems are caused by snow avalanches. Avalanche disturbances primarily affect subalpine forests (i.e., forests closest to the upper tree line; Bebi et al. 2009). Individual trees may be damaged or killed by avalanches. Forest stands disturbed by snow avalanches consist of smaller and shorter trees, shade intolerant species, lower stem densities, and greater structural diversity than stands not affected by snow avalanches. The C balance of these forests may differ from those of forests not disturbed by snow avalanches but these disturbances have received little research attention compared to other major types of forest disturbances (Bebi et al. 2009). The structure of forests and, in particular, tree biodiversity and C storage may also be altered by erosion, siltation, debris and sand avalanches, glaciers and floods but also by earthquakes and volcanic eruptions. There is, however, insufficient quantitative information about the forest area affected globally by these potentially stand-replacing disturbances and their effects on the forest ecosystem C balance (FAO 2006a).

In some forests, ice storms can be the primary cause of tree mortality (Galik and Jackson 2009). For example, substantial ice damage occurs in forests along a belt from Texas to New England in the U.S. (Irland 2000). Forest stands affected by ice storms may also be increasingly susceptible to diseases (Bragg et al. 2003). About 10% of total U.S. annual forest C sequestration can be lost by single ice storms (McCarthy et al. 2006).

Biotic Factors

Herbivory by mammals and insects, and diseases are integral natural components of forests that normally cause little damage and have only small impact on tree growth and vigour (FAO 2006a). However, biotic agents can adversely affect tree growth, and the yield of wood and non-wood products (FAO 2009). In 2000, approximately 4.7 million hectare of the global forest area were adversely affected by diseases, and 35.7 million hectare by insects. On average 1.4% of the forest area of each country was adversely affected by insects and another 1.4% affected by diseases (FAO 2007b). However, only 20 countries reported the area of forest affected by forest fires, insect pests, diseases and other disturbances. Furthermore, the discovery of previously unrecorded pest problems in new locations, countries and regions requires continuous updating (FAO 2009).

During outbreaks, insect species may reach damaging numbers, extent their spatial distribution, and cause stand-replacing disturbances (Fig. 3.3). Pest outbreaks are cyclical, occurring every 7–10 years (FAO 2009). By killing trees, epidemic outbreaks by bark beetles may greatly alter forest ecosystems and their C balance over large areas (Table 3.2; Waring and Running 2007). The impact of insects on forest C dynamics is, however, not well documented (Kurz et al. 2008a). For example, by the end of 2006 mountain pine beetle (*Dendroctonus ponderosae* Hopkins) infestations on an area of 130,000 km^2 in British Columbia resulted in the loss of 435 million cubic metre timber. As a result of such large insect outbreaks, Canada's managed forests turned recently from a C sink into a C source (Kurz et al. 2008b). Recently, infestations by bark beetles and defoliating insects in forest ecosystems in North America have been more intense than at any time in recorded history (Solomon and Freer-Smith 2007). Warmth is permitting beetles to invade northern tree populations that have been previously inaccessible by virtue of too-short growing season. Furthermore, gaps created by wind-throw may facilitate bark beetle outbreaks by harboring beetle communities (Bouget and Duelli 2004). Thus, damage from bark beetles is usually highly correlated with wind damage

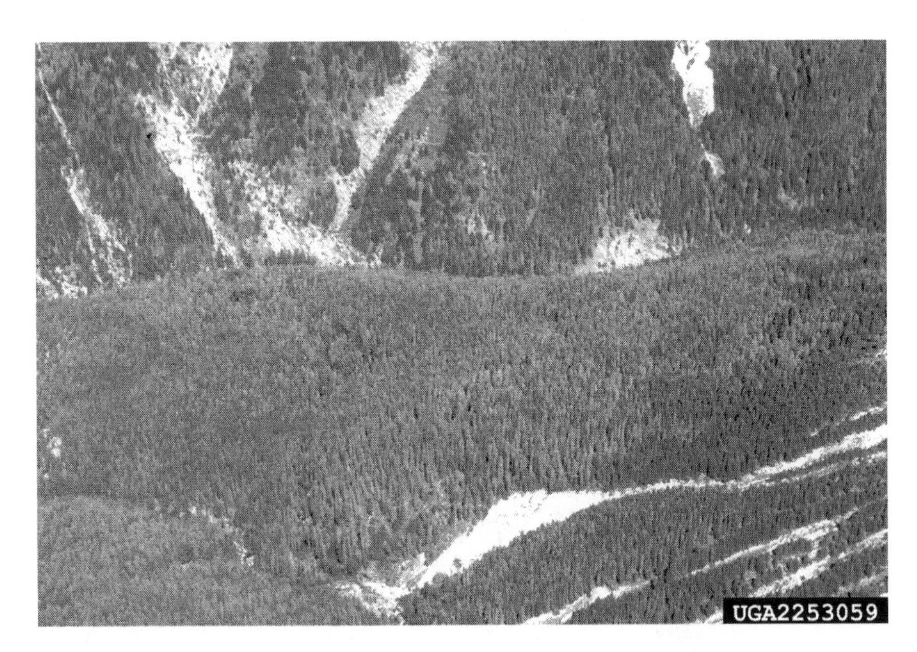

Fig. 3.3 Forest damage by insect outbreak (Mountain pine beetle, Jerald E. Dewey, USDA Forest Service, Bugwood.org, http://creativecommons.org/licenses/by/3.0/us/)

(Schelhaas et al. 2003). Data about pests associated with natural forests in the tropics are, however, scanty (FAO 2007b).

Aside outbreaks by indigenous insect pests, the outbreak of invasive species entering forest ecosystems can have stand-replacing potential. Examples of invasive species are the Asian longhorn beetle (*Anoplophora glabripennis* Cano) and the emerald ash borer (*Agrilus planipennis* Fairmaire) (FAO 2007b). In North America, more than 25 million ash trees (*Fraxinus* L.) were killed by the emerald ash borer since 2002. Extremely destructive is the cypress aphid (*Cinara cupressi* (Buckton)) in Eastern and Southern Africa and South America (FAO 2006a). Other invasive species affecting forests in Asia and Europe are the pine wood nematode (*Bursaphelenchus xylophilis*) and sirex woodwasps (*Sirex noctilio* F.) in the Southern Hemisphere (Seppälä et al. 2009). Introduced diseases can also kill vigorous trees, and may be major disturbances in mono-specific stands. Examples from North America are the Dutch elm disease (*Ophiostoma ulmi*) killing the American elm (*Ulmus americana* L.), the white pine blister rust affecting five-needles pines, and the Chestnut blight (*Cryophonectria parasitica*) that brought the American chestnut (*Castanea dentata* (Marsh.) Borkh.) close to extinction (Oliver and Larson 1996; Waring and Running 2007). Aside defoliating insects and stem-killing bark beetles, root pathogens can kill all dominant trees and, thus, stand replacement may occur (Waring and Running 2007). However, impacts of pathogen and insect disturbances on SOC pools over landscape scales are rather unknown (Overby et al. 2003).

3.1.1.2 Minor Disturbances

In contrast to major disturbances, minor or small-scale disturbances result only in the mortality of individual tree parts, trees or small groups of trees (Fahey et al. 2005). Thus, C pools and fluxes in the forested landscape are affected to a lesser extent compared to stand-replacing major disturbances (Table 3.2).

- Non-stand replacing fires, for example, are associated with low fire intensity (Binkley et al. 1997). Ground fires, in particular, are largely flameless and burn slowly through thick surface accumulations of OM (Kimmins 2004). Otherwise, surface fires move quickly through a forest area but consume litter and kill the aboveground parts of understory trees and shrubs. Thus, non-stand replacing fires produce relatively low immediate C release (Binkley et al. 1997). The subsequent C uptake by the vegetation, however, may be increased as pathogens in the forest floor may have been killed and nutrients remobilized. Vegetation growth may also be promoted as VOCs that inhibit decomposition may be consumed by wildfires (Waring and Running 2007). Furthermore, ground and surface fires directly emit smaller amounts of gaseous C forms than crown fires. If they burn slowly through the forest soil, however, they may produce more severely or partly charred necromass compared to crown fires (Knicker 2007). Thus, the nature and stability of SOM may be more affected by ground and surface fires than by crown fires. A wildfire-insect synergism is frequently important in North American boreal and temperate forests (Solomon and Freer-Smith 2007).

In particular, trees damaged but not killed by fire attract large numbers of pathogenic insects.

- Aside stand-replacing winds, intense winds on small areas within a forest stand or less intense winds may also affect the forest ecosystem C flux and storage. Tornadoes are the most intense storms and create gaps in forests by uprooting trees (Oliver and Larson 1996). On the other hand, blowdown of single trees or of small groups of trees occurs through less intense winds (Fig. 3.4). Windthrow gaps created by winds promote the gap-phase dynamics in forest communities (Ulanova 2000). Small- and broad-leaved trees in boreal coniferous forests, for example, may depend on the development of gaps. On the other hand, trees surrounding the gap grow more rapidly as they extend branches into the sunlit space. The vegetation structure in the gap may change depending on the gap size (Ulanova 2000). Less strong winds have also the potential to break individual tree parts. Snow and ice may also cause the breakage of tree branches and stems by their strong mechanical forces (Waring and Running 2007). Hail may separate leaves and needles from trees and break tree branches. The increased litter input and loss of photosynthetic tissue by these climatic events are associated with a reduction in GPP and enhanced C losses by higher amounts of decomposing plant residues on the forest floor.

Fig. 3.4 Trees uprooted by wind (European beech, Haruta Ovidiu, University of Oradea, Bugwood.org, http://creativecommons.org/licenses/by/3.0/us/)

- Minor disturbances in the forest C balance are caused by defoliating insects and stem-killing bark beetles as long as they don't reach epidemic proportions (Oliver and Larson 1996). Some native defoliating insect species allow surviving trees to maintain moderate growth efficiencies and the forest ecosystem to produce near maximum NPP (Waring and Running 2007).
- Browsing animals are other examples of biotic factors causing minor disturbances in forests. The forest vegetation is well adapted to herbivory by native animal browsers. Extremely high animal populations may, however, result in the reduction in biodiversity and long-term site productivity (Waring and Running 2007). Furthermore, the introduction of non-native animals can result in overbrowsing of understory trees and shrubs as is, for example, obvious in the extensive damage in New Zealand forests by the Common Brushtail Possum (*Trichosurus vulpecula* (Kerr 1792)) introduced from Australia.
- Forest diseases can be less severe or minor disturbances for the forest ecosystem C pool (Kimmins 2004). Less resistant to diseases are individual trees which are weakened by climatic changes, fires, overcrowding, or other conditions (Oliver and Larson 1996). At low level, endemic diseases have, thus, the potential to kill individual trees and consume a fraction of the forest NPP (Binkley et al. 1997). However, only for a few examples the fraction consumed by diseases has been estimated. Mortality specifically increases the amount and substrate quality of leaf and fine-root detritus (Waring and Running 2007). The microclimate for decomposition and mineralization is improved by the reduction in LAI. Thus, C is lost from the forest ecosystem pool mainly by heterotrophic respiration.
- Drought is another minor forest disturbance that is frequently involved with death of individual tree parts or trees or small clusters of trees (Bréda et al. 2006). Hydraulic constraints within the soil-plant-atmosphere continuum and deficits in the C balance, i.e., deficiency in stored reserves, are discussed as possible reasons for drought-related mortality of temperate forest trees. Periodic droughts and heat waves coupled with insect outbreaks are, in particular, the most important forest ecosystem disturbance in North America (Potter et al. 2005). On the other hand, large areas of tropical forests are influenced by severe drought episodes (Nepstad et al. 2007). The drought-related mortality of tropical trees alters forest structure, composition, C content, and flammability. Multi-year severe drought can substantially reduce Amazon forest C pools (Brando et al. 2008). The C sink in mature global forests may also be substantially threatened by ACC due to changed regimes of drought and heat waves (Chapters 2 and 4; Nabuurs et al. 2007).

3.2 The Natural Successional Cycle of Forest Stand Development and Carbon Sequestration

Aside natural disturbances, changes in forest structure and growth occur also during undisturbed stand development (Waring and Running 2007). Both natural disturbances and stand development affect, in particular, the C cycle in primary forests where past or present human activities are missing (FAO 2006a). Patterns of species dominance and changes in stand structure occur as a result of interactions among trees (Oliver and Larson 1996). The changes in forest structure and growth cause changes in above- and belowground C pools. Four idealized stages are as follows: (i) initiation, (ii) stem exclusion, (iii) understory reinitiation, and (iv) the old-growth phase. These stages can be distinguished for characterizing forest stand development in even-aged stands. Functional changes are related to alterations in foliage mass or leaf area, and the accumulation of woody biomass can be described by an asymptotic curve (Landsberg and Gower 1997). Forest succession is accompanied by changes in plant and soil C pools, and C fluxes (NPP, R_h, NEP). Successional stage and site history determine in part global patterns in NEP among boreal, temperate and tropical forests (Luyssaert et al. 2007). C partitioning to aboveground wood production increases with the stand age but decreases in the total belowground C flux (Litton et al. 2007). However, changes with age in partitioning within a site are less than 15% of GPP.

Primary or secondary succession occurs after disturbances (Chapin et al. 2002). Primary succession after severe disturbances leaves little or no OM or organisms within an area. Examples include the development of forest communities on previously non-vegetated areas on rock or exposed parent material, deposited substrate or the hydric environment (Landsberg and Gower 1997; Kimmins 2004). Globally more important and a faster process is the secondary succession (i.e., reforestation of an area after the destruction of the existing stand by disturbance). In the tropics, however, the conversion of primary to secondary forests can result in large losses in tree diversity (Stokstad 2008). Few estimates on the net change in C storage exist for similar forest types at different stages of stand development. Thus, generalizations are not possible (Landsberg and Gower 1997). Specifically, studies of succession use the method of substituting space for time or chronosequence because of the length of time required to actually observe successional vegetation changes in a single forest stand (Johnson and Miyanishi 2008). The validity of this chronosequence methodology is, however, questionable. Also, whether the net C change approaches zero at the end of succession in steady-state or old-growth forests is under discussion (Schulze et al. 2000). Specifically, boreal and temperate old-growth forests in the Northern Hemisphere accumulate C for centuries. For example, boreal and temperate old-growth forests sequester 1.3 Pg C $year^{-1}$, and contain large quantities of C (Luyssaert et al. 2008). Otherwise, boreal, temperate and tropical forest ecosystems not subjected to catastrophic disturbances over thousands of years may decline through increasing lack of P and reduced performance of the decomposer community (Wardle et al. 2004).

The functional changes in an even-aged forest stand during secondary succession are described in the following paragraph (Fig. 3.5). In contrast to primary succession, many of the initial colonizers are already present immediately after disturbance (Chapin et al. 2002). Also, the initial C pools and fluxes are much larger in secondary than in primary succession. The patterns of successional changes in dissolved and particulate C fluxes, and C fluxes other than CO_2 are, however, not well documented.

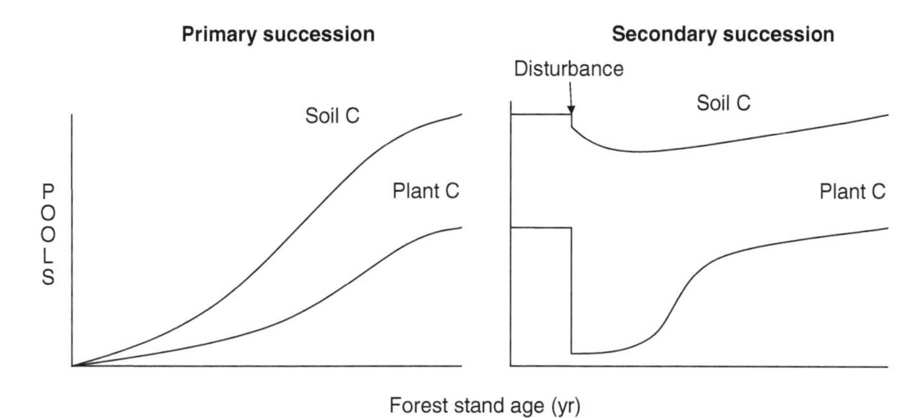

Fig. 3.5 Idealized patterns of changes in soil and plant carbon pools during primary and secondary succession. In early primary succession, plant and soil carbon accumulates slowly as NPP from colonizing vegetation is slowly but faster increasing than R_a. After disturbance in early secondary succession, soil carbon declines as losses by R_a exceed NPP from regrowing vegetation. In late succession, plant and soil carbon may not approach steady state but continue to increase slowly (Luyssaert et al. 2008). Other dissolved, gaseous and particulate C losses are assumed to be negligible (Modified from Chapin et al. 2002)

3.2.1 Initiation Stage

The stand-level C at the initiation stage of forest development consists mainly of detritus C and SOC as large trees may have been killed by major disturbances (Oliver and Larson 1996). Carbon is lost from the stand in gaseous, liquid and solid form, and the CO_2 release from decomposition of wood residues and SOC decomposition exceeds the CO_2 uptake of regrowing vegetation in the young stand (Schulze et al. 2000). This period dominated by C loss from decomposition can last for several decades (Janisch and Harmon 2002). Depending on the disturbance, however, forest floor herbs, shrubs, and advance regeneration may have survived and, thus, contribute to the recovery of vegetation C (Fig. 3.6). The understory biomass, in particular, is greatest in this stage, and due to its rapid turnover, understory detritus is important to C cycling (Landsberg and Gower 1997). The competition among forest vegetation is low as growth resources are abundant. Tree growth and, thus, photosynthetic fixation of CO_2 is, however, slow and NEP and NECB may be negative as the stand loses more C than is gained by re-growth (Chapin et al. 2002; Apps et al. 2006; Smithwick et al. 2007). Thus, ecosystem C pools decline in early secondary succession (Chapin et al. 2002).

The re-establishment of forest vegetation occurs from seeds, spores, or other regenerating parts of plants which are available to germinate or sprout (Waring and Running 2007). As the leaf area develop, forest growth and the C pool stored in vegetation increase (Ryan et al. 1997). Forest stand regeneration increasingly sequesters C as live wood and eventually exceeds C loss from heterotrophic respiration. Forest floor and SOC begin to accumulate (Waring and Running 2007).

Fig. 3.6 Secondary succession (Trees colonizing uncultivated fields and meadows, Tomasz Kuran, http://en.wikipedia.org/wiki/GNU_Free_Documentation_License)

In even-aged stands, leaf area and fine roots increase to full site occupancy by the canopy and the root system (Maguire et al. 2005). At crown closure, the productivity of crown, bole and roots reaches its maximum (Waring and Running 2007). Fine-litter accumulations on the forest floor approach an equilibrium level. Eventually, NPP and NECB become positive as the vegetation recovers and the canopy closes (Landsberg and Gower 1997; Chapin et al. 2002; Hyvönen et al. 2007).

3.2.2 Stem Exclusion Stage

As the forest stand ages, C accumulates in woody plant biomass but also increasingly in detritus and soil (Landsberg and Gower 1997). During this development stage, the competition among tree species and individual trees becomes particularly intense (Waring and Running 2007). Some trees in single- and mixed-species stands outgrow others (Oliver and Larson 1996). Thus, inter- and intraspecific competition for resources results in increased tree mortality and stem exclusion. Fewer but larger trees survive at the cost of smaller trees, a process called self-thinning of stands (Schulze et al. 2005). The transition of an entire stand form the initiation to the stem exclusion stage may take several decades (Oliver and Larson 1996). The LAI of the overstory and foliage mass reaches its maximum in this stage (Waring and Running 2007). The maximum of foliage mass is most rapidly reached in tropical forests whereas it is relatively slowly reached in boreal forests (Landsberg and Gower 1997). The tallest dominant trees usually produce the most growth per unit of leaf area. In some tree species leaf area or mass peaks early, followed by a gradual decline to a level slightly less than the peak (Maguire et al. 2005). Other tree species show a gradual increase of leaf area or mass to a stable maximum but no subsequent decline. On the shaded forest floor, understory biomass decreases and dead leaves, twigs and stems accumulate (Oliver and Larson 1996).

3.2.3 Understory Reinitiation Stage

Stochastic events (e.g., lightning, storms or diseases) during aging of the forest stand result in a more open canopy (Schulze et al. 2005). With increasing stand development, sunlight can increasingly penetrate to the forest floor as tree growth slows down and gaps created in even-aged stands by tree mortality cannot be filled completely by branch extension (Waring and Running 2007). Thus, understory vegetation suppressed in the stem exclusion stage is able to reinitiate but growth is slow because of the small amount of light reaching the forest floor. The re-initiation stage may last for several decades (Oliver and Larson 1996). For C sequestration in forest ecosystems, accumulation of woody debris on the forest floor from tree mortality is important (Waring and Running 2007). CWD is an important component of the long-term C storage in forests (Janisch and Harmon 2002).

3.2.4 Old-Growth Stage

For many tree species, the age at which the mean annual increment culminates and subsequently declines is quite early relative to their potential maximum age (Bauhus et al. 2009). Thus, in the old-growth stage of even-aged forests, growth and biomass accumulation decline further (Ryan et al. 1997). Although the forest stand growth peaks as leaf area reached its maximum, the total live biomass of dominant trees reaches its maximum in the early old-growth stage (Fig. 3.7; Ryan et al. 1997). Maximum total live biomass values of 400, 2,500 and 690 Mg ha^{-1} have been reported for boreal, temperate and tropical forests, respectively (summarized in Waring and Running 2007). The possible upper limit for tree biomass C accumulation in boreal and temperate forests based on forest C-flux estimates is about 200 and 700 Mg C ha^{-1}, respectively (Luyssaert et al. 2008). Similar forest C-flux based estimates for tropical forests are not possible as data are scanty (Section 4.3.1). Diameter growth of dominant trees continues during the old-growth stage but only little if any height growth occurs. The maintenance respiration in stems and branches is up to 50% reduced as sapwood volume is reduced by the larger proportion of trees adapted to grow in shaded conditions. Slowly-decomposing bolewood is a large fraction of total litterfall (Vitousek et al. 1988).

Fig. 3.7 Old-growth forest (European beech, Snežana Trifunović, http://en.wikipedia.org/wiki/GNU_Free_Documentation_License)

Thus, wood C inputs into the soil may be higher than foliage C inputs and a large amount of CWD in various decay classes is present (Waring and Running 2007). The canopy is relatively open with foliage in many layers and diverse understories (Oliver and Larson 1996). In late secondary succession, forest stands either approach a C balance of zero or continue to accumulate C at a slow rate (Chapin et al. 2002). The NECB of old-growth stands is, however, not exactly zero as fluctuations in local weather and finer-scale disturbances can occur (Smithwick et al. 2007).

The commonly accepted patterns of live biomass accumulation and NPP in relation to stand age are based on small-scale ecological studies of homogenous stands (Hudiburg et al. 2009). However, patterns of growth in all types of stands in a forest landscape may not follow idealized trends. Specifically, the C balance of undisturbed, pristine forests is not in equilibrium as previously hypothesized but these forests continue to sequester C (Bolin et al. 2000; Suchanek et al. 2004). Thus, pristine old-growth forests are important components of the global terrestrial C budget (Schulze 2006). For example, the top 20-cm of soils in old-growth forests over 400 years stand age in southern China accumulated atmospheric CO_2 at an unexpectedly high rate (Zhou et al. 2006). An old-growth forest in Washington state in the U.S. continues to sequester atmospheric CO_2 at 500 years stand age (Ingerson and Loya 2008). Biomass in 300 and 600 years old forests in the Pacific Northwest in the U.S. is still increasing (Hudiburg et al. 2009). The net C balance of forests 800 years of age is usually positive (Luyssaert et al. 2008). Thus, contrary to previous suggestions centuries old forests may not be C neutral but continue to be significant atmospheric CO_2 sinks (Carey et al. 2001). Specifically, centuries old boreal and temperate forests in the Northern Hemisphere may provide 10% of the global NEP (Luyssaert et al. 2008). These forests accumulate 0.4 Mg C ha^{-1} $year^{-1}$ in stem biomass, 0.7 Mg C ha^{-1} $year^{-1}$ in CWD, and 1.3 Mg C ha^{-1} $year^{-1}$ of the sequestered C is contained in roots and soil. Increasing tree C storage (i.e., 0.49 Mg C ha^{-1} $year^{-1}$) was also observed in tropical old-growth forests in Africa, America and Asia (Lewis et al. 2009). This can be explained by slow succession in tropical old-growth forests taking hundreds or thousands of years (Muller-Landau 2009). Otherwise, increasing atmospheric CO_2 concentrations are discussed as explanation for increasing tree C in these forests (Lewis et al. 2009).

3.3 Forest Management and Carbon Sequestration

Globally, two-thirds of all forests show clearly visible indications of human activity and, thus, ecological process may be disturbed (FAO 2006a). Forest management, in particular, affects the C cycle of natural forests where the species composition of the forest overstory has been largely determined by nature (Binkley et al. 1997). Assisted natural regeneration, planting or seeding are management activities associated with the establishment of semi-natural forests (FAO 2006a). Forest plantations, on the other hand, are managed forests with an overstory largely determined by human activities. Plantations supply a large proportion of the timber products globally with implications for forest ecosystem C (Landsberg and Gower 1997). The quantitatively most important product extracted from forests is wood destined for industrial use (FAO 2006a). Quantitative important non-wood forest products are food (e.g., berries, mushrooms, edible plants, bushmeat) and fodder (Seppälä et al. 2009).

Traditional forest management strategies in natural, semi-natural and plantation forests have the intention to maximize forest productivity and yield, and to sustain both in the long term (Nyland 2002; Jarvis et al. 2005; Jandl et al. 2007b). For example, for centuries forestry in Western Europe focused on charcoal, then the interest shifted to timber, and later from high-density hardwoods to conifers and longer rotations as the demand for high-quality large timber increased (Schoene and Netto 2005). Harvesting and litter raking, however, depleted SOC and aboveground biomass C in most European forests until the mid twentieth century (Ciais et al. 2008b). Between 1950 and 2000, the net biomass C sink was controlled by foresters and harvesting rates. Thus, environmental conditions in combination with the type of silviculture can efficiently sequester C (Ciais et al. 2008b).

Because of natural disturbances and harvesting, most forests are not at maximum C storage and potentially store more C after forest management changes (Dixon et al. 1994). Thus, aside meeting the demand for wood, a novel intention of forest management is the increase of the forest C pool by sequestration but the knowledge about forest production and forest growth is still incomplete (Andersson et al. 2000).

A quantitative understanding of how forest management simultaneously enhances wood production and C sequestration is lacking for most forest types (Gough et al. 2008). In particular, the potential for conserving C through changes in management of boreal and temperate forests has not been estimated (Ciccarese et al. 2005). Thus, it is unclear how the forest C sink and reservoir can be managed to mitigate atmospheric CO_2 buildup (Canadell and Raupach 2008). Management strategies to increase forest C storage in one area may have unintended negative or positive consequences on C storage elsewhere (Magnani et al. 2009).

The forest C stores carry the risk to return to the atmosphere by disturbances such as fire and insect outbreaks, exacerbated by climate extremes and ACC. Adaptive forest management strategies which simultaneously achieve risk reduction and C storage objectives under a changing climate are needed (Galik and Jackson 2009). Adaptation is defined as 'adjustment in natural or human systems in response to actual or expected climate stimuli or their effects which moderates harm or exploits beneficial opportunities' (Seppälä et al. 2009). Possible approaches are reactive adaptation after a ACC-related disturbance occurs, and planned adaptation involving redefining forestry goals and practices in advance in view of ACC-related risks and uncertainties (Bernier and Schoene 2009). Also, biophysical factors such as reflectivity, evaporation, and surface roughness of forests can alter temperatures much more than C sequestration does but this is rarely acknowledged (Jackson et al. 2008). Specifically, by using more reflective and deciduous species, forest management can increase the climate benefit of some forest projects.

Forest management contributes to the variability in NEP among boreal, temperate and tropical forests (Luyssaert et al. 2007). Specific management activities for increasing C storage include practices aimed at increasing NPP by fertilization, planting genetically improved fast growing trees (Section 6.1), and activities that enhance growth without causing increases in decomposition (Birdsey et al. 2007). To increase C sequestration in the forested landscape, the fraction allocated to trunks, large branches and large roots must be increased as these are tree C stores for several decades or centuries (Beedlow et al. 2004). Trees must also live longer to increase C sequestration. The increase in C density at both the stand and landscape scale is another strategy to mitigate C emissions through forestry activities (Canadell and Raupach 2008). Another activity is the expansion of the use of forest products to sustainably replace fossil-fuel CO_2 emissions (Section 6.1).

Forest management should be aimed at increasing the SOC pool as it often exceeds the amount of C stored in trees (Jarvis and Linder 2007). Increasing SOC pools with long turnover times, and minimizing erosion and oxidation of SOC during and following the logging activities increase C sequestration in the forested landscape (Beedlow et al. 2004). Increasing SOM turnover time and stabilization capacity in the deeper mineral soil by management activities is challenging but required to strengthen C sequestration in forest ecosystems (Diochon and Kellman 2009). Forest soil types that sequester substantial amounts of C must be protected (Section 6.3). Other activities for increasing C storage are aimed at minimizing the time forests are a C source to the atmosphere by low impact harvesting, utilization of logging residues, incorporation of logging residues into soil, thinning to capture mortality and fertilization (Lal 2005). However, little research about effects of specific management activities on forest C storage have been published (Birdsey et al. 2007). Furthermore, the assumption that the climate will remain relatively stable throughout a forest's life is challenged by the widely accepted ACC projections. Thus, forest management activities must be adapted to ACC (Bernier and Schoene 2009; Spittlehouse and Stewart 2003).

Carbon storage in forests and wood products (especially in landfills) must be included in management strategies to maximize forest-sector contributions to reduce atmospheric CO_2 concentrations (Martin et al. 2001; Hennigar et al. 2008). This may also include wood burial to prevent decomposition (Zeng 2008). For regional C budgets the lateral C fluxes with wood products, emission and oxidation of reduced C species (non-CO_2 gaseous species such as CO, CH_4, terpenes, isoprenes; Section 2.3.1.3), and by river erosion and transport must be accounted for (Eq. (3.1); Ciais et al. 2008a).

$$\text{Forest Ecosystem } CO_2 \text{ Sink} = \text{Forest Ecosystem C Accumulation} + \text{Lateral C Flux} \tag{3.1}$$

The magnitude of the C flux by lateral transport in Europe, for example, is comparable to the C accumulation in European forests. If management keeps the standing stock of forest constant over time and the stock of wood products is increasing, a net removal of CO_2 from the atmosphere is observed (Tonn and Marland 2007). The emissions associated with management activities and forest products are largely offset by increasing C pools in forest products (Miner and Perez-Garcia 2007).

Aside natural disturbances and successional changes, forest management (i.e., any manipulation aimed at achieving specified objectives) has the potential to alter C sequestration (Landsberg and Gower 1997). In addition to natural disturbance events, patterns of forest harvest, regeneration and growth control the C balance on managed forest land and respond to management practices (Krankina and Harmon 2006). In particular, the C balance of temperate and boreal forests is directly controlled through forest management (Magnani et al. 2007). Furthermore, the large forest C pool in the soil is also affected by direct influences of forest management activities on the C flow into the soil (Jandl et al. 2007a). Partitioning of direct human impacts by forest management on C sequestration from indirect impacts (i.e., increasing temperature, increasing atmospheric CO_2 concentration, increasing N fertilization) is, however, difficult (Vetter et al. 2005).

3.3.1 Management Activities in Natural Forests

The tree species composition of natural forests is primarily determined by nature (Binkley et al. 1997). In addition, regeneration also occurs by management activities using the same or similar tree species as those removed by logging. Silviculture practices focussing on wood production commonly result in production cycles of

Table 3.3 Qualitative effects of forest management and silvicultural activities on forest carbon pools

Management activity	Soil C	Biomass C	Ecosystem C
Lowering water table	–	+	Variable
Decreased regeneration delay	–	+	+
Change species composition	?	+	+
Density, thinning	–	–	–
Pruning	?	–	?
Fertilization	?	+	+
Suppression and prevention forest fires	+	Variable	Variable
Biotic factors	?	+	+
Timber harvest	Variable	Variable	Variable
Increased rotation length	+	+	+

25–150 years but succession cy may continue for several hundred or a thousand years between stand-replacing disturbances (Bauhus et al. 2009). Thus, managed forests often cover only between 10 and 40% of the potential stand development period.

Strategies for the management of natural forests which increase the C density (MgC ha^{-1}) and on-site adaptation of specific forest management activities to ACC are discussed in the following section (Table 3.3; Spittlehouse and Stewart 2003; Apps et al. 2006; Krankina and Harmon 2006). The effects of management activities on C budgets of a forest stand through a complete management or rotation cycle are, however, poorly known (Jarvis et al. 2005). Chronosequence studies at the scale of forest stands provide some answers. Furthermore, the impacts of a management activity on C pools and fluxes may be clear at the stand scale, but may be different at the landscape scale (García-Quijano et al. 2008; Nabuurs et al. 2008). Mitigation options at the landscape scale may include: (i) maximizing stocks at low risk sites, (ii) maintaining lower stocks or reducing the risks at high risks sites, and (iii) maximizing biomass production through changes in existing stands (Nabuurs et al. 2008). Fossil fuel C emissions associated with management operations must also be taken into account for assessing the net C sequestration in forests (Markewitz 2006).

3.3.1.1 Fire Management

Carbon sequestration in natural forest ecosystems depends on the successional stage of individual forest stands in the landscape and how they are affected by the landscape disturbance regime (Smithwick et al. 2007). As discussed previously (Section 3.1.1.1), fires of both natural and anthropogenic origin, play an important role in the life cycle of many forests. Specifically, non-stand replacing fires are associated with low immediate C releases but large C releases occur by stand-replacing fires (Binkley et al. 1997). The prevention and suppression of forest fires is, thus, a potentially attractive forest management option to reduce C losses.

Fire prevention is aimed at reducing the fuel load through preventive burning or other fuel reduction methods such as controlled grazing (FAO 2007a). Prevention methods reduce fire severity and extent, and potential C losses. Thinning and removing excess debris reduce the possibility that intense, stand-destroying crown fires release large amounts of C from forest ecosystems (Richards et al. 2006). Fuel-reduction treatments may offer climate benefits by increasing C storage and by increasing surface reflectance (Hurteau et al. 2008). The disruption of the continuity of the fuel within (open forest) and between stands (fire breaks, variation in stand characteristics) are important measures to decrease forest fire risks (Nabuurs et al. 2008). However, the long-term impacts of fire suppression on succession are probably more significant. Large-scale studies comparing effects of prescribed fires with those of mechanical treatments that are intended to reduce fire risk and restore resiliency are scanty (McIver et al. 2009). For example, a large-scale comparison of effectiveness and ecological consequences of commonly used fuel reduction treatments in the U.S. indicated that treatment effects on the SOC content in the top 15-cm are negligible (Boerner et al. 2009). In many countries, however, forest fire management focus on suppression or extinguishing rather than prevention. Fire suppression in U.S. forests during the twentieth century is thought to have increased the forest biomass (Canadell and Raupach 2008). Inventory data over 60 years, however, indicate that aboveground C pools in western U.S. forests decreased by fire suppression (Fellows and Goulden 2008). Specifically, stem density increased due to increases in the number of small trees but fire suppression led to a net loss in the number of large trees which contained a disproportionate large amount of biomass C.

The forest ecosystems where fires are natural occurrence are adapted to them, and depend on them for nutrient recycling, regeneration, removal of pests and disease vectors and other ecological functions (Waring and Running 2007). The landscape of boreal forests, for example, is often an irregular patchwork mosaic created by wildfires (Stocks 2004). Thus, fire suppression in boreal forests potentially results in a more uniform mosaic of uneven-aged stands which also includes tree species not adapted to fire (Soja et al. 2007). Fire suppression over long periods can, thus, lead to large changes in forest species composition (Landsberg and Gower 1997). Houghton et al. (1999) suggested that fire suppression in North America reduced the overall forest C sink. Specifically, fire suppression since the 1940s resulted in a shift from dominance by surface fires to increasing frequency of stand-replacing crown fires (Solomon and Freer-Smith 2007). Thus, over 10 million hectare of forests in the U.S. are listed under elevated fire hazard condition (McIver et al. 2009). On the other hand, longer fire intervals following fire suppression may result in higher landscape C stores as the progressive accumulation of C continues (Smithwick et al. 2007).

Prescribed or controlled burning is a forest management activity which increases subsequent C uptake by certain forest types (Binkley et al. 1997). In the U.S., for example, prescribed burning is a tool to restore and maintain forest ecosystems aside being a tool for fire prevention (FAO 2007a). Large CO_2 emissions, however, occur in the Southeastern U.S. from prescribed burns in forested areas (Wiedinmyer and Neff 2007).

Depending on regeneration success and nutrient loss, prescribed burning may decrease, increase or have no effect on the size of the forest biomass C pool and the forest ecosystem C pool (Hyvönen et al. 2007). The optimal management for fire and C sequestration, however, differs among regions (Boerner et al. 2008).

Potential changes in forest fire regimes by future warmer and drier climate require adaptive actions in fire management (Spittlehouse and Stewart 2003). As future forest fire increases may overwhelm the ability to respond to all occurrences, protecting areas with high economic or social value while allowing fire to run its course in other areas are adaptive forest management activities. Reducing risks and extent of fire disturbance through alterations in forest structure, and increasing the use of prescribed burning contributes to adaptive actions. Developing forest landscapes where fuels are managed to control the spread of wildfire and enhancing recovery after fire disturbances are other adaptive forest management activities to deal with potential changes in forest fire regimes (Spittlehouse and Stewart 2003).

3.3.1.2 Management of Herbivores and Diseases

Forest insects, diseases and browsing animals can have large effects on the C cycle through reducing leaf area and by killing or removing trees, and by affecting forest succession and age-class structure. Where browsing animals are native, however, the vegetation is likely to be well adapted to herbivory (Waring and Running 2007). Only when animal populations reach levels where they negatively impact forest ecosystem function and structure, forest management activities must limit their population size. Loss of C from outbreaks of insect defoliators can be reduced by the application of biocides or pesticides. However, this may also reduce the buildup of pathogens in the insect population which would otherwise control the population. Furthermore, the reduction of native defoliating insects may affect stand development as they reduce competition among trees (Waring and Running 2007). Thinning and fertilization are management options which may increase tree growth efficiencies during outbreaks of budworms and bark beetles. To reduce the risk of subsequent contagious disturbance by insects, fire or pathogens, the removal of dead or dying trees is a management option which potentially increases the net C storage (Binkley et al. 1997). The salvaged tree biomass can be used to replace fossil fuel although moving the biomass into decomposition-free permanent storage optimizes forest C storage (Janisch and Harmon 2002; Eriksson et al. 2007). Against damages from root pathogens, the removal of stumps or their treatment with fungicides is another management option. The rotation of resistant tree species is, however, recommended when root pathogens are well established.

For protection against increased disturbances by insects and diseases due to ACC, several adaptive forest management activities are recommended (Spittlehouse and Stewart 2003). Adaptive actions include increasing stand vigor and lowering the susceptibility by partial cutting or thinning, removing infected trees through sanitation cuts, and decreasing the period a stand is vulnerable to insects and diseases by shortening the rotation length and facilitating a change to more suitable species. Furthermore, using insecticides and fungicides, and genetically improved pest-resistant genotypes may protect forests against increased insect and disease disturbance (Spittlehouse and Stewart 2003).

3.3.1.3 Silvicultural Management

Young forests have a large C sink capacity but old forests have a high C density (Jandl et al. 2007b). Thus, short rotation lengths maximize aboveground biomass production but not C storage. If management activities are aimed to increase the terrestrial C sink, prolonging rotations will generally contribute to C sequestration (Binkley et al. 1997). Longer harvesting cycles may increase the forest C density (Canadell and Raupach 2008). The increase in the rotation period of natural forests may allow a progressive accumulation of C (Hyvönen et al. 2007). Long rotation periods are favorable for C sequestration as the sum of biomass C and SOC peaks in mature forest ecosystems. Increased rotation lengths increase biomass C and ecosystem C pools, and C density at the landscape level (Krankina and Harmon 2006; Hyvönen et al. 2007; Hudiburg et al. 2009). Also, total C storage and average NEP in a forest landscape increases with the length of the rotation interval but NECB does not increase (Euskirchen et al. 2002; Jarvis and Linder 2007; Smithwick et al. 2007).

Otherwise, increased rotation lengths mean that the total amount of wood to be harvested must be found in older forests (Nabuurs et al. 2008). Thus, the average age of the forest at the landscape level may be reduced. Site-specific conditions may, however, also prohibit the prolongation of rotation periods as risk of disturbances may also increase (Jandl et al. 2007b). Very long rotation lengths do not necessarily maximize the total C balance (Jandl et al. 2007a). Furthermore, increasing on-site forest C stores versus increasing off-site C stores in forest products, and fossil C emissions must be considered when rotation periods are compared (Liski et al. 2001; Krankina and Harmon 2006). The provisions of forest-derived products (i.e., timber, biomass for energy) are particularly attractive in temperate forests (Canadell and Raupach 2008).

The density of natural forests is actively managed during planting by initial spacing and during stand development by thinning (i.e., removing a proportion of the trees and leaving the residues on the ground) to reduce competition-induced tree mortality and to avoid natural disturbances (Jarvis et al. 2005; Jandl et al. 2007b). For the production of knot-free wood, branches are sometimes removed from trees by pruning (Smith et al. 1997). Pruning is most advantageous for conifers whereas most hardwoods prune well naturally. Pruning potentially reduces tree NPP and stem volume growth (Maguire et al. 2005). Otherwise, the pruned branches contribute to the detrital C pool. For all spacing densities in conifers, the sum of standing volume and cumulative mortality is the same (Maguire et al. 2005). Conifer stands with wider spacing, however, attain the maximum net yield at later stand ages.

Thinning is one of the most important silvicultural activities, and holds the most direct promise to increase C sequestration in forests and provide forest resilience to increasing synergistic stresses (Solomon and Freer-Smith 2007; Bravo et al. 2008a). Thinning affects C sequestration by reducing the amount of tree biomass and OM, and by stimulating microbial decomposition as more solar radiation and throughfall precipitation reaches the forest floor. The reduction in forest density increases in the amount of soil moisture and nutrients available to the remaining trees (Solomon and Freer-Smith 2007). Whether an optimum relationship exists between thinning intensity and tree production is unclear (Nabuurs et al. 2008). The forest floor C pools, however, decrease with increase in thinning intensity but the effects of thinning on the mineral SOC pool are less known (Jandl et al. 2007a). Overall, thinned stands hold less than the maximum C density in vegetation, detritus and soil. Otherwise, thinning activities improve forest stand stability and its longevity. This may result in higher landscape C sink (Smithwick et al. 2007). Experimental evidence indicates that thinning causes only short-term decreases in NEP as GPP is more reduced than total ecosystem respiration, but soon NEP reaches levels observed before the thinning occurred (Jarvis and Linder 2007).

The C density in natural forests can be increased by increasing the ecosystem productivity through fertilization (Chapter 5; Hyvönen et al. 2007). Nitrogenous fertilizers alone or in combination with other nutrients are most commonly applied (Binkley et al. 1997). The stimulated tree growth increases also C inputs to the soil through litterfall and rhizodeposition but decreases in root biomass have also been observed under experimental N addition (Jandl et al. 2007a). Litter decomposition is stimulated by fertilization as litter nutrient contents are increased but the fraction of stable SOM may also increase. Thus, in the long-term the SOC pool in the forest soil profile may increase (Hyvönen et al. 2007). Site-specific responses of SOC sequestration to N fertilization, however, do not allow generalizations of fertilization effects on SOC (Johnson and Curtis 2001). In addition, on fertile soils the stimulating effect of fertilization on growth is reduced as other factors may limit growth (Apps et al. 2006).

Lowering the water table in peat soils is a forest management option which increases the C density in the short term (Apps et al. 2006). Over a rotation period, however, SOC pools decrease, biomass C pools increase but effects of drainage on forest ecosystem C pools are variable depending on C loss from soil and C gain in biomass (Hyvönen et al. 2007). In particular, C stores in the vegetation often increase following forestry drainage through increased NPP but direct measurements of SOC balances following drainage are scanty (Jandl et al. 2007a). For example, drainage was the reason for very small or negative NEP of a mixed Scots pine (*Pinus sylvestris* L.)/Norway spruce (*Picea abies* (L.) H.Karst.) forest (Lindroth et al. 1998). Also, CH_4 emissions from drained peat are reduced but CO_2 emissions increased (Binkley et al. 1997).

In the face of future ACC, silvicultural management must be aimed at managing the declining and disturbed stands. Adaptive activities include pre-commercial thinning or selectively removing suppressed, damaged or poor quality trees to increase growth conditions for remaining trees (Spittlehouse and Stewart 2003). Silvicultural management can also assist in reducing the vulnerability of forest to future disturbances by altering tree density, species composition and forest structure. When current advanced regeneration is unacceptable as source of the future forest, underplanting suitable other species or genotypes is recommended. The establishment of better-adapted forests can be accelerated by reducing the rotation age followed by planting (Spittlehouse and Stewart 2003).

3.3.1.4 Management of Tree Species and Genotypes

Tree species differ widely in growth pattern over time, specific achievable stand density, rooting depth and pattern, specific wood density, life span, vulnerability to disturbance, decomposition of dead wood and their dependency on climate (Jandl et al. 2007b; Purves and Pacala 2008). Most of these factors are relevant to the forest C cycle. Thus, the species composition of natural forests can be managed to increase C density by selecting species and genotypes (Apps et al. 2006). Provenances of the same tree species, for example, may vary substantially in NPP (Binkley et al. 1997). As the growth rate of many coniferous species is higher over longer periods than that of many deciduous species, switching to conifers from broadleaves increases SOC, biomass and ecosystem C pools over a rotation period (Hyvönen et al. 2007). Differences in forest floor and soil C pools among conifer and broadleaf forests may, thus, be relevant to forest C inventory (Schulp et al. 2008b).

Mixing complementary species that occupy different ecological niches is another forest management option to increase biomass production compared to pure stands of each species (Jandl et al. 2007a). Examples of such diverse stands include mixtures of shade-tolerant and shade-intolerant tree species, and those with N-fixing or litter decomposition accelerating species (Maguire et al. 2005). A suitable mixture may increase the dry biomass production up to 30% in boreal and temperate, and up to 50% in subtropical forests (Pretzsch 2005). Enrichment planting below the canopy can increase NEP when the growing site is not fully utilized (Binkley et al. 1997). However, increasing tree diversity can also result in decreased productivity (Larjavaara 2008).

Otherwise, tree species effects on the SOC pool are not consistent (Gleixner et al. 2005). It remains unclear whether tree biodiversity can influence biogeochemical cycles significantly (Mund and Schulze 2005). In particular, while shallow rooting species tend to accumulate C in the forest floor, deep-rooting species may transfer C into the mineral soil. For C sequestration in natural forests, species that increase the stabilized SOC pool in the mineral soil would be preferable choices but data are insufficient to identify functionally important tree species which increase the terrestrial C sink (Jandl et al. 2007b). Aside from the SOC pool, tree species effects on stand stability needs also to be considered when comparing the suitability of species for long-term C sequestration in forests. There is, for example, substantial evidence that mixed-species stands are more resistant against biotic and abiotic disturbances (Knoke et al. 2008). Avoiding wind damage by selecting species that are less sensitive to wind-throw, for example, has landscape level effects on C pools and fluxes (Nabuurs et al. 2008).

For maintaining genetic diversity and resilience, and to adapt to ACC-induced alterations in area extent and new species assemblages, several adaptive management activities are recommended (Spittlehouse and Stewart 2003). One possibility is to identify and plant trees species, genotypes and provenances which are resilient to ACC occurring during the stand development (Solomon and Freer-Smith 2007). Further adaptive actions include breeding and planting specific genotypes with higher resistance against disturbances by pests and climate stresses and extremes (Bravo et al. 2008b). Locating seed orchards in climate regimes for seed production of species adapted to a future climate is also important. Planting a mixture of provenances at a forest site is another activity useful to adapt forests to ACC (Spittlehouse and Stewart 2003).

3.3.1.5 Management of Forest Regeneration

Disturbances in natural forests may cause a regeneration delay because viable seed sources are not available and the site conditions not suitable for the recovery of forest vegetation (Binkley et al. 1997). Soil nutrients required for re-growth may have been extracted by timber harvest, which may further be intensified on steep slopes through accelerated erosion and runoff (Waring and Running 2007). By appropriate choices of harvesting timing and method, and mitigation techniques such as in-planting, site preparation and nutrient supplements forest management can accelerate full stand occupancy with tree species. Site preparation techniques include manual, mechanical and chemical methods, and prescribed burning (Jandl et al. 2007a). Often, decomposition is stimulated by these techniques by exposing the mineral soil and removal or mixing of the forest floor. Low-intensive site preparation may, however, cause only small losses from the SOC pool over a rotation period but may increase biomass production, and biomass C and ecosystem C pools (Hyvönen et al. 2007). Planting itself is an additional disturbance that may stimulate further decomposition and C loss (Jarvis et al. 2005). In general, the reduction in regeneration delay results only in a small one-time gain in C density (Apps et al. 2006).

Adaptive management activities are required to reduce the susceptibility of forest regeneration to ACC (Spittlehouse and Stewart 2003). Activities include promoting drought-tolerant genotypes, planting trees in their future ranges, planting provenances growing adequately under a wide variety of conditions and planting a range of provenances at a site. Vegetation management must be aimed at controlling undesirable plant species which compete with commercial trees in a changing climate (Spittlehouse and Stewart 2003).

3.3.1.6 Management of Forest Operations

Timber harvest involves the mechanical extraction of biomass and affects the C balance of forest ecosystems (Fig. 3.8; Krankina and Harmon 2006). The NPP is not maintained, and NEP may become zero immediately and negative for thereafter

Fig. 3.8 Timber harvest (Poplar clearcut, Doug Page, USDI Bureau of Land Management, Bugwood.org, http://creativecommons.org/licenses/by/3.0/us/)

(Jarvis and Linder 2007). Impacts of logging methods on the C density and C sequestration are, however, poorly documented (Binkley et al. 1997; Apps et al. 2006). Reducing harvest rates increases the land-based C storage in the forest biomass (Hudiburg et al. 2009). At the landscape scale, effects of harvesting methods on the NECB are probably small as the same amount of wood is cut anyway (Smithwick et al. 2007; Nabuurs et al. 2008).

The most severe harvest disturbance is clear-cutting as most of the trees within a stand are removed. Thus, canopy photosynthesis and auto- and heterotrophic components of ecosystem respiration are affected (Kowalski et al. 2004). For example, clear-cutting of coniferous forests leads to decline in total ecosystem respiration, reduced GPP, and converts mature forest C sinks into sources. Furthermore, after clear-cutting the deep SOC below 20-cm depth may be destabilized and C lost due to increased rates of decomposition aside alterations of SOM in the shallow mineral soil (Diochon and Kellman 2009). Clear-cut stands, however, are not always large C sources (Kowalski et al. 2004). In seed-tree and shelterwood systems, most of the trees are also removed. Selective logging, on the other hand, has effects similar to thinning and is a less severe disturbance as the forest cover is maintained while the stand is gradually replaced.

C losses associated with harvesting can be reduced through modifications of the harvesting methods. Low-impact harvesting, for example, is aimed at reducing damage to residual trees and soil structure (Binkley et al. 1997). Otherwise, highly mechanized forest harvest usually increases soil disturbance by disrupting the forest floor and organic soil horizons, and stimulating C emissions into the atmosphere through enhanced oxidation of SOM (Jarvis et al. 2005; Jandl et al. 2007b). Mechanical incorporation of harvest residues may, thus, cause SOC losses but the net effects of residue incorporation on SOC accrual in the mineral soil depend on decomposition rate and clay mineralogy (Busse et al. 2009). On the other hand, the changes in the SOC pool depend also on the residue removal by the harvesting method (Hyvönen et al. 2007). For example, 10 years after complete forest floor removal, the SOC concentrations to 20-cm depth at plots of the North American long-term soil productivity experiment were reduced but standing forest biomass was generally not affected (Powers et al. 2005). The 10-year production was less on severely compacted plots if understory was present. It is, however, possible that these trends may change with stand development. On average, little or no harvesting effects on SOC were observed but sawlog harvesting (i.e., residues are left on site) increased SOC in coniferous forests whereas whole-tree harvesting (i.e., removal of all aboveground residues) caused small decreases (Johnson and Curtis 2001). Losses of nutrients with the harvested biomass may cause losses in subsequent forest production although the full successional effects may sometimes be different than expected (Nabuurs et al. 2008). Positive effects by leaving C on site through managing harvest residues last only for a few years. Forest harvest has probably much smaller effect on forest floor and SOC pools than was predicted earlier from the 'Covington Curve' as OM is mixed or moves into the mineral soil (Fig. 3.9; Yanai et al. 2003). In summary, the effects of harvesting methods on the SOC, biomass and ecosystem C pools in managed natural forests over a rotation period are highly variable (Hyvönen et al. 2007).

Forest operations need to be adapted to the projected ACC (Spittlehouse and Stewart 2003). For example, warmer winters may reduce the opportunities for winter logging when it relies on frozen surfaces for site access. Otherwise, heavy rainfall

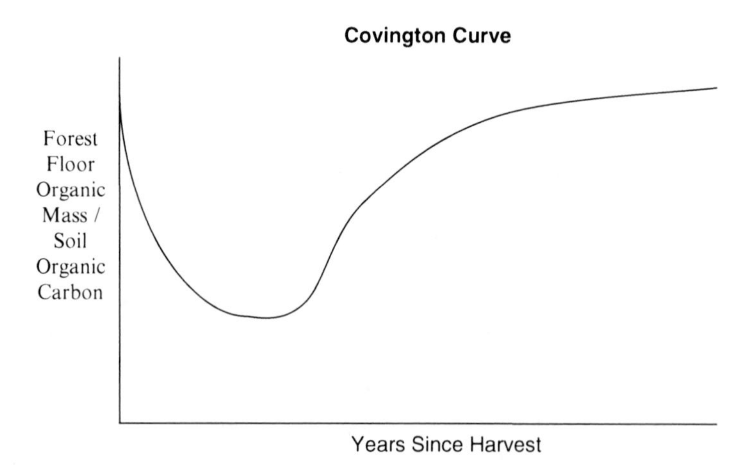

Fig. 3.9 Schematic Covington curve showing harvest effects on forest floor organic mass and soil organic carbon pool

may restrict harvesting due to the risk of landslides. Salvage logging needs to be increased in a future climate to remove timber from fire- or insect-disturbed stands. Protecting primary forests and their diversity of functional groups is another adaptive forestry activity as established forests are often able to survive extensive periods of unfavorable climates (Spittlehouse and Stewart 2003).

3.3.2 Management Activities in Forest Plantations

Globally, forest plantations provide a large proportion of timber products (Fig. 3.10; Landsberg and Gower 1997). Half of the productive planted forests are used for saw- or veneer-logs, but the area planted for pulpwood/fibre, non-wood products and bioenergy is increasing (Section 6.1; FAO 2006b). Management of plantations aims to capture the whole inherent site productivity or further enhance it (Powers 1999). Thus, management of forest plantations is much more intensive than that of natural forests described in Section 3.3.1. NEP of a new plantation forest is negative for several years after establishment as disturbance increases C losses by heterotrophic respiration (i.e., OM oxidation and mineralization), and the NPP of the young trees is low (Jarvis and Linder 2007). The period of net C loss may last between 5 and 15 years. The NEP stabilizes for a number of years once the canopy closes. For fast-growing species in favorable environments, NEP may reach up to 8 Mg C ha^{-1} $year^{-1}$, and 3 to 6 Mg C ha^{-1} $year^{-1}$ in temperate forests but NEP is more

Fig. 3.10 Forest plantation (*Pinus radiata* and *Eucalyptus nitens*, photo credit: Michael Ryan)

variable and considerably lower in boreal forests (Gower et al. 2001; Griffiths and Jarvis 2005; Hyvönen et al. 2007; Magnani et al. 2007).

Forest plantations may differ in their biomass C and SOC pools from those of natural forests they replace. While differences in SOC pools show no general trend, plantations have a small time-averaged biomass as they are felled near the time of maximum mean annual volume increment but before the forest reaches its maximum volume or C mass (Cannell 1999). Up to two-thirds of C stored in trees is 'lost' when an area of mature, old-growth forest is replaced by the same area of harvested plantations with the same growth characteristics. Thus, plantation forestry represents no net benefit over the conservation of mature forests for C sequestration (Stoy et al. 2008). Also, albedo and hydrological effects must be taken into account to assess the C sequestration benefits of extra-tropical plantations whereas hydrological effects of tropical forest plantations must be accounted for (Schaeffer et al. 2006).

For productive purposes, a relative restricted range of tree species are established in plantations, and growth rates and yields are high for most species. The dominant tree species in plantations are *Pinus* spp. growing on 20% and *Eucalyptus* spp. growing on 10% of the global plantation area (FAO 2001). For many commercially important tree species, however, data on total root biomass, fine root biomass, fine root surface area and root depth distribution are lacking (Jose et al. 2006). These data would be useful in the selection of functionally important species for belowground C sequestration. Invasive species may also be introduced by non-native tree plantations. This may cause major impacts on biodiversity, ecosystem processes, and the production of ecosystem goods and services (Seppälä et al. 2009).

Management in plantations can be optimized to achieve a specified product in the minimum possible time. Forest plantations are established on prepared land, uniform and genetically improved seedlings are grown at standardized spacings, weeds are controlled and fertilizers are used to improve tree nutrition. The management of competing vegetation, primarily using herbicides, may result in increases in wood volume (Wagner et al. 2006). However, vegetation control with herbicide application often causes a reduction in SOC storage (Sartori et al. 2007). Weed control by soil scarification may also result in SOC losses by accelerated decomposition and erosion (Paul et al. 2002). The soil under forest plantations is readily affected by management through fertilization and techniques that minimize soil compaction or erosion (Powers 1999).

Further increases in productivity are expected through plantations of genetically modified trees, and positive growth interactions are observed when genetically improved trees and effective vegetation control are combined (Section 6.1; Rousseau et al. 2005). Potentially important to forest C sequestration are genetic transformations in forest trees such as wood modifications including increases in quantity and quality, reductions in the prolonged juvenile phase, resistance to viral, fungal and bacterial pathogens, and tolerance to cold, heat, drought, and other abiotic stresses (Hoenicka and Fladung 2006). However, genetic improvement cannot compensate losses caused by unfavorable climate and poor soil quality. Sustainable productivity of planted forests requires that forest management maintains NPP without any decline in rate. However, this has not been conclusively demonstrated (Powers 1999). Also, the potential of mixed-species plantations for C sequestration is probably higher than that of monocultures as they more completely utilize a site (Nichols et al. 2006). However, only a small fraction of industrial plantations are polycultures, and operational-scale demonstrations of mixed-species plantations are lacking. One reason is that forest management of mixed-species plantations is more complicated than that of monocultures (Larjavaara 2008).

3.4 Effects of Peatland, Mining and Urban Land Uses on Forest Carbon Sequestration

Carbon sequestration in forest ecosystems may be adversely affected by a reduction in forest area through deforestation and conversion to other land uses such as agriculture and infrastructure, and through natural disasters. Deforestation and, in particular, the conversion of forests to agricultural land continues at an alarmingly high rate, affecting about 13 million hectare $year^{-1}$ (FAO 2006a). Specifically, about 6 million hectare of primary forests are either completely lost or modified each year, which may severely affect tree diversity and cause net changes in forest C storage. Other industrial and urban activities such as mining, drilling for oil and gas, installing pipeline, developing roads etc. also affect C sequestration in forest ecosystems (Lal et al. 2003). Examples of some major effects of forest land-use in peatlands, mining and urban areas are discussed in the following Section.

3.4.1 Forested Peatlands

Peatlands are defined as terrestrial environments where NPP exceeds OM decomposition over the long term, leading to the substantial accumulation of deposits of incompletely decomposed OM or peat (Wieder et al. 2006; Keddy et al. 2009). However, a range of definitions of peatlands and peat exist, and hamper comparisons among studies (Schils et al. 2008). Peatlands represent a disproportional large terrestrial C stock as they cover only about 3% of the global land surface but contain between 20% and 30% of the global soil C stock (Moore 2002; Strak 2008). In particular, northern peatlands play an important role in the global C cycle (Fig. 3.11; Vasander and Kettunen 2006). About one third of world's soil carbon (i.e., 270–450 Pg C) is stored in northern peatlands but massive peat deposits also occur in tropical peatlands which account for 68% of the global peat area (Page and Banks 2007; Limpens et al. 2008; Vitt and Wieder 2006; Beilman et al. 2009). For example, about 55 Pg C are stored in Indonesia's peatlands and peatlands of equatorial Asia contain about 70 Pg C (Jaenicke et al. 2008; van der Werf et al. 2008). The total C stocks and fluxes associated with Amazonian peatlands may be of global significance but is less well studied (Lähteenoja et al. 2009). Permafrost peatlands, on the other

Fig. 3.11 Peatland forest (pine trees in *Sphagnum spp.* L. bog, Paul Bolstad, University of Minnesota, Bugwood.org, http://creativecommons.org/licenses/by/3.0/us/)

hand, contain about 277 Pg C (Schuur et al. 2008). Furthermore, wetlands (i.e., mostly peatland soils) in North America store an estimated 36 Pg C in the SOC pool (Bridgham et al. 2006). Otherwise, boreal and tropical wetlands are one of the dominant emission source of CH_4 (Rigby et al. 2008). Methane bursts from wetlands occurred, for example, during a past sudden global warming transition (Petrenko et al. 2009). In particular, peatlands are expected to be sensitive to ACC (Smith et al. 2008). For example, over the last 2,000 years peatlands in the southern West Siberian Lowland region were mainly responsible for C accumulation in this region but future warming will shift long-term C sequestration northward (Beilman et al. 2009). Permafrost thaw, in particular, may promote a boost in peat C sequestration. The current generation of coupled carbon-climate models, however, neglects peats (Doney and Schimel 2007).

Disturbances by anthropogenic activities impact the C balance of wetland and peatland ecosystems. Land-use conversions, in particular, result in an increase in net GHG emissions from wetlands (Roulet 2000). Since 1800 about one third of the global wetland area has been lost through land-use change, and wetlands currently occupy about 5.96×10^6 km^2 of the terrestrial global surface (Bridgham 2007). The majority of peatlands are currently not used. In contrast, the natural vegetation of tropical lowland peatlands, the peat swamp forest, is intensively exploited in Southeast Asia as a timber resource and cleared for the establishment of oil palm (*Elaeis guineensis*) plantations for biofuel production (Page 2004; Jaenicke et al. 2008). Replacing peatland with oil-palm monocultures will accelerate both ACC and biodiversity loss (Danielsen et al. 2009). Furthermore, peatlands in equatorial Asia are converted to establish rice fields (van der Werf et al. 2008).

About 28% of the total European peatland area has been drained for forestry (Schils et al. 2008). For example, large peatland areas in Fennoscandia, especially Finland, are used for timber/wood production (Paavilainen and Päivänen 1995). The role of forested wet- or peatlands is typically not separated from that of the upland forests despite their disproportional influence on terrestrial C storage due to the huge amounts of C stored in peat (Trettin and Jurgensen 2003). Globally, peatlands are a moderate source of atmospheric C of about 150 Tg C year^{-1} whereas freshwater mineral-soil wetlands and estuarine wetlands are sinks of about 39 and 43 Tg C year^{-1}, respectively (Bridgham 2007). The C cycle of peatlands is discussed in the following section with a focus on forested peatlands.

The C cycle in peatlands differs from those in other ecosystems, but the C sink-source relationship is not known for tropical peatlands (Page 2004). Adaptation of the plants in peatland forests to periods of limited soil oxygen because of hydric soil conditions can enhance C storage (Trettin and Jurgensen 2003). Thus, ecohydrological changes are the primary driving force of changes in the peatland C cycle (Vasander and Kettunen 2006). The CO_2 emissions from tropical peatland ecosystems, for example, are strongly controlled by groundwater level (Hirano et al. 2009). The extracellular enzyme activity and transport and associated accumulation of decomposition products in peat control OM decomposition and methanogenesis (Limpens et al. 2008).

The NPP of boreal peatlands is lower than that in many other ecosystems, but productivity varies widely among wetland forest types. There are, however, few NPP data available for tropical peatland vegetation (Page 2004). The decay rates of OM in peatlands are low due to anoxic conditions. Thus, peat accumulates when the OM production rate exceeds the decay rate. The C remaining after respiratory losses is transformed into plant biomass, in particular, in belowground plant parts but direct measurements of belowground productivity, especially in wetland forests are scanty (Trettin and Jurgensen 2003). Stump and coarse-root biomass, and mycorrhizal fungi are important wetland soil components. *Sphagnum* mosses, sedges, and shrubs but also tree foliage and fine roots are litter sources in peatlands (Laine et al. 2006). Shrubs and herbs are, in particular, important OM sources in northern wetlands (Trettin and Jurgensen 2003). Dead plant matter and root exudates are the substrate for CO_2 and CH_4 production. Carbon dioxide is also dissolved in the soil-water column aside its emission as gas. Other C losses occur through surface DOC flows and downward leaching of C in the peat profile. Little research data are, however, available on DOC losses from forested wetlands (Trettin and Jurgensen 2003). During a wet year, C sequestration in peatlands may be positive, but is often negative during a subsequently dry year. Peatlands are, however, the most important single CH_4 source, globally (Vasander and Kettunen 2006).

Undisturbed peatlands are net C sinks (Limpens et al. 2008). Carbon accumulation rates for tropical peats, for example, may exceed values attributed to temperate and boreal peats by a factor of 3–6 (Neuzil 1997). However, direct measurements of peat C balances are rare (Laine et al. 2006). There are very few examples of

NECB of peatlands where all fluxes have been measured at the same time (Limpens et al. 2008). For example, the NECB of northern peatlands ranges from 10 to 30 g C m^{-2} $year^{-1}$, with the largest term for NEE ranging between 0 and 100 g C m^{-2} $year^{-1}$, and losses by CH_4 and dissolved C (DC) fluxes ranging between 0 and 12, and between 2 and 70 g C m^{-2} $year^{-1}$, respectively. The high DC export fluxes were measured, in particular, in disturbed or extensively used peatlands. Thus, any human-induced perturbation of the ecohydrological conditions may affect the peatland C cycle. In northern peatlands, for example, the feedback between water table and peat depth increases the sensitivity of peat decomposition to temperature, and intensifies the loss of SOC in a changing climate (Ise et al. 2008). Otherwise, the temperature response of CO_2 emissions from tropical peatland soils is probably similar to those in temperate forests (Hirano et al. 2009). However, predictions of future rates of peatland C sequestration are uncertain (Belyea and Malmer 2004).

Peat accumulation may be altered by animal husbandry (Turetsky and St. Louis 2006). Sheeps and other animals are grazing in peatlands, and remove aboveground biomass, alter microtopography, soil structure and microbial activity, and increase soil erosion. Furthermore, C sequestration in peatlands can also be reduced by fires. Land-clearing activities by burning forested peatlands, in particular, in Southeast Asia and rotational burning to increase plant diversity and improve grazing value potentially release C to the atmosphere and affect SOC storage (Page et al. 2002). Thus, Indonesia's peatlands turned from a CO_2 sink into a CO_2 source (Jaenicke et al. 2008). Larger areas of peatland becoming vulnerable to fire in drought years in combination with higher rates of forest loss caused strong nonlinear relation between drought and fire emissions in southern Borneo (van der Werf et al. 2008). Fire may result, in particular, in diminished NPP for a certain period of time (Limpens et al. 2008). Otherwise, if severe fires melt permafrost in northern peatlands, more permanent vegetation changes may occur and eventually increase C pools.

A major threat to the C store in peatlands is flooding by the creation of reservoirs for hydroelectric power generation, water supply, recreation, and aquaculture (Turetsky and St. Louis 2006). Specifically, large amounts of accumulated peat are decomposed in flooded peatlands and C emitted as CO_2 and CH_4 (Roulet 2000). Furthermore, the C sink is severely altered as terrestrial plants are killed by flooding. On the other hand, drainage also affects the biogeochemistry and vegetation in peatlands (Limpens et al. 2008). Compared to undisturbed peatlands, manipulating the water table for forestry, agriculture, and peat extraction may release more CO_2 and N_2O and consume more CH_4 (Turetsky and St. Louis 2006). Peat extraction is another serious perturbation of the peatland C cycle. About 50% of extracted peat is used as fuel source, globally. Extracted peat is also used as fertilizer and as raw material for chemical products. Peat extraction or mining involves drainage and removal of surface vegetation which result in increased CO_2 emissions but decreased CH_4 emissions.

Saturated conditions inhibit tree root growth in pristine peatlands. Thus, for commercial forestry operations large areas of peatlands have been drained (Turetsky and St. Louis 2006). Most profitable, in particular, are forestry plantations on nutrient-rich peatland sites (Laine et al. 2006). Otherwise, on nutrient-poor sites, P and K have to be applied to sustain forest growth. Thus, the soil nutrient regime becomes the primary edaphic constraint to tree growth aside temperature. The increased tree establishment and production in drained peatlands has the potential to offset C releases from peat associated with drainage. On organic soils, for example, increasing site productivity can also increase C storage both above and below ground (Trettin and Jurgensen 2003). Thus, forest drainage may have a negative radiative forcing for up to 500 years after drainage, i.e., a cooling effect on the global climate (Laine et al. 2006). However, soil emissions of CO_2, CH_4 and N_2O in forested peatlands decrease or offset the C sink in forest biomass (Schils et al. 2008). Thus, afforestation of peat soils is not considered an effective means of C sequestration. Drainage and tillage, and the application of mineral soils and fertilizers are required activities to prepare peatlands for agricultural operations (Turetsky and St. Louis 2006). Rapid declines in productivity may, however, occur as peat degradation in agricultural sites accelerates.

In summary, forested peatlands are large terrestrial C stores because anoxic conditions slow microbial decomposition of OM stored in peat. Ecohydrological changes are, thus, the primary natural driving force of changes in the peatland C cycle. Aerobic peat decomposition, in particular, may have a greater feedback to global warming on a century timescale than anaerobic decomposition. Furthermore, peatlands may release large amounts of CH_4 into the atmosphere. Direct measurements of peat C balances are, however, rare. Land-use changes in the past have caused significant losses in the global peatland area but the majority of peatlands are currently not used. Tropical peat swamp forests are, however, exploited as a timber resource. Deforestation and conversion to other land uses, in combination with drought and fire in tropical peatlands, are responsible for drastic losses of C to the atmosphere. A strong nonlinearity between fire emissions and drought in southern Borneo may further strengthen the positive feedback of reduced NPP in the tropics in response to warming and drought. Also, replacing tropical peatland with oil-palm monocultures accelerates ACC. Otherwise, global warming appears to be a major driver of the net C balance of northern latitude peatlands.

3.4.2 Mining Activities in Forests

Mining is defined as the extraction of material from the ground in order to recover one or more component parts of the mined material (Lottermoser 2007). Mining is always associated with mineral processing at mine sites, and sometimes accompanied by the metallurgical extraction of commodities. Major impacts of mining on land include the construction of access roads, infrastructure, survey lines, drill sites, and exploration tracks, creation of large voids, piles of wastes, and tailing dams, destruction or disturbance of natural habitats, and the release of contaminants into surrounding ecosystems (Lottermoser 2007). Surface mining, in particular, removes the vegetation, canopy cover and surface soil to get at the mineral resource being

Fig. 3.12 Forest regrowth after surface mining for coal (Photo credit: Johannes Fasolt)

mined from an ore body, vein or (coal) seam (Fig. 3.12; Bradshaw 2000). Thus, large drastic direct and indirect reductions in biomass C and soil C pools occur by the principal activities of the mining industry (Lal et al. 2004). Mines account, however, only for a small part of the land area of a country. For example, less than 1% of the total area of South Africa or the U.S. are disturbed by the mining industry whereas about 0.06% of the Australian land mass are occupied by mining activities (Lottermoser 2007).

The post-mining land surface consists of skeletal subsoil and rock materials. Distinct changes in topography, hydrology and stability of a landscape may occur (Lottermoser 2007). Because the post-mining land surface often contains little C, the degraded land has a large potential for C sequestration. For the successful rehabilitation of a mine site for terrestrial C sequestration, problematic mine wastes must be isolated safely for long periods of time to reduce transport of contaminants into the environment. Most important, suitable landforms need to be identified to reduce erosion, a suitable plant growth medium needs to be developed and a vegetation cover must be established to increase C storage. Many modern mines, particularly in industrialized nations, are designed to have minimum environmental impacts but unregulated mining activities still occur in developing nations and former communist states (Lottermoser 2007). There are, however, no global standards for environmentally compatible or responsible mining (Miranda et al. 2003).

About 10% of active mines and 20% of exploratory mine sites are located in areas of high conservation values (Miranda et al. 2003). Clusters of mines occur, in particular, in the boreal forest of North America. To promote the restoration of a mined site into a forest, reclamation practices are required although it may not strictly be possible to restore it to the original forest ecosystem (Koch and Hobbs 2007). One goal of restoration activities is the establishment of self-sustaining native forest ecosystems (Bell 2001). Planted forests are also established on rehabilitated degraded lands (FAO 2006b). Rehabilitation and reforestation of land degraded by mining are potential strategies for ACC mitigation through C sequestration (Harris et al. 2006). Forest land reclamation, in particular, involves restoration of land that was forested before mining to restore a productive forestry post-mining land use (Torbert and Burger 2000). The construction of a deep, non-compacted, nontoxic mine soil, and the use of a noncompetitive ground cover are some requirements for the restoration of productive forest land. The soil microorganisms, in particular, play a key role for the successful establishment of higher plants on reclaimed sites (Young et al. 2005).

The mined soils under forest can be more productive than native soils (Torbert and Burger 2000). High productivity can be achieved by selecting appropriate overburden material, preventing compaction, establishing tree-compatible ground cover, and using proper tree handling and planting techniques. The potential sequestration on mined soil reclaimed to forest plantation in seven east-central U.S. states was estimated to be between 0.7 and 2.2 Tg C year^{-1} (Sperow 2006). However, these estimates are highly uncertain as the correction for any coal C impurities in the soil is required to account for the C sequestration potential of forests established on reclaimed coal mine soils (Lorenz and Lal 2007). For example, coal C contributed up to 91% of C in 0–5 cm depth at reclaimed forest sites in Germany, and comparable values have been reported for sub-soil layers at sites in Ohio (Rumpel et al. 2003; Ussiri and Lal 2008). Accurate SOC quantification is required as the knowledge of SOC pools may adequately integrate all ecosystem components, and, thus, can be used to determine the level of forest ecosystem restoration (Koch and Hobbs 2007).

Without any regulation of reclamation activities, great disparities between pre- and post-mining forest C pools are generally observed. For example, after 60 years of forest development on mined land in the Midwestern and Appalachian coalfields of the U.S., the C sequestration potential was less restored at sites with higher original forest site quality (Amichev et al. 2008). In contrast to the tree and litter C pools, the SOC pools were much lower on the mined sites indicating the potential for sequestering up to 100 Mg C ha^{-1} in the soil. Thirty-five years after reclamation in the eastern USA, however, the forest vegetation composition still differed substantially from reference sites indicating that only the less common forest species slowly colonize reclaimed sites (Holl 2002).

In summary, surface mining drastically affects the C sequestration in forest ecosystems by removing the canopy cover, forest floor and the soil. Suitable reclamation practices are required to return the post-mining landscape to a forest and restore its

C storage. Specifically, the restoration of pre-mining C sequestration rates may take decades or centuries. Globally, however, the forest area affected by mining activities is probably small.

3.4.3 Urbanization and Forest Ecosystems

The current rate of urbanization is unprecedented in human history. Only 29% of people lived in cities in 1950, compared with the projection of 70% of the population living in urban centers by 2050 (Anonymous 2008). The urban population occupies less than 3% of the global terrestrial surface but 78% of C emissions from fossil fuel burning and cement manufacturing, and 76% of wood used for industrial purposes is attributed to urban areas (Brown 2001). Otherwise, per capita GHG emissions from cities are often lower than the average for the countries in which they are located (Dodman 2009). However, urbanization drastically affects local and global biogeochemical cycles and climate (Grimm et al. 2008). Urban areas produce major disturbances of the C cycle through land use change, climate modification, and atmospheric pollution (Trusilova and Churkina 2008). For example, urban regions in Europe emit 50 g C m^2 $year^{-1}$ whereas European forests take up 70 g C m^2 $year^{-1}$ (Ciais et al. 2008a). The modeled net effect of urbanization is an increase in C sequestration in Europe (Trusilova and Churkina 2008). Thus, cities are a crucial front in the fight against ACC (Huq et al. 2007).

The global extent of land classified as urban and built-up area is rather small compared to the area covered by forest (256,332 vs. 32,457,759 km^2; Loveland et al. 2000). However, the global forest area converted to urban land and its consequences to forest NPP are not well known (DeFries et al. 1999; FAO 2006a). Predictions how forest areas will grow or shrink as direct results of human land-use decisions are extremely uncertain (Field et al. 2007). In particular, studies on the impacts of urbanization on the structure and function of forested wetlands are missing despite their importance to C sequestration (Faulkner 2004). Urbanization effects on forests are seldom included in coupled biosphere-atmosphere models to simulate the ACC (Bonan 2008). The C cycle science on measuring C pools and fluxes has focused on natural ecosystems and not on settlements (Pataki 2007; Lorenz and Lal 2009). For example, for North America the net balance of C loss during conversion of natural to urban or suburban land cover is highly uncertain. Also, in a recent study on effects of land use changes on the future C sequestration in Europe no C emission or sequestration was allocated to built-up areas (Schulp et al. 2008a). In the 1990s, however, urbanization in Europe caused a loss of 3.3 Tg C $year^{-1}$ from the terrestrial environment (Zaehle et al. 2007). In the following section, effects of urbanization on terrestrial C storage in the U.S. are discussed in more detail.

Urban areas in the conterminous U.S. may store between 23 and 42 kg C m^{-2}, and about 18 Pg C were stored in human settlements (urban and exurban areas) in 2000 (Churkina et al. 2009). Most of the urbanization in the U.S. has taken place on lands with higher NPP, i.e., on productive agricultural land (Imhoff et al. 2004). However, housing growth is particularly high in forests (Radeloff et al. 2005). Between 1997 and 2001, a loss of about 16,000 km^2 of forestland to urban development was observed (Lubowski et al. 2006). Much of newly developed land had been forested, i.e., 40% of the land developed in the 1980s, and 46% of the land converted by urban sprawl between 1997 and 2001 (Duryea and Vince 2005). Furthermore, many forests are no longer remote from cities but surrounded and penetrated by development and indirectly affected by urbanization. Specifically, in many northern U.S. cities urbanization occurred in mostly forested areas and this caused an overall annual NPP loss (Imhoff et al. 2004). In the southeastern U.S, the amount of C release attributable to urbanization increased over time (Zhang et al. 2008). The majority of new urban and developed land is projected to come from forestland and, thus, the C storage potential of terrestrial ecosystems in the U.S. may be reduced (USGCRP 2003). In particular, significant amounts of U.S. forestland may be transformed by urbanization, i.e., between 2000 and 2050 about 5% of forestland outside of urban areas may be directly subsumed by urban growth (Nowak and Walton 2005). The increase in developed land area in the Chesapeake Bay region, for example, is predicted to consume 14% of forest land in the region by 2030 (Goetz et al. 2004). In the southeastern U.S., on the other hand, most newly urbanized land has been converted from forestland but total forest area changed little as cropland area was converted to forest (Zhang et al. 2008). Soils of residential lawns may have higher C densities than many forest soils they replaced (Pouyat et al. 2006). In general, substantial losses of SOC in temperate regions in the U.S. may occur through urban development whereas urban conversions in arid climates have the potential to increase the SOC storage.

Many of the areas with large urban expansion in the U.S. are heavily forested (Nowak et al. 2005). Otherwise, in urban areas trees and forests may also contribute to the vegetation cover and total C storage but the abundance of planted trees in urban areas in many countries other than the U.S. is less well documented (Fig. 3.13; Long and Nair 1999). For example, in Baltimore, MD, soils of urban forest remnants had more than 70% higher SOC densities to 1-m depth than rural forest soils but reasons for this difference were not clear (Pouyat et al. 2009). About 3% of the total tree cover in the continental U.S. is located in urban areas (Nowak et al. 2001). Urban trees show different growth rates compared to rural trees because of the relatively open structure of urban forests and the proximity to impervious areas (Nowak and Crane 2002; Quigley 2004). Thus, the accuracy of urban tree biomass estimates is low as allometric relationships developed outside of urban environments cannot be used to estimate urban forest C storage (McHale et al. 2009). Per hectare, however, the C storage in rural forests is higher as urban areas contain less tree coverage. Otherwise, aside reducing atmospheric CO_2 by directly storing photosynthetically fixed C, utilization of biomass from urban trees and wood waste can be a source of bio-based fuels for power and heat generation, and reduces fossil fuel consumption (Mead 2005; MacFarlane 2009). Advanced combustion of wood from urban trees, for example, offers environmental benefits as renewable energy source in the U.S. (Richter et al. 2009).

Urban trees can also offset CO_2 emissions from power plants and minimize its effect by lowering temperatures and shading buildings in summer, and by blocking winds in winter (Heisler 1986). Tree planting influences urban air temperatures by altered albedo, shading, and latent heat flux (Pataki et al. 2009). Thus, large benefits of urban tree planting in terms of ACC are rather effects of afforestation on surface

Fig. 3.13 Urban forest park (Joseph LaForest, University of Georgia, Bugwood.org, http://creativecommons.org/licenses/by/3.0/us/)

energy balance than those from direct C sequestration. Urban forestry projects that reduce energy use benefit the climate (Jackson et al. 2008). Promoting tree planting outside forests, especially in urban areas is an adaptive forest management practice to mitigate ACC (Bravo et al. 2008b). Thus, ACC mitigation using urban forestry includes increasing C density in settlements, using wood from urban trees as renewable energy source, and accentuating indirect effects such as reducing heating and cooling energy use in buildings and changing the albedo of paved parking lots and roads (Nabuurs et al. 2007; Richter et al. 2009). Planting of trees in arid urban areas, however, can also increase ozone production due to increased BVOC emissions from trees (Papiez et al. 2009). Thus, low BVOC-emitting trees should be planted.

In summary, as a result of urbanization forests may be converted to urban land uses. The effects on the terrestrial C pools are, however, variable. Specifically, urban forests have the potential to directly and indirectly reduce C emissions from cities. As urban population continues to grow at unprecedented levels, both urban trees and urban forests are important ACC mitigation options.

3.5 Conclusions

Carbon sequestration in forest ecosystems can be affected by natural disturbances and forest management activities. Temporal changes in vegetation, detritus and soil C pools and C fluxes occur, in addition, during primary and secondary succession.

Wildfires, wind-throw, hurricanes, herbivore outbreaks, landslides, floods, glacial advances, and volcanic eruptions are examples of natural disturbances. The primary forests which occupy one-third of the global forest area are, in particular, affected by natural disturbances. Best studies are the effects of wildfires on NPP and NECB in boreal forests. Short-term C losses from soil and biomass burning occur but fire-affected boreal forest stands may be a net C sink in the long-term. Less well documented are effects of other minor and major disturbances on forest C. In general, however, forest ecosystem C pools decrease after disturbances. On two thirds of the global forest area, forest management activities are aimed at maximizing the forest productivity and yield in a sustainable way. Effects of management practices on C sequestration in managed forest ecosystems are, however, not well documented. For example, lowering the water table in peatland forests, decreasing regeneration delay, choice of tree species, fertilization, suppression and prevention of forest fires and increasing the rotation length are forest management practices which potentially result in increased ecosystem C pools. Temporal changes in soil, detritus and vegetation C pools occur in forest stands during stand development. After disturbances removed all products of ecosystem processes, primary succession occurs and this is characterized by very low initial C pools which slowly recover by vegetation re-growth. Secondary succession, on the other hand, occurs after disturbances on previously vegetated sites. Initially, soil C pools are higher than during primary succession and plant C pools recover faster as plant parts for reproduction are abundant. If plant and soil C, and the forest ecosystem C pool approach a steady state at the end of succession cy needs a careful appraisal as old-growth forests continue to be net C sinks. For a comprehensive assessment of effects of disturbance, management activities and succession on the NECB, however, data on all gaseous, liquid and particulate C fluxes in forest ecosystems are required but yet not available. Forested peatlands accumulate large amounts of C as peat over long periods of time. Thus, deforestation of peatlands in the tropics and global warming in northern latitude peatlands will cause significant C losses. Restoring surface mined areas back to forest land use is a ACC mitigation option by establishing C sequestration in vegetation, detritus and soil. Worldwide cities are growing rapidly. Urban areas are major sources of C emissions but urban forestry can also contribute to terrestrial C sequestration.

3.6 Review Questions

1. What are the most severe natural disturbances in each of the major global forest biomes? What effects do they have on C sequestration processes?
2. How may the severity and frequency of disturbances be altered by ACC and interact with C sequestration processes in each of the major global forest biomes?
3. Forest fires and insect outbreaks are major natural disturbances. Discuss activities for their management and how they can be optimized to reduce alterations of forest C pools.

4. Imagine a barren and C poor landscape which is naturally re-vegetating. What may be the reasons that C accumulation in plants and soil never approaches equilibrium? In an idealized situation, can C accumulate in forest soils over the entire interglacial period?
5. Classify forest management and silvicultural activities for their C footprint by comparing C costs associated with the activities, i.e., CO_2 emissions from fossil fuel burning versus C sequestration in plants and soil.
6. The tree species composition can be actively managed. What favorable properties should a functional important tree species have for C sequestration? Can you recommend tree species for the major forest biomes?
7. Compare C sequestration processes and perturbations in forested peatlands in northern and tropical regions.
8. What are appropriate reclamation activities to restore the C sequestration potential of a surface mined area by forest land use?
9. Urban trees and urban forests can reduce C emissions from cities. What can urban planning including housing development and infrastructure contribute to ACC mitigation?
10. Compare and contrast C cycling processes in tropical and temperate forests and peatlands.

References

Amichev BY, Burger JA, Rodrigue JA (2008) Carbon sequestration by forests and soils on mined land in the Midwestern and Appalachian coalfields of the U.S. For Ecol Manag 246:1959–1969

Andersson FO, Ågren GI, Führer E (2000) Sustainable tree biomass production. For Ecol Manag 132:51–62

Andreae MO (2004) Assessment of global emissions from vegetation fires. Int For Fire News 31:112–121

Anonymous (2008) Turning blight into bloom. Nature 455:137

Apps MJ, Bernier P, Bhatti JS (2006) Forests in the global carbon cycle: implications of climate change. In: Bhatti JS, Lal R, Apps MJ, Price MA (eds) Climate change and managed ecosystems. Taylor & Francis, Boca Raton, FL, pp 175–200

Bauhus J, Puettmann K, Messier C (2009) Silviculture for old-growth attributes. For Ecol Manag 258:525–537

Bebi P, Kulakowski D, Rixen C (2009) Snow avalanche disturbances in forest ecosystems-state of research and implications for management. For Ecol Manag 257:1883–1892

Beedlow PA, Tingey DT, Phillips DL, Hogsett WE, Olszyk DM (2004) Rising atmospheric CO_2 and carbon sequestration in forests. Front Ecol Environ 2:315–322

Beilman DW, MacDonald GM, Smith LC, Reimer PJ (2009) Carbon accumulation in peatlands of West Siberia over the last 2000 years. Global Biogeochem Cy 23, GB1012, doi:10.1029/2007GB003112

Bell LC (2001) Establishment of native ecosystems after mining – Australian experience across diverse biogeographic zones. Ecol Eng 17:179–186

Belyea LR, Malmer N (2004) Carbon sequestration in peatland: patterns and mechanisms of response to climate change. Glob Change Biol 10:1043–1052

Bernier P, Schoene D (2009) Adapting forests and their management to climate change: an overview. Unasylva 60:5–11

Binkley CS, Apps MJ, Dixon RK, Kauppi PE, Nilsson LO (1997) Sequestering carbon in natural forests. Crit Rev Environ Sci Tech 27:S23–S45

Birdsey RA, Jenkins JC, Johnston M, Huber-Sanwald E (2007) Principles of forest management for enhancing carbon sequestration. In: King AW, Dilling L, Zimmerman GP, Fairman DM, Houghton RA, Marland G, Rose AZ, Wilbanks TJ (eds) The first state of the carbon cycle report (SOCCR) – the North American carbon budget and implications for the global carbon cycle. Global Change Research Information Office, Washington, DC, pp 175–176

Boerner REJ, Huang J, Hart SC (2008) Fire, thinning, and the carbon economy: effects of fire and fire surrogate treatments on estimated carbon storage and sequestration rate. For Ecol Manag 255:3081–3097

Boerner REJ, Huang J, Hart SC (2009) Impacts of fire and fire surrogate treatments on forest soil properties: a meta-analytical approach. Ecol Appl 19:338–358

Bolin B, Sukumar R, Ciais P, Cramer W, Jarvis P, Kheshgi H, Nobre C, Semenov S, Steffen W (2000) Global perspective. In: Watson RT, Noble IR, Bolin B, Ravindranath NH, Verardo DJ, Dokken DJ (eds) Land use, land-use change, and forestry. Cambridge University Press, Cambridge, pp 23–52

Bonan GB (2008) Forests and climate change: forcings, feedbacks, and the climate benefits of forests. Science 320:1444–1449

Bond TC, Streets DG, Yarber KF, Nelson SM, Woo J-H, Klimont Z (2004) A technology-based global inventory of black and organic carbon emissions from combustion. J Geophys Res 109:D14203. doi:10.1029/2003JD003697

Bormann BT, Homann PS, Darbyshire RL, Morrissette BA (2008) Intense forest wildfire sharply reduces mineral soil C and N: the first direct evidence. Can J For Res 38:2771–2783

Bouget C, Duelli P (2004) The effects of wind-throw on forest insect communities: a literature review. Biol Conserv 118:281–299

Bowman DMJS, Balch JK, Artaxo P, Bond WJ, Carlson JM, Cochrane MA, D'Antonio CMD, DeFries R, Doyle JC, Harrison SP, Johnston FH, Keeley JE, Krawchuk MA, Kull CA, Marston JB, Moritz MA, Prentice IC, Roos CI, Scott AC, Swetnam TW, van Der Werf GR, Pyne SP (2009) Fire in the Earth system. Science 324:481–484

Bradshaw A (2000) The use of natural processes in reclamation – advantages and difficulties. Landscape Urban Plan 51:89–100

Bragg DC, Shelton MG, Zeide B (2003) Impacts and management implications of ice storms on forests in the southern United States. For Ecol Manag 186:99–123

Brando PM, Nepstad DC, Davidson EA, Trumbore SE, Ray D, Camargo P (2008) Drought effects on litterfall, wood production and belowground carbon cycling in an Amazon forest: results of a throughfall reduction experiment. Phil Trans R Soc B 363:1839–1848

Bravo F, del Río M, Bravo-Oviedo A, Del Peso C, Montero G (2008a) Forest management strategies and carbon sequestration. In: Bravo F, LeMay V, Jandl G, von Gadow K (eds) Managing forest ecosystems: the challenge of climate change. Springer, New York, pp 179–194

Bravo F, Jandl R, Gadow KV, LeMay V (2008b) Introduction. In: Bravo F, LeMay V, Jandl R, von Gadow (eds) Managing forest ecosystems: the challenge of climate change. Springer, New York, pp 3–11

Bréda N, Huc R, Granier A, Dreyer E (2006) Temperate forest trees and stands under severe drought: a review of ecophysiological responses, adaptation processes and long-term consequences. Ann For Sci 63:625–644

Bridgham SD (2007) Wetlands. In: King W, Dilling L, Zimmermann GP, Fairman DM, Houghton RA, Marland G, Rose AZ, Wilbanks TJ (eds) The first state of the carbon cycle report. U.S. Climate change science program, Washington, DC, pp 139–148

Bridgham SD, Megonigal JP, Keller JK, Bliss NB, Trettin C (2006) The carbon balance of North American wetlands. Wetlands 26:889–916

Brown LR (2001) Eco-economy: building an economy for the earth. Norton, New York

Busse MD, Sanchez FG, Ratcliff AW, Butnor JR, Carter EA, Powers RF (2009) Soil carbon sequestration and changes in fungal and bacterial biomass following incorporation of forest residues. Soil Biol Biochem 41:220–227

Canadell JG, Raupach MR (2008) Managing forests for climate change mitigation. Science 320:1456–1457

Cannell MGR (1999) Environmental impacts of forest monocultures: water use, acidification, wildlife conservation, and carbon storage. New Forest 17:239–262

Carey EV, Sala A, Keane R, Callaway RM (2001) Are old forests underestimated as global carbon sinks? Glob Change Biol 7:339–344

Chambers JQ, Fisher JI, Zeng H, Chapman EL, Baker DB, Hurtt GC (2007) Hurricane Katrina's carbon footprint on U.S. gulf coast forests. Science 318:1107

Chapin FS III, Matson PA, Mooney HA (2002) Principles of terrestrial ecosystem ecology. Springer, New York

Chapin FS III, Woodwell GM, Randerson JT, Rastetter EB, Lovett GM, Baldocchi DD, Clark DA, Harmon ME, Schimel DS, Valentini R, Wirth C, Aber JD, Cole JJ, Goulden ML, Harden JW, Heimann M, Howarth RW, Matson PA, McGuire AD, Melillo JM, Mooney HA, Neff JC, Houghton RA, Pace ML, Ryan MG, Running SW, Sala OE, Schlesinger WH, Schulze E-D (2006) Reconciling carbon-cycle concepts, terminology, and methods. Ecosystems 9:1041–1050

Cheng C-H, Lehmann J (2009) Ageing of black carbon along a temperature gradient. Chemosphere 75:1021–1027

Churkina G, Brown D, Keoleian G (2009) Carbon stored in human settlements: the conterminous US. Glob Change Biol (in press), doi: 10.1111/j.1365-2486.2009.02002.x

Ciais P, Borges AV, Abril G, Meybeck M, Folberth G, Hauglustaine D, Janssens IA (2008a) The impact of lateral carbon fluxes on the European carbon balance. Biogeosciences 5:1259–1271

Ciais P, Schelhaas M-J, Zaehle S, Piao SL, Cescatti A, Liski J, Luyssaert S, Le Maire G, Schulze E-D, Bouriaud O, Freibauer A, Valentini R, Nabuurs G-J (2008b) Carbon accumulation in European forests. Nat Geosci 1:425–429

Ciccarese L, Brown S, Schlamadinger B (2005). Carbon sequestration through restoration of temperate and boreal forests. In: Stanturf JA, Madsen P (eds) Restoration of boreal and temperate forests. CRC Press, Boca Raton, FL, pp 111–120

Czimczik CI, Masiello CA (2007) Controls on black carbon storage in soils. Global Biogeochem Cy 21, GB3005, doi:10.1029/2006GB002798

Danielsen F, Beukema H, Burgess ND, Parish F, Brühl CA, Donald PF, Murdiyarso D, Phalan B, Reijnders L, Struebig M, Fitzherbert EB (2009) Biofuel plantations on forested lands: double jeopardy for biodiversity and climate. Conserv Biol 23:348–358

DeFries R, Field C, Fung I, Collatz GJ, Bounoua L (1999) Combined satellite data and biogeochemical models to estimate global effects of human-induced land cover change on carbon emissions and primary productivity. Global Biogeochem Cy 13:803–815

Diffenbaugh NS, Trapp RJ (2008) Does global warming influence tornado activity? EOS Trans Am Geophys Union 89:553–554

Diochon AC, Kellman L (2009) Physical fractionation of soil organic matter: destabilization of deep soil carbon following harvesting of a temperate coniferous forest. J Geophys Res 114:G01016. doi:10.1029/2008JG000844

Dixon RK, Brown S, Houghton RA, Solomon AM, Trexler MC, Wisniewski J (1994) Carbon pools and flux of global forest ecosystems. Science 263:185–190

Dodman D (2009) Blaming cities for climate change? An analysis of urban greenhouse gas emission inventories. Environ Urban 21:185–201

Doney SC, Schimel DS (2007) Carbon and climate system coupling on timescales from the Precambrian to the Anthropocene. Annu Rev Environ Resour 32:14.1–14.36

Duryea ML, Vince SW (2005) Introduction: the city is moving to our frontier's doorstep. In: Vince SW, Duryea ML, Macie EA, Hermansen LA (eds) Forests at the wildland-urban interface: conservation and management. CRC Press, Boca Raton, FL, pp 3–13

Eriksson E, Gillespie AR, Gustavsson L, Langvall O, Olsson M, Sathre R, Stendahl J (2007) Integrated carbon analysis of forest management practices and wood substitution. Can J For Res 37:671–681

Euskirchen ES, Chen J, Li H, Gustafson EJ, Crow TR (2002) Modeling landscape net ecosystem productivity (LandNEP) under alternative management regimes. Ecol Model 154:75–91
Fahey TJ, Siccama TG, Driscoll CT, Likens GE, Campbell J, Johnson CE, Battles JJ, Aber JD, Cole JJ, Fisk MC, Groffman PM, Hamburg SP, Holmes RT, Schwarz PA, Yanai RD (2005) The biogeochemistry of carbon at Hubbard Brook. Biogeochemistry 75:109–176
FAO (Food and Agricultural Organization of the United Nations) (2001) Global forest resources assessment 2000-main report. FAO Forestry paper 140. FAO, Rome
FAO (Food and Agricultural Organization of the United Nations) (2006a) Global forest resources assessment 2005. Progress towards sustainable forest management. FAO Forestry paper 147. FAO, Rome
FAO (Food and Agricultural Organization of the United Nations) (2006b) Global planted forests thematic study: results and analysis. Planted forests and trees working paper 38. FAO, Rome
FAO (Food and Agricultural Organization of the United Nations) (2007a) Fire management – global assessment 2006. FAO Forestry paper 151. FAO, Rome
FAO (Food and Agricultural Organization of the United Nations) (2007b) State of the world's forests 2007. FAO, Rome
FAO (Food and Agricultural Organization of the United Nations) (2009) Global review of forest pests and diseases. FAO Forestry paper 156. FAO, Rome
Faulkner S (2004) Urbanization impacts on the structure and function of forested wetlands. Urban Ecosyst 7:89–106
Fellows AW, Goulden ML (2008) Has fire suppression increased the amount of carbon stored in western U.S. forests? Geophys Res Lett 35, L12404, doi:10.1029/2008GL033965
Field CB, Lobell DB, Peters HA, Chiariello NR (2007) Feedbacks of terrestrial ecosystems to climate change. Annu Rev Environ Resour 32:7.1–7.29
Fisher JI, Hurtt GC, Thomas RQ, Chambers JQ (2008) Clustered disturbances lead to bias in large-scale estimates based on forest sample plots. Ecol Lett 11:554–563
Galik CS, Jackson RB (2009) Risks to forest carbon offset projects in a changing climate. For Ecol Manage 257:2209–2216
García-Quijano JF, Deckmyn G, Ceulemans R, van Orshoven J, Muys B (2008) Scaling from stand to landscape scale of climate change mitigation by afforestation and forest management: a modeling approach. Clim Change 86:397–424
Gleixner G, Kramer C, Hahn V, Sachse D (2005) The effect of biodiversity on carbon storage in soils. In: Scherer-Lorenzen M, Körner C, Schulze E-D (eds) Forest diversity and function. Ecological studies, Vol. 176. Springer, Berlin, pp 165–183
Goetz SJ, Jantz CA, Prince SD, Smith AJ, Wright R, Varlyguin D (2004) Integrated analysis of ecosystem interactions with land use change: the Chesapeake Bay watershed. In: De Fries RS, Asner GP, Houghton RA (eds) Ecosystems and land use change. Geophysical Monograph Series. American Geophysical Union, Washington, DC, pp 263–275
Gough CM, Vogel CS, Schmid HP, Curtis PS (2008) Controls on annual forest carbon storage: lessons from the past and predictions for the future. BioScience 58:609–622
Goward SN, Masek JG, Cohen W, Moisen G, Collatz GJ, Healy S, Houghton RA, Huang C, Kennedy R, Law B, Powell S, Turner D, Wulder MA (2008) Forest disturbance and North American carbon flux. EOS Trans Am Geophys Union 89:105–106
Gower ST, Krankina O, Olson RJ, Apps M, Linder S, Wang C (2001) Net primary production and carbon allocation patterns of boreal forest ecosystems. Ecol Appl 11:1395–1411
Griffiths H, Jarvis PG (2005) The carbon balance of forest biomes. Taylor & Francis, Oxon, UK
Grimm NB, Faeth SH, Golubiewski NE, Redman CL, Wu J, Bai X, Briggs JM (2008) Global change and the ecology of cities. Science 319:756–760
Harmon ME (2001) Carbon sequestration in forests. J For 99:24–29
Harris JA, Hobbs RJ, Higgs E, Aronson J (2006) Ecological restoration and global climate change. Restor Ecol 14:170–176
Heisler GM (1986) Energy savings with trees. J Arboricult 12:113–125
Helms JA (ed) (1998) The dictionary of forestry. Society of American Forestry, Bethesda, MD

Hennigar CR, MacLean DA, Amos-Binks LJ (2008) A novel approach to optimize management strategies for carbon stored in both forests and wood products. For Ecol Manag 256:786–797

Hirano T, Jauhiainen J, Inoue T, Takahashi H (2009) Controls on the carbon balance of tropical peatlands. Ecosystems DOI: 10.1007/s10021-008-9209-1

Hoenicka H, Fladung M (2006) Biosafety in *Populus* spp. and other forest trees: from non-native species to taxa derived from traditional breeding and genetic engineering. Trees 20:131–144

Holl KD (2002) Long-term vegetation recovery on reclaimed coal surface mines in the eastern USA. J Appl Ecol 39:960–970

Houghton RA, Hackler JL, Lawrence KT (1999) The U.S. carbon budget: contributions from land use change. Science 295:574–578

Hudiburg T, Law B, Turner DP, Campbell J, Donato D, Duane M (2009) Carbon dynamics of Oregon and Northern California forests and potential land-based carbon storage. Ecol Appl 19:163–180

Hungate BA, Naiman RJ, Apps M, Cole JJ, Moldan B, Satake K, Stewart JWB, Victoria R, Vitousek PM (2003) Disturbance and element interactions. In: Melillo JM, Field CB, Moldan B (eds) Interactions of the major biogeochemical cy: global change and human impacts. Island Press Washington, DC, pp 47–62

Huq S, Kovats S, Reid H, Satterthwaite D (2007) Editorial: reducing risks to cities from disasters and climate change. Environ Urban 19:3–15

Hurteau MD, Koch GW, Hungate BA (2008) Carbon protection and fire risk reduction: toward a full accounting of forest carbon offsets. Front Ecol Environ 6:493–498

Hyvönen R, Ågren GI, Linder S, Persson T, Cotrufo MF, Ekblad A, Freeman M, Grelle A, Janssens IA, Jarvis PG, Kellomäki S, Lindroth A, Loustau D, Lundmark T, Norby RJ, Oren R, Pilegaard K, Ryan MG, Sigurdsson BD, Strömgren M, Oijen M, Wallin G (2007) The likely impact of elevated [CO_2], nitrogen deposition, increased temperature and management on carbon sequestration in temperate and boreal forest ecosystems: a literature review. New Phytol 173:463–480

Imhoff ML, Bounoua L, DeFries R, Lawrence WT, Stutzer D, Tucker CJ, Ricketts T (2004) The consequences of urban land transformation on net primary productivity in the United States. Remote Sens Environ 89:434–443

Ingerson A, Loya W (2008) Measuring forest carbon: strengths and weaknesses of available tools. Science and policy brief. The Wilderness Society, Washington, DC

Irland LC (2000) Ice storms and forest impacts. Sci Total Environ 262:231–242

Ise T, Dunn AL, Wofsy SC, Moorcroft PR (2008) High sensitivity of peat decomposition to climate change through water-table feedback. Nat Geosci 1:763–766

Jackson RB, Randerson JT, Canadell JG, Anderson RG, Avissar R, Baldocchi DD, Bonan GB, Caldeira K, Diffenbaugh NS, Field CB, Hungate BA, Jobbágy EG, Kueppers LM, Nosetto MD, Pataki DE (2008) Protecting climate with forests. Environ Res Lett 3:044006

Jaenicke J, Rieley JO, Mott C, Kimman P, Siegert F (2008) Determination of the amount of carbon stored in Indonesian peatlands. Geoderma 147:151–158

Jandl R, Lindner M, Vesterdahl L, Bauwens B, Baritz R, Hagedorn F, Johnson DW, Minkkinen K, Byrne KA (2007a) How strongly can forest management influence soil carbon sequestration? Geoderma 137:253–268

Jandl R, Vesterdal L, Olsson M, Bens O, Badeck F, Rock J (2007b) Carbon sequestration and forest management. CAB Rev Perspect Agr Vet Sci Nutr Nat Res,doi:10.1079/PAVSNNR20072017

Janisch JE, Harmon ME (2002) Successional changes in live and dead wood carbon stores: implications for net ecosystem productivity. Tree Physiol 22:77–89

Jarvis PG, Ibrom A, Linder S (2005) 'Carbon forestry': managing forests to conserve carbon. In: Griffiths H, Jarvis PG (eds) The carbon balance of forest biomes. Taylor & Francis, Oxon, UK, pp 331–349

Jarvis PG, Linder S (2007) Forests remove carbon dioxide from the atmosphere: spruce forest tales! In: Freer-Smith PH, Broadmeadow MSJ, Lynch JM (eds) Forestry and climate change. CAB International, Wallingford, UK, pp 60–72

Johnson DW, Curtis PS (2001) Effects of forest management on soil C and N storage: meta analysis. For Ecol Manag 140:227–238
Johnson EA, Miyanishi K (2008) Testing the assumptions of chronosequences in succession. Ecol Lett 11:419–431
Jose S, Williams R, Zamora D (2006) Belowground ecological interactions in mixed-species forest plantations. For Ecol Manag 233:231–239
Kashian DM, Romme WH, Tinker DB, Turner MG, Ryan MG (2006) Carbon storage on landscapes with stand-replacing fires. BioScience 56:598–606
Keddy PA, Fraser LH, Solomeshch AI, Junk WJ, Campbell DR, Arroyo MTK, Alho CJR (2009) Wet and wonderful: the world's largest wetlands are conservation priorities. BioScience 59:39–51
Kimmins JP (2004) Forest ecology. Prentice Hall, Upper Saddle River, NJ
Kirilenko AP, Sedjo RA (2007) Climate change impacts on forestry. Proc Natl Acad Sci U S A 104:19697–19702
Knicker H (2007) How does fire affect the nature and stability of soil organic nitrogen and carbon? A review. Biogeochemistry 85:91–118
Knoke T, Ammer C, Stimm B, Mosandl R (2008) Admixing broadleaved to coniferous tree species: a review on yield, ecological stability and economics. Eur J For Res 127:89–101
Koch JM, Hobbs RJ (2007) Synthesis: is Alcoa successfully restoring a jarrah forest ecosystem after bauxite mining in Western Australia? Restor Ecol 15:S137–S144
Körner C (2006) Plant CO_2 responses: an issue of definition, time and resource supply. New Phytol 172:393–411
Kowalski AS, Loustau D, Berbigier P, Manca G, Tedeschi V, Borghetti M, Valentini R, Kolari P, Berninger F, Rannik Ü, Hari P, Rayment M, Mencuccini M, Moncrieff J, Grace J (2004) Paired comparisons of carbon exchange between undisturbed and regenerating stands in four managed forests in Europe. Glob Change Biol 10:1707–1723
Krankina ON, Harmon ME (2006) Forest management strategies for carbon storage. In: Matrazzo D (ed) Forests, carbon and climate change – a synthesis of science findings. Oregon Forest Resources Institute, Portland, OR, pp 79–91
Krawchuk MA, Moritz MA, Parisien M-A, Van Dorn J, Hayhoe K (2009) Global pyrogeography: the current and future distribution of wildfire. PloS ONE 4(4):e5102. doi:10.1371/journal.pone.0005102
Kurz WA, Dymond CC, Stinson G, Rampley GJ, Neilson ET, Carroll AL, Ebata T, Safranyik L (2008a) Mountain pine beetle and forest carbon feedback to climate change. Nature 452: 987–990
Kurz WA, Stinson G, Rampley GJ, Dymond CC, Neilson ET (2008b) Risk of natural disturbances makes future contribution of Canada's forest to the global carbon cycle highly uncertain. Proc Natl Acad Sci U S A 105:1551–1555
Lähteenoja O, Ruokolainen K, Schulman L, Oinonen M (2009) Amazonian peatlands: an ignored C sink and potential source. Glob Change Biol 15:2311–2320
Laine J, Laiho R, Minkkinen K, Vasander H (2006) Forestry and boreal peatlands. In: Wieder RK, Vitt DH (eds) Boreal peatland ecosystems. Ecological studies, Vol. 188. Springer-Verlag, Berlin, pp.331–357
Lal R (2005) Forest soils and carbon sequestration. For Ecol Manag 220:242–258
Lal R, Kimble JM, Birdsey RA, Heath LS (2003). Research and development priorities for carbon sequestration in forest soils. In: Kimble JM, Heath LS, Birdsey RA, Lal R (eds) The potential of U.S. forest soils to sequester carbon and mitigate the greenhouse effect. Lewis Publishers, Boca Raton, FL, pp 409–420
Lal R, Sobecki TM, Jivari T, Kimble JM (2004) Soil degradation by mining and other disturbance. In: Lal R, Sobecki TM, Jivari T, Kimble JM (eds) Soil degradation in the United States – extent, severity, and trends. CRC Press, Boca Raton, FL, pp 163–171
Landsberg JJ, Gower ST (1997) Applications of physiological ecology to forest management. Academic, San Diego, CA
Larjavaara M (2008) A review on benefits and disadvantages of tree diversity. Open For Sci J 1:24–26

Lehmann J, Sohi S (2008) Comment on "fire-derived charcoal causes loss of forest humus". Science 321:1295c

Lewis SL, Lopez-Gonzalez G, Sonké B, Affum-Baffoe K, Baker TR, Ojo LO, Phillips OL, Reitsma JM, White L, Comiskey JA, M-N DK, Ewango CEN, Feldpausch TR, Hamilton AC, Gloor M, Hart T, Hladik A, Lloyd J, Lovett JC, Makana J-R, Malhi Y, Mbago FM, Ndangalasi HJ, Peacock J, S-H PK, Sheil D, Sunderland T, Swaine MD, Taplin J, Taylor D, Thomas SC, Votere R, Wöll H (2009) Increasing carbon storage in intact African tropical forests. Nature 457:1003–1007

Limpens J, Berendse F, Blodau C, Canadell JG, Freeman C, Holden J, Roulet N, Rydin H, Schaepman-Strub G (2008) Peatlands and the carbon cycle: from local processes to global implications – a synthesis. Biogeosciences 5:1475–1491

Lindroth A, Grelle A, Morén AS (1998) Long-term measurements of boreal forest carbon balance reveal large temperature sensitivity. Glob Change Biol 4:443–450

Lindroth A, Lagergren F, Grelle A, Klemedtsson L, Langvall O, Weslien P, Tuulik J (2009) Storms can cause Europe-wide reduction in forest carbon sink. Glob Change Biol 15:346–355

Liski J, Pussinen A, Pingoud K, Mäkipää R, Karjalainen T (2001) Which rotation length is favourable to carbon sequestration? Can J For Res 31:2004–2013

Litton CM, Raich JW, Ryan MG (2007) Review: carbon allocation in forest ecosystems. Glob Change Biol 13:2089–2109

Long AJ, Nair PKR (1999) Trees outside forests: agro-, community-, and urban forestry. New Forest 17:145–174

Lorenz K, Lal R (2007) Stabilization of organic carbon in chemically separated pools in reclaimed coal mine soils in Ohio. Geoderma 141:294–301

Lorenz K, Lal R (2009) Biogeochemical C and N cy in urban soils. Environ Int 35:1–8

Lottermoser BG (2007) Mine wastes: characterization, treatment and environmental impacts. Springer-Verlag, Berlin

Loveland TR, Reed BC, Brown JF, Ohlen DO, Zhu Z, Yang L, Merchant J (2000) Global land cover characteristics database (GLCCD) version 2.0. http://edc2.usgs.gov/glcc/globe_int.php, accessed August 25, 2009

Lubowski RN, Vesterby M, Bucholtz S, Baez A, Roberts MJ (2006) Major uses of land in the United States, 2002. Economic information bulletin no. 14. United States Department of Agriculture, Economic Research Service

Luyssaert S, Inglima I, Jung M, Richardson AD, Reichstein M, Papale D, Piao SL, Schulze E-D, Wingate L, Matteucci G, Aragao L, Aubinet M, Beer C, Bernhofer C, Black KG, Bonal D, Bonnefond J-M, Chambers J, Ciais P, Cook B, Davis KJ, Dolman AJ, Gielen B, Goulden M, Grace J, Granier A, Grelle A, Griffis T, Grünwald T, Guidolotti G, Hanson PJ, Harding R, Hollinger DY, Hutyra LR, Kolari P, Kruijt B, Kutsch W, Lagergren F, Laurila T, Law BE, Le Maire G, Lindroth A, Loustau D, Malhi Y, Mateus J, Migliavacca M, Misson L, Montagnani L, Moncrieff J, Moors E, Munger JW, Nikinmaa E, Ollinger SV, Pita G, Rebmann C, Roupsard O, Saigusa N, Sanz MJ, Seufert G, Sierra C, Smith M-L, Tang J, Valentini R, Vesala T, Janssens IA (2007) The CO_2-balance of boreal, temperate and tropical forests derived from a global database Glob Change Biol 13:2509–2537

Luyssaert S, Schulze E-D, Börner A, Knohl A, Hessenmöller D, Law BE, Ciais P, Grace J (2008) Old-growth forests as global carbon sinks. Nature 455:213–215

MacFarlane DW (2009) Potential availability of urban wood biomass in Michigan: implications for energy production, carbon sequestration and sustainable forest management in the U.S.A. Biomass Bioenerg 33:628–634

Magnani F, Dewar RC, Borghetti M (2009) Leakage and spillover effects of forest management on carbon storage: theoretical insights from a simple model. Tellus 61B:385–393

Magnani F, Mencuccini M, Borghetti M, Berbigier P, Berninger F, Delzon S, Grelle A, Hari P, Jarvis PG, Kolari P, Kowalski AS, Lankreijer H, Law BE, Lindroth A, Loustau D, Manca G, Moncrieff JB, Rayment M, Tedeschi V, Valentini R, Grace J (2007) The human footprint in the carbon cycle of temperate and boreal forests. Nature 447:848–852

Maguire DA, Osawa A, Batista JLF (2005) Primary production, yield and carbon dynamics. In: Andersson F (ed) Ecosystems of the world 6. Coniferous forests. Elsevier, Amsterdam, The Netherlands, pp 339–383
Markewitz D (2006) Fossil fuel carbon emissions from silviculture: impacts on net carbon sequestration in forests. For Ecol Manag 236:153–161
Martin PH, Nabuurs G-J, Aubinet M, Karjalainen T, Vine EL, Kinsman J, Heath LS (2001) Carbon sinks in temperate forests. Annu Rev Energy Environ 26:435–465
McCarthy HR, Oren R, Kim H-S, Johnsen KH, Maier C, Pritchard SG, Davis MA (2006) Interaction of ice storms and management practices on current carbon sequestration in forests with potential mitigation under future CO_2 atmosphere. J Geophys Res 111:D15103. doi:10.1029/2005JD006428
McHale MR, Burke IC, Lefsky MA, Peper PJ, McPherson EG (2009) Urban forest biomass estimates: is it important to use allometric relationships developed specifically for urban trees? Urban Ecosyst 12:95–113
McIver J, Youngblood A, Stephens SL (2009) The national fire and fire surrogate study: ecological consequences of fuel reduction methods in seasonally dry forests. Ecol Appl 19:283–284
Mead DJ (2005) Forests for energy and the role of planted trees. Crit Rev Plant Sci 24:407–421
Miner R, Perez-Garcia J (2007) The greenhouse gas and carbon profile of the global forest products industry. Forest Prod J 57:80–90
Miranda M, Burris P, Bincang JF, Shearman P, Briones JO, La Viña A, Menard A (2003) Mining and critical ecosystems: mapping the risks. World Resources Institute, Washington, DC
Moore PD (2002) The future of cool temperate bogs. Environ Conserv 29:3–20
Muller-Landau HC (2009) Sink in the African jungle. Nature 457:969–970
Mund M, Schulze E-D (2005) Silviculture and its interaction with biodiversity and the carbon balance of forest soils. In: Scherer-Lorenzen M, Körner C, Schulze E-D (eds) Forest diversity and function. Ecological studies, Vol. 176. Springer, Berlin, pp 185–210
Nabuurs GJ, Masera O, Andrasko K, Benitez-Ponce P, Boer R, Dutschke M, Elsiddig E, Ford-Robertson J, Frumhoff P, Karjalainen T, Krankina O, Kurz WA, Matsumoto M, Oyhantcabal W, Ravindranath NH, Sanz Sanchez MJ, Zhang X (2007): Forestry. In: Metz B, Davidson OR, Bosch PR, Dave R, Meyer LA (eds) Climate change 2007: mitigation. Contribution of working group III to the fourth assessment report of the intergovernmental panel on climate change. Cambridge University Press, Cambridge, UK/New York, NY, pp 541–584
Nabuurs G-J, Thürig E, Heidema N, Armolaitis K, Biber P, Cienciala E, Kaufmann E, Mäkipää R, Nilsen P, Petritsch R, Pristova T, Rock J, Schelhaas M-J, Sievanen R, Somogyi Z, Vallet P (2008) Hotspots of the European forests carbon cycle. For Ecol Manag 256:194–200
Nelson BW, Kapos V, Adams JB, Oliveira WJ, Braun OPG, do Amaral IL (1994) Forest disturbance by large scale blowdowns in the Brazilian Amazon. Ecology 75:853–885
Nepstad DC, Tohver IM, Ray D, Moutinho P, Cardinot G (2007) Mortality of large trees and lianas following experimental drought in an Amazon forest. Ecology 88:2259–2269
Neuzil SG (1997) Onset and rate of peat and carbon accumulation in four domed ombrogenous peat deposits, Indonesia. In: Rieley JO, Page SE (eds) Biodiversity and sustainability of tropical peatlands. Samara Publishing, Cardigan, UK, pp 55–72
Nichols JD, Bristow M, Vanclay JK (2006) Mixed-species plantations: prospects and challenges. For Ecol Manag 233:383–390
Nowak DJ, Crane DE (2002) Carbon storage and sequestration by urban trees in the USA. Environ Pollut 116:381–389
Nowak DJ, Noble MH, Sisinni SM, Dwyer JF (2001) Assessing the U.S. urban forest resource. J For 99:37–42
Nowak DJ, Walton JT (2005) Projected urban growth (2000–2050) and its estimated impact on the US forest resource. J For 103:383–389
Nowak DJ, Walton JT, Dwyer JF, Kaya LG, Myeong S (2005) The increasing influence of urban environments on US forest management. J For 103:377–382

Nyland RD (2002) Silviculture: concepts and applications. McGraw-Hill, New York
Oliver CD, Larson BC (1996) Forest stand dynamics. Wiley, New York
Overby ST, Hart SC, Neary DG (2003) Impacts of natural disturbance on soil carbon dynamics in forest ecosystems. In: Kimble JM, Heath LS, Birdsey RA, Lal R (eds) The potential of U.S. forest soils to sequester carbon and to mitigate the greenhouse effect. CRC Press, Boca Raton, FL, pp 159–172
Page SE (2004) The natural resource functions of tropical peatlands. In: Mansor M, Ali A, Rieley J, Ahmad AH, Mansor A (eds) Tropical peat swamps-safe-guarding a global natural resource. Proceedings of the international conference and workshop on tropical peat swamps. Penerbit Universiti Sains Malaysia, Pulau Pinang, Malaysia, pp 153–161
Page SE, Banks C (2007) Tropical peatlands: distribution, extent and carbon storage – uncertainties and knowledge gaps. Peatlands Int 2:26–27
Page SE, Siegert F, Rieley JO, Boehm H-D V, Jaya A, Limin S (2002) The amount of carbon released from peat and forest fires in Indonesia during 1997. Nature 420:61–65
Papiez MR, Potosnak MJ, Goliff WS, Guenther AB, Matsunaga SN, Stockwell WR (2009) The impacts of reactive terpene emissions on air quality in Las Vegas, Nevada. Atmos Environ 43:4109–4123
Pataki DE (2007) Human settlements and the North American carbon cycle. In: King W, Dilling L, Zimmermann GP, Fairman DM, Houghton RA, Marland G, Rose AZ, Wilbanks TJ (eds) The first state of the carbon cycle report. U.S. Climate change science program, Washington, DC, pp 149–156
Pataki DE, Emmi PC, Forster CB, Mills JI, Pardyjak ER, Peterson TR, Thompson JD, Dudley-Murphy E (2009) An integrated approach to improving fossil fuel emissions scenarios with urban ecosystem studies. Ecol Complex 6:1–14
Paul KI, Polglase PJ, Nyakuengama JG, Khanna PK (2002) Change in soil carbon following afforestation. For Ecol Manag 168:241–257
Paavilainen E, Päivänen J (1995) Peatland forestry. Ecological studies, Vol. 111. Springer-Verlag, Berlin
Petrenko VV, Smith AM, Brook EJ, Lowe D, Riedel K, Brailsford G, Hua Q, Schaefer H, Reeh N, Weiss RF, Etheridge D, Severinghaus JP (2009) $^{14}CH_4$ measurements in Greenland ice: investigating last glacial termination CH_4 sources. Science 324:506–508
Potter C, Gross P, Klooster S, Fladeland M, Genovese V (2008) Storage of carbon in U.S. forests predicted from satellite data, ecosystem modeling, and inventory summaries. Clim Change 90:269–282
Potter C, Tan P-N, Kumar V, Kucharik C, Klooster S, Genovese V, Cohen W, Healey S (2005) Recent history of large-scale ecosystem disturbances in North America derived from the AVHRR satellite record. Ecosystems 8:808–824
Pouyat RV, Yesilonis ID, Golubiewski NE (2009) A comparison of soil organic carbon stocks between residential turf grass and native soil. Urban Ecosyst 12:45–62
Pouyat RV, Yesilonis ID, Nowak DJ (2006) Carbon storage by urban soils in the United States. J Environ Qual 35:1566–1575
Powers RF (1999) On the sustainable productivity of planted forests. New For 17:263–306
Powers RF, Scott DA, Sanchez FG, Voldseth RA, Page-Dumroese DS, Elioff JD, Stone DM (2005) The North American long-term soil productivity experiment: findings from the first decade of research. For Ecol Manag 220:31–50
Pregitzer KS, Euskirchen ES (2004) Carbon cycling and storage in world forests: biome patterns related to forest age. Glob Change Biol 10:2052–2077
Pretzsch H (2005) Diversity and productivity in forests: evidence from long-term experimental plots. In: Scherer-Lorenzen M, Körner C, Schulze E-D (eds) Forest diversity and function. Ecological studies, Vol. 176. Springer, Berlin, pp 41–64
Purves D, Pacala S (2008) Predictive models of forest dynamics. Science 320:1452–1453
Quigley MF (2004) Street trees and rural conspecifics: will long-lived trees reach full size in urban conditions? Urban Ecosyst 7:29–39
Radeloff VC, Hammer RB, Stewart SI, Fried JS, Holcomb SS, McKeefry JF (2005) The wildland–urban interface in the United States. Ecol Appl 15:799–805

Rauch JN, Pacyna JM (2009) Earth's global Ag, Al, Cr, Cu, Fe, Ni, Pb, and Zn cy. Global Biogeochem Cy 23, GB2001, doi:10.1029/2008GB003376

Ren D, Wang J, Fu R, Karoly DJ, Hong Y, Leslie LM, Fu C, Huang G (2009) Mudslide-caused ecosystem degradation following Wenchuan earthquake 2008. Geophys Res Lett 36:L05401. doi:10.1029/2008GL036702

Richards KR, Sampson RN, Brown S (2006) Agricultural & forestlands: U.S. carbon policy strategies. Pew Center on Global Climate Change

Richter deB D Jr, Jenkins DH, Karakash JT, Knight J, McCreery LR, Nemestothy KP (2009) Wood energy in America. Science 323:1432–1433

Rigby M, Prinn RG, Fraser PJ, Simmonds PG, Langenfelds RL, Huang J, Cunnold DM, Steele LP, Krummel PB, Weiss RF, O'Doherty S, Salameh PK, Wang HJ, Harth CM, Mühle J, Porter LW (2008) Renewed growth of atmospheric methane. Geophys Res Lett 35:L22805. doi:10.1029/2008GL036037

Roulet NT (2000) Peatlands, carbon storage, greenhouse gases, and the Kyoto Protocol: prospects and significance for Canada. Wetlands 20:605–615

Rousseau R, Kaczmarek D, Martin J (2005) A review of the biological, social, and regulatory constraints to intensive plantation culture. S J Appl For 29:105–109

Rumpel C, Balesdent J, Grootes P, Weber E, Kögel-Knabner I (2003) Quantification of lignite- and vegetation-derived soil carbon using ^{14}C activity measurements in a forested chronosequence. Geoderma 112:155–166

Running SW (2008) Ecosystem disturbance, carbon, and climate. Science 321:652–653

Ryan MG, Binkley D, Fownes JH (1997) Age-related decline in forest productivity: pattern and process. In: Begon M, Fitter AH (eds) Advances in ecological research Vol. 27. San Diego, CA, pp 214–262

Sartori F, Markewitz D, Borders BE (2007) Soil carbon storage and nitrogen and phosphorous availability in loblolly pine plantations over 4 to 16 years of herbicide and fertilizer treatments. Biogeochemistry 84:13–30

Schaeffer M, Eickhout B, Hoogwijk M, Strengers B, van Vuuren D, Leemans R, Opsteegh T (2006) CO_2 and albedo climate impacts of extratropical carbon and biomass plantations. Global Biogeochem Cy 20, GB2020, doi:10.1029/2005GB002581

Schelhaas M-J, Nabuurs G-J, Schuck A (2003) Natural disturbances in the European forests in the 19th and 20th centuries. Glob Change Biol 9:1620–1633

Schils R, Kuikman P, Liski J, van Oijen M, Smith P, Webb J, Alm J, Somogyi Z, van den Akker J, Billett M, Emmett B, Evans C, Lindner M, Palosuo T, Bellamy P, Jandl R, Hiederer R (2008) Review of existing information on the interrelations between soil and climate change (CLIMSOIL). http://ec.europa.eu/environment/soil/pdf/climsoil_report_dec_2008.pdf, Accessed August 25, 2009

Schoene D, Netto M (2005) The Kyoto Protocol: what does it mean for forests and forestry? Unasylva 56:3–11

Schulp CJE, Nabuurs G-J, Verburg PH (2008a) Future carbon sequestration in Europe – effects of land use change. Ag Ecosyst Environ 127:251–264

Schulp CJE, Nabuurs G-J, Verburg PH, de Waal RW (2008b) Effect of tree species on carbon stocks in forest floor and mineral soil and implications for soil carbon inventories. For Ecol Manag 256:482–490

Schulze E-D (2006) Biological control of the terrestrial carbon sink. Biogeosciences 3:147–166

Schulze E-D, Beck E, Müller-Hohenstein K (2005) Plant ecology. Springer, Berlin

Schulze E-D, Lloyd J, Kelliher FM, Wirth C, Rebmann C, Lühker B, Mund M, Knohl A, Milyukova IM, Schulze W, Ziegler W, Varlagin AB, Sogachev AF, Valentini R, Dore S, Grigoriev S, Kolle O, Panfyorov MI, Tchebakova N, Vygodskaya NN (1999) Productivity of forests in the Eurosiberian boreal region and their potential to act as a carbon sink – a synthesis. Glob Change Biol 5:703–722

Schulze E-D, Valentini R, Sanz M-J (2002) The long way from Kyoto to Marrakesh: implications of the Kyoto Protocol negotiations for global ecology. Glob Change Biol 8:505–518

Schulze E-D, Wirth C, Heimann M (2000) Managing forests after Kyoto. Science 289:2058–2059

Schuur EAG, Bockheim J, Canadell JG, Euskirchen E, Field CB, Goryachkin SV, Hagemann S, Kuhry P, Lafleur PM, Lee H, Mazhitova G, Nelson FE, Rinke A, Romanovsky VE, Shiklomanov N, Tarnocai C, Venevsky S, Vogel JG, Zimov SA (2008) Vulnerability of permafrost carbon to climate change: implications for the global carbon cycle. Bioscience 58:701–714

Seppälä R, Buck A, Katila P (eds) (2009) Adaptation of forests and people to climate change. A global assessment report. International Union of Forest Research Organizations (IUFRO), Helsinki, Finland

Shorohova E, Kuuluvainen T, Kangur A, Jõgiste K (2009) Natural stand structures, disturbance regimes and successional dynamics in the Eurasian boreal forests: a review with special reference to Russian studies. Ann For Sci 66:200–220

Smith DM, Larson BC, Kelty MJ, Ashton PMS (1997) The practice of silviculture: applied forest ecology. Wiley, New York

Smith P, Fang C, Dawson JJC, Moncrieff JB (2008) Impact of global warming on soil organic carbon. Adv Agron 97:1–43

Smithwick EAH, Harmon ME, Domingo JB (2007) Changing temporal patterns of forest carbon stores and net ecosystem carbon balance: the stand to landscape transformation. Landscape Ecol 22:77–94

Soja AJ, Tchebakova NM, French NHF, Flannigan MD, Shugart HH, Stocks BJ, Sukhinin AI, Parfenova EI, Chapin FSIII, Stackhouse PW Jr (2007) Climate-induced boreal forest change: predictions versus current observations. Global Planet Change 56:274–296

Solomon AM, Freer-Smith PH (2007) Forest responses to global change in North America: interacting forces define a research agenda. In: Freer-Smith PH, Broadmeadow MSJ, Lynch JM (eds) Forestry and climate change. CAB International, Wallingford, UK, pp 151–159

Sperow M (2006) Carbon sequestration potential in reclaimed mine sites in seven east-central states. J Environ Qual 35:1428–1438

Spittlehouse DL, Stewart RB (2003) Adaptation to climate change in forest management. BC J Ecosyst Manag 4:1–11

Stocks BJ (2004) Forest fires in the looreal zone: climate change and carbon implications. International Forest Fire News 31:122–131

Stokstad E (2008) A second chance for rainforest biodiversity. Science 320:436–438

Stoy PC, Katul GG, Siqueira MBS, Juang J-Y, Novick KA, McCarthy HR, Oishi AC, Oren R (2008) Role of vegetation in determining carbon sequestration along ecological succession in the southeastern United States. Glob Change Biol 14:1409–1427

Strak M (ed) (2008) Peatlands and climate change. International Peat Society, Jyväskylä, Finnland

Suchanek TH, Mooney HA, Franklin JF, Gucinski H, Ustin SL (2004) Carbon dynamics of an old-growth forest. Ecosystems 7:421–426

Tonn B, Marland G (2007) Carbon sequestration in wood products: a method for attribution to multiple parties. Environ Sci Policy 10:162–168

Torbert JL, Burger JA (2000) Forest land reclamation. In: Barnhisel RJ, Darmody RG, Lee Daniels W (eds) Reclamation of drastically disturbed lands, Agronomy Monograph no. 41, ASA-CSSA-SSSA, Madison, WI, pp 371–398

Trettin CC, Jurgensen MF (2003) Carbon cycling in wetland forest soils. In: Kimble JM, Heath LS, Birdsey RA, Lal R (eds) The potential of U.S. forest soils to sequester carbon and mitigate the greenhouse effect. Lewis Publishers, Boca Raton, FL, pp 311–331

Trusilova K, Churkina G (2008) The response of the terrestrial biosphere to urbanization: land cover conversion, climate, and urban pollution. Biogeosciences 5:1505–1515

Turetsky MR, St. Louis VL (2006) Disturbance in boreal peatlands. In: Wieder RK, Vitt DH (eds) Boreal peatland ecosystems. Ecological studies, Vol. 188. Springer-Verlag, Berlin, pp 359–379

Ulanova NG (2000) The effects of wind-throw on forests at different spatial scales: a review. For Ecol Manag 135:155–167

United States Global Change Research Program (USGCRP) (2003) Land use / land cover change – USGCRP fiscal year 2003 accomplishments. Washington, D.C., http://www.usgcrp.gov/usgcrp/ProgramElements/recent/landFY2003.htm, accessed August 25, 2009

Ussiri DAN, Lal R (2008) Method for determining coal carbon in the reclaimed minesoils contaminated with coal. Soil Sci Soc Am J 72:231–237

van der Werf GR, Dempewolf J, Trigg SN, Randerson JT, Kasibhatla PS, Giglio L, Murdiyarso D, Peters W, Morton DC, Collatz GJ, Dolman AJ, DeFries RS (2008) Climate regulation of fire emissions and deforestation in equatorial Asia. Proc Natl Acad Sci USA 105:20350–20355

Vasander H, Kettunen A (2006) Carbon in boreal peatlands. In: Wieder RK, Vitt DH (eds) Boreal peatland ecosystems. Ecological studies, Vol. 188. Springer-Verlag, Berlin, pp 165–194

Vetter M, Wirth C, Böttcher H, Churkina G, Schulze E-D, Wutzler T, Weber G (2005) Partitioning direct and indirect human-induced effects on carbon sequestration of managed coniferous forests using model simulations and forest inventories. Glob Change Biol 11:810–827

Vitousek PM, Fahey TJ, Johnson DW, Swift MJ (1988) Element interactions in forest ecosystems: succession, allometry and input-output budgets. Biogeochemistry 5:7–34

Vitt DH, Wieder RK (2006) Boreal peatland ecosystems: our carbon heritage. In: Wieder RK, Vitt DH (eds) Boreal peatland ecosystems. Ecological studies, Vol. 188. Springer-Verlag, Berlin, pp 425–429

Wagner RG, Little KM, Richardson B, McNabb K (2006) The role of vegetation management for enhancing productivity of the world's forests. Forestry 79:57–79

Wardle DA, Nilsson M-C, Zackrisson O (2008) Fire-derived charcoal causes loss of forest humus. Science 320:629

Wardle DA, Walker LR, Bardgett RD (2004) Ecosystem properties and forest decline in contrasting long-term chronosequences. Science 305:509–513

Waring RW, Running SW (2007) Forest ecosystems – analysis at multiple scales. Elsevier Academic Press, Burlington, MA

Wieder RK, Vitt DH, Benscoter BW (2006) Peatlands and the boreal forest. In: Wieder RK, Vitt DH (eds) Boreal peatland ecosystems. Ecological studies, Vol. 188. Springer-Verlag, Berlin, pp 1–8

Wiedinmyer C, Neff JC (2007) Estimates of CO_2 from fires in the United States: implications for carbon management. Carbon Bal Manag 2:10

Yanai RD, Currie WS, Goodale CL (2003) Soil carbon dynamics after forest harvest: an ecosystem paradigm reconsidered. Ecosystems 6:197–212

Young TP, Petersen DA, Clary JJ (2005) The ecology of restoration: historical links, emerging issues and unexplored realms. Ecol Lett 8:662–673

Zaehle S, Bondeau A, Carter T, Cramer W, Erhard M, Prentice I, Reginster I, Rounsevell M, Sitch S, Smith B, Smith P, Sykes M (2007) Projected changes in terrestrial carbon storage in Europe under climate and land-use change, 1990–2100. Ecosystems 10:380–401

Zeng N (2008) Carbon sequestration via wood burial. Carbon Bal Manag 3:1

Zeng H, Chambers JQ, Negrón-Juárez RI, Hurtt GC, Baker DB, Powell MD (2009) Impacts of tropical cyclones on U.S. forest tree mortality and carbon flux from 1851 to 2000. Proc Natl Acad Sci USA 106:7888–7892

Zhang C, Tian H, Pan S, Liu M, Lockaby G, Schilling EB, Stanturf J (2008) Effects of forest regrowth and urbanization on ecosystem carbon storage in a rural-urban gradient in the southeastern United States. Ecosystems 11:1211–1222

Zhou G, Liu S, Li Z, Zhang D, Tang X, Zhou C, Yan J, Mo J (2006) Old-growth forests can accumulate carbon in soils. Nature 314:1417

Chapter 4
Carbon Dynamics and Pools in Major Forest Biomes of the World

Forests covered an estimated area of 3,952 million hectare in 2005, or about 30% of the world's land area, and contained an estimated 283 Pg C in biomass, 38 Pg C in dead wood and 317 Pg C in litter and soils to 30-cm depth (FAO 2006). Through natural processes described in Chapter 2, C continuously enters and leaves the forest ecosystems. Natural habitat conditions, determined by factors such as temperature, moisture availability and frequency of disturbance, influence the C budget of a forest ecosystem. The major global forest biomes are the boreal, temperate and tropical forest biomes. Biomes are broad vegetation types defined by climate, life-form and ecophysiology (Woodward et al. 2004). As the climatic and environmental variables differ widely among biomes, the C dynamics and pools must be discussed separately for major global forest biomes. Otherwise, the biogeophysical processes in boreal, temperate and tropical forests influence the climate more immediately than does the C cycle (Bonan 2008). In particular, the negative climate forcing through C sequestration can be diminished or enhanced through biogeophysical feedbacks such as evaporative cooling, albedo changes, and changes in surface roughness, and the feedbacks differ among boreal, temperate and tropical forests.

Estimates of the area extent of the boreal forest biome range between 950 and 1,570 million hectare depending on its definition. Estimates of the size of the C pool range between 78 and 143 Pg C for the vegetation and 338 Pg C in soils to 1-m depth (Saugier et al. 2001; Robinson 2007; Nieder and Benbi 2008). In the temperate forest biome, between 73 and 159 Pg C are stored in vegetation and between 153 and 195 Pg C in soils to 1-m depth on an area with an estimated size between 920 and 1,600 million hectare. Estimates of the area covered by the tropical forest biome range between 1,450 and 2,200 million hectare. Between 206 and 389 Pg C are stored in vegetation, and between 214 and 435 Pg C in soils to 1-m depth in the tropical forest biome. Thus, except for tropical forests, the largest amount of C in forest biomes is stored belowground in the soil (Reichstein 2007). However, data on the forest area and area trends, and C pools are not reliable and have a high degree of uncertainty (Mather 2005). Specifically, data on the C pools in above- and belowground biomass, dead wood, litter and SOC to 1-m and the maximum soil depth are not available for the entire forest biomes (FAO 2006). For example,

K. Lorenz and R. Lal, *Carbon Sequestration in Forest Ecosystems*,
DOI 10.1007/978-90-481-3266-9_4,

data on SOC pools for 51% of the global forest area are lacking. Lack of data on key C fluxes such as root biomass production restricts also the ability to build integrated C budgets for forest biomes. For example, significantly more C is probably stored in roots than hitherto estimated (Robinson 2007). The climatic changes in the last decades have a generally positive impact on forest productivity on sites where water is not strongly limiting (Boisvenue and Running 2006). Projected changes in CO_2 concentrations and climate may increase forest plant diversity capacity (Woodward and Kelly 2008). However, the ACC-induced modifications of frequency and intensity of wildfires, outbreaks of insects and pathogens, and extreme events such as high winds, may be more important than direct effects of elevated CO_2 and higher temperatures (Kirilenko and Sedjo 2007). Also, current and projected tropospheric ozone levels may reduce the C pools in the forest biomes but the understanding of ozone effects on forest ecosystems in a changing climate is rudimentary (Wittig et al. 2009). In the following chapter, C dynamics and pools for major global forest biomes are compared with the focus on GPP, R_a, NPP, R_h, NEP, vegetation, detritus and soil C, and how ACC may affect them. Furthermore, a few examples for C exchange processes associated with natural disturbances and other components of the NECB are also discussed.

4.1 Boreal Forests

The boreal biome is located primarily between 50°N and 60°N in the Northern Hemisphere, south of the Polar biome (Reich and Eswaran 2006). It can be subdivided into a semiarid component covered mainly by shrubs (355 million hectare), and the humid boreal biome covered mainly by trees (945 million hectare). The area extent has been estimated to be as high as 1,470 million hectare depending on the definition of boreal forest (Table 4.1; Jarvis et al. 2001). Specifically, much of the region between 46°N and 66°N is covered by the boreal forest (Jarvis et al. 2001), the world's second largest forest biome (Landsberg and Gower 1997). About 48% of the world's relatively undisturbed forest lies in the boreal region (Bryant et al. 1997).

The boreal climate is particularly harsh to tree growth. The climate is characterized by long, very cold winters and short, mild summers with wide annual and daily temperature variations (Table 4.1; Jarvis et al. 2001). Otherwise, long summer daylength in the high latitudes may help to compensate for the short growing season (Taggart and Cross 2009). Probably due to the climatic conditions and the short geological history after glacial retreat 18,000 years ago, tree species diversity in this biome is relatively poor, and the boreal forest is dominated by conifers of the family *Pinaceae* (Landsberg and Gower 1997). The boreal forest biome may be the youngest of the major forest biomes (Taggart and Cross 2009). Important tree genera include *Abies*, *Alnus*, *Betula*, *Larix*, *Picea*, *Populus*, *Salix*, and *Tsuga* (Black et al. 2005). Dominant species in boreal parts of North America are *Picea mariana* (Mill.) BSP and *Picea glauca* (Moench) Voss whereas *Pinus sylvestris* L. covers

Table 4.1 Characteristics of the major global forest biomes

	Boreal	Temperate	Tropical
Location	50°N to 60°N	25°N to 55°N Counterparts in the Southern Hemisphere	23.5°S to 23.5°N
Estimated area	9.5–14.7 × 10^8 ha	10.4–14.2 × 10^8 ha	14.5–22.0 × 10^8 ha
Climate	Short, moist and moderately warm summers (less than 50 frost-free days in summer)	Warm summers (up to 30 °C) Four to six frost-free months	Temperature varies between 20°C and 25°C
	Long, cold and dry winters (permafrost)	Cold winters (up to −30°C)	Winter absent
	Precipitation 400–1,000 mm, mainly as snow	Precipitation 500–1,500 mm, evenly distributed throughout the year	Precipitation exceeding 2,000 mm, evenly distributed throughout the year
Soils	Thin, nutrient-poor, acidic	Fertile, enriched with litter	Nutrient-poor, acidic, heavily leached
Vegetation	Mostly cold-tolerant evergreen conifers with needle-like leaves	Mostly broadleaf deciduous and needleleaf evergreen trees, three to four tree species per square kilometer	Mostly evergreen trees, up to 100 tree species per square kilometer
	Canopy permits low light penetration and limits understory vegetation	Canopy moderately dense, allows light to penetrate, resulting in well-developed and richly diversified understory vegetation	Multilayered canopy allows little light penetration
	Growing season length 130 days	Growing season length 140–200 days	Highly diverse flora
Natural disturbance	Fire, insect outbreaks		El Niño Southern Oscillation

Fig. 4.1 Boreal forest (*Picea mariana*, photo credit: L.B. Brubaker)

65% of boreal forests in Scandinavia (Gower et al. 2001). *Larix sibirica* Ledeb and *Larix gmelinii* (Rupr.) Kuzen are the dominant tree species in Siberia.

Boreal forest trees are characterized by low LAI and spiral canopies (Fig. 4.1; Landsberg and Gower 1997). Average LAI range from 2.6 for boreal deciduous broadleaf forests to 2.7 for boreal evergreen needleleaf forests (Asner et al. 2003). Low fluxes of mineral N, in particular, limit boreal forest productivity (Jarvis et al. 2001). In the understory, ericaceous shrubs dominate, and soils of xeric sites are dominated by lichens (*Cladina* spp.) whereas bryophytes such as feathermoss (*Pleurozium* spp.) and sphagnum (*Sphagnum* spp.) predominate in soils of hydric sites (Gower et al. 2003). Wildfire is the dominant natural driver of ecological processes, and maintains age structure, species composition and the floristic diversity of boreal forests (Soja et al. 2007). For example, large areas in Canada and Russia have been changed from coniferous species into *Betula* and *Populus* spp. by repeated fires (Jarvis et al. 2001). On average, 1.59 million hectare of boreal forest in North America has burned annually since 1920. Estimates for the total forest area burned in Siberia, however, are highly uncertain (Soja et al. 2007). Other common disturbances in boreal forests are insect outbreaks, windthrows and fluctuations in moisture regimes (Shorohova et al. 2009).

Soils in the boreal forests are poorly developed and tend to be nutrient-poor (Table 4.1; Gower et al. 2003). Spodosols occupy 20.6% of the boreal forest area, and are characterized by pH around 4, a thick litter layer, and leaching of organic acids into deeper soil horizons (Jarvis et al. 2001). A range of other soils also present in the boreal forest biome include Alfisols, Inceptisols, Mollisols, and Histosols, particularly in depressions (Reich and Eswaran 2006). In colder areas of the boreal biome, soils are deeply frozen by permafrost, defined as subsurface earth materials remaining below 0°C for two consecutive years (Schuur et al. 2008). This buried frozen soil layer impedes root development and soil water drainage (Camill 2005). Permafrost covers 25% of the land surface in the Northern Hemisphere. Large amounts of peat (undecomposed mosses and *Sphagnum*) may also accumulate in poorly drained boreal soils (Gower et al. 2003).

4.1.1 Carbon Dynamics and Pools

Boreal forests play a significant role in the global C cycle (Jarvis et al. 2001). The boreal forest is both a major depository and at times a major source of C (Ruckstuhl et al. 2008; Taggart and Cross 2009). Boreal forests are covered by extensive areas of deep organic soils and represent one of the largest reservoirs of organic C on the globe (Aber and Melillo 2001; FAO 2006). In particular, ca. 48% of the total forest SOC pool to 1-m depth is stored in boreal forest soils (Saugier et al. 2001). The ACC may affect the boreal terrestrial environment and, thus, drastically alter the global C cycle. However, measurements of sources and sinks of C in boreal forests are hampered by data uncertainties as observed for the forests of Russia (Houghton et al. 2007). Thus, the understanding of the environmental and ecological processes in boreal forests is rudimentary (Ruckstuhl et al. 2008).

According to a comprehensive global database assembled by Luyssaert et al. (2007) based on measurements of C pools and fluxes, the gross uptake of C by photosynthesis or GPP among boreal forest types ranges from 773 g C m^{-2} $year^{-1}$ for semi-arid evergreen forests to 1,201 g C m^{-2} $year^{-1}$ for semi-arid deciduous forests. Comparable moderate values for GPP have been reported by others (Table 4.2; Jarvis et al. 2001; Litton et al. 2007). In general, half of GPP in forests is used for synthesis and maintenance of living cells (Waring and Running 2007). The other half accounts for the NPP (e.g., build-up of plant tissues). In boreal forests, however, only 25–30% of GPP ultimatively accounts for the NPP (Trumbore 2006). Thus, a large fraction of C fixed in boreal forests is respired quickly. Specifically, estimates for respiratory losses by R_a among boreal forest types range between 489 g C m^{-2} $year^{-1}$ for humid evergreen and 755 g C m^{-2} $year^{-1}$ for semi-arid deciduous forests (Table 4.2; Luyssaert et al. 2007). Thus, boreal forests partition substantially higher amounts of C to R_a than the estimated average of 57% for forest ecosystems (Litton et al. 2007). The aboveground R_a for boreal tree species range between 134 and 291 g C m^{-2} $year^{-1}$. In comparison, the belowground R_a estimates range between 151 and 382 g C m^{-2} $year^{-1}$ (Litton et al. 2007).

The C remaining after autotrophic losses is allocated as NPP to foliage, wood, and roots, and can be estimated based on direct measurements of the main components (i.e., stem, branch, foliage, reproductive tissue, roots, herbivory; Gower et al. 2001). NPP can also be estimated on the basis of the CO_2 gradient between vegetation and atmosphere by using the eddy covariance technique (Baldocchi 2003). Eddy covariance-based flux estimates of NEP, GPP, and Re, however, require independent measurements to be validated (Luyssaert et al. 2009). The eddy covariance

Table 4.2 Carbon dynamics and estimated pools for major global forest biomes (references, see text)

	Boreal	Temperate	Tropical
GPP (g C m^{-2} $year^{-1}$)	556–1,201	1,228–1,762	3,551
R_a (g C m^{-2} $year^{-1}$)	489–755	498–951	2,323
NPP (g C m^{-2} $year^{-1}$)	238–539	354–801	170–2,170
R_h (g C m^{-2} $year^{-1}$)	150–381	280–970	460–877
NEP (g C m^{-2} $year^{-1}$)	30–178	133–398	360–408
Plant C (Pg)	78–143	73–159	206–389
Soil C to 1- and 3-m depth (Pg)	338 (0–1 m)[a]	153–195 (0–1 m)	214–435 (0–1 m)
	150 (0–3 m)[b]	262 (0–3 m)	692 (0–3 m)
	625[c]		
C sink (Pg $year^{-1}$)	0.49–0.7	0.37	0.72–1.3

[a] Saugier et al. (2001)
[b] Jobbágy and Jackson (2000)
[c] Soil depth not specified, including peat (Kasischke 2000)

method is most valuable for obtaining NEE over regional scales (Smith et al. 2008). For boreal forest types, estimated NPP values range between 271 (humid evergreen biome) and 539 g C m^{-2} $year^{-1}$ (semi-arid deciduous biome) (Table 4.2; Luyssaert et al. 2007). Other studies have reported mean NPP values in boreal forests ranging between 200 and 1,500 g C m^{-2} $year^{-1}$ (summarized in Nieder and Benbi 2008). Pregitzer and Euskirchen (2004) estimated a NPP mean of 280 g C m^{-2} $year^{-1}$, and the peak NPP was observed in the 71–120 years age class.

The NPP values among major boreal tree species range between 218 and 302 g C m^{-2} $year^{-1}$ for *P. mariana*, between 371 and 431 g C m^{-2} $year^{-1}$ for *P. glauca*, between 221 and 429 g C m^{-2} $year^{-1}$ for *Pinus banksiana* Lamb., between 215 and 912 g C m^{-2} $year^{-1}$ for *P. sylvestris*, between 394 g C m^{-2} $year^{-1}$ and 658 g C m^{-2} $year^{-1}$ for *Populus tremuloides* Michx., and NPP was estimated to be 323 g C m^{-2} $year^{-1}$ for *L. gmelinii* (Gower et al. 2001; Litton et al. 2007). In boreal humid evergreen and semi-arid deciduous forests, the major fraction of NPP (205 and 304 g C m^{-2} $year^{-1}$, respectively) is allocated to the production of wood (Luyssaert et al. 2007). In contrast, in boreal semi-arid evergreen forests the major fraction of NPP (157 g C m^{-2} $year^{-1}$) is allocated to the production of roots. In general, aboveground NPP is consistently larger for deciduous than for evergreen boreal forests (Gower et al. 2001). Specifically, the fraction of NPP allocated to root production in *P. mariana* and *P. banksiana* is greater than the fraction allocated to wood or foliage (Gower et al. 2001; Litton et al. 2007). In contrast, the fraction of NPP allocated to wood production in *P. tremuloides* and *L. gmelinii* is greater than the fraction allocated to foliage or roots whereas NPP fractions are variable among tree components in *P. sylvestris*.

However, it is likely that NPP is strongly underestimated by most inventories as they may not completely account for fine root production, root exudates, mycorrhiza, reproductive organs, herbivory, tree mortality, and emissions of BVOCs and CH_4 (Gower et al. 2001; Ciais et al. 2005a; Luyssaert et al. 2007). In particular, roots may represent 30–60% of NPP, and root C in boreal forests may be larger than that previously estimated (Jarvis et al. 2001; Robinson 2007). About 17% of roots in boreal forests may occur below 30-cm depth but the knowledge about root biomass distribution is inadequate in most terrestrial biomes (Jackson et al. 1996; Schenk and Jackson 2005). The maximum rooting depths in boreal forests may reach 2.0 m (Canadell et al. 1996).

Often missing in credible data on NPP are the NPP for understory and bryophyte vegetation. For example, the productivity of bryophytes in boreal forests can be equal to or exceed that of the stem growth of trees (Gower et al. 2001). Non-CO_2 and non-respiratory CO_2 losses from terrestrial ecosystems are generally estimated to be 11% of the NPP (Randerson et al. 2002). In boreal peatlands, in particular, fluxes of DOC, DIC and CH_4 may be as much as 5–10% of the net ecosystem-atmosphere CO_2 flux. Furthermore, boreal forests take up 2.6 kg CH_4 ha^{-1} year^{-1} and the total flux is estimated to be 4.5 Tg year^{-1} (Dutaur and Verchot 2007). Also, streams draining boreal forests may transport significant amounts of soluble C into adjacent ecosystems (Randerson et al. 2002). Boreal forests emit BVOCs, and boreal plant species among the genera *Populus* and *Salix*, and *Sphagnum* are significant isoprene emitters (Guenther et al. 2006). For example, monoterpene concentrations in northern European boreal forests range between 1 ppb and 10 ppm (Tunved et al. 2006).

Estimates of respiratory losses by R_h among boreal forests range between 247 (semi-arid evergreen) and 381 g C m^{-2} year^{-1} (humid evergreen) (Table 4.2; Luyssaert et al. 2007). Pregitzer and Euskirchen (2004) reported the average rates for R_h in boreal forests ranging from 150 to 350 g C m^{-2} year^{-1}. Respiratory losses by R_a and R_h can also be considered as total ecosystem respiration (R_e). The R_e among boreal forest types is the highest for deciduous semi-arid forests (1,029 g C m^{-2} year^{-1}), followed by the humid evergreen (824 g C m^{-2} year^{-1}) and the lowest for semi-arid evergreen forests (734 g C m^{-2} year^{-1}) (Luyssaert et al. 2007).

The observed C uptake of boreal ecosystems or the net ecosystem production (NEP) is estimated to be 43 g C m^{-2} year^{-1} for the semi-arid evergreen, 130 g C m^{-2} year^{-1} for the humid evergreen and 178 g C m^{-2} year^{-1} for the semi-arid deciduous forests (Table 4.2; Luyssaert et al. 2007). The CO_2-balance for the boreal forest types is, however, not closed. Substantial correction (ranging from 14% to 45% of GPP) is needed to close it (Luyssaert et al. 2007). Specifically, the decomposition of historical C through land-use change or ecosystem perturbation contributes to the CO_2-balance of the boreal biome (Luyssaert et al. 2007). Pregitzer and Euskirchen (2004) reported the mean NEP values of 30 g C m^{-2} year^{-1} for the boreal forests. Valid data on respiratory processes and lateral C fluxes (i.e., advection, BVOC, DOC) and the SOC pool are, however, required to improve the estimates of C budgets for the boreal forest biome. These data are also needed for the boreal tree species. For example, *L. sibirica* has twice as much capacity to preserve SOC compared to *Betula pendula* Roth and *P. sylvestris* (Menyailo et al. 2002).

Estimates of the aboveground biomass among boreal forest types range between 5,761 g C m^{-2} for humid evergreen, 4,766 g C m^{-2} for semi-arid evergreen and 7,609 g C m^{-2} for deciduous semi-arid forests (Luyssaert et al. 2007). Estimates for the belowground biomass-C range from 1,352 g C m^{-2} (semi-arid deciduous) to

1,604 g C m^{-2} (semi-arid evergreen). Assuming that plant C is 50% of dry mass, Saugier et al. (2001) reported the aboveground biomass in boreal forests to be 3,050 g C m^{-2} compared with 1,100 g C m^{-2} stored in the belowground plant biomass. The proportion of the boreal forest C stored in CWD is small (790 g C m^{-2}; Pregitzer and Euskirchen 2004). Furthermore, the biomass of epiphytic lichens is very low and can amount to 50 g C m^{-2} on *P. banksiana* (calculated from Maguire et al. 2005, assuming organic dry matter to contain 50% C). Maximum biomass in the boreal forests at the old-growth stage is 400 Mg ha^{-1}, although the mean is 40 Mg ha^{-1} (Waring and Running 2007). The average biomass for boreal forests in Russia range between 38 and 133 Mg ha^{-1} (Houghton et al. 2007). On average, 95 Mg C ha^{-1} are fixed in the standing biomass (Schlesinger 1997). In the boreal forest biome, the aboveground plant biomass stores ca. 42 Pg C whereas ca. 15 Pg C are stored in the belowground plant biomass (Saugier et al. 2001). The root C pool, however, is probably underestimated according to assumptions by Robinson (2007), and 25 Pg C are probably stored in roots. Other studies reported that up to 143 Pg C are stored in the boreal forest biomass (summarized in Nieder and Benbi 2008).

The largest amount of boreal forest C is stored in the soil, and is estimated to be 338 Pg C to 1-m depth (Table 4.2; Saugier et al. 2001). About 34% of the SOC pool in boreal forests to 3-m depth, however, is probably stored below 1-m depth but almost no SOC data are available for deeper soil layers making SOC pool estimations highly uncertain (Jobbágy and Jackson 2000). Also, for the top 30-cm of soil large data gaps exist for the major boreal forests with typically large amounts of SOC pool (FAO 2006). For example, Kasischke (2000) estimated a soil C pool of 625 Pg C in boreal forests which includes peat. The total soil C in the northern circumpolar permafrost zone was estimated to be 1,672 Pg (Schuur et al. 2008). This could still be an underestimate as deep soils were not adequately included because of data scarcity. Due to cryogenic (freeze-thaw) mixing and sediment deposition, organic C pools can be large at depth. In permafrost peatlands up to 277 Pg C are probably stored to several meters depth. In boreal peatlands, in particular, about one third of world's soil carbon may be stored (Limpens et al. 2008). Therefore, the current estimates of the SOC pool for the boreal forest biome are highly uncertain. The long-term C accumulation rates in boreal soils are highly variable, ranging between 0.8 and 11.7 g C m^{-2} $year^{-1}$ (Schlesinger 1990). The CWD accounts for only 5% of the total boreal ecosystem C (Pregitzer and Euskirchen 2004). Total C sink of the boreal forest biome (i.e., in plants and soil) is estimated to range between 0.49 and 0.7 Pg C $year^{-1}$ (Table 4.2; Apps et al. 1993; Robinson 2007).

Wildfires strongly influence the C cycle in boreal forests (Schulze et al. 1999; Stocks 2004). Fire is, for example, the dominant driver of the central Canadian boreal forest C balance, and enhances deciduous tree and moss production at the expense of coniferous trees (Bond-Lamberty et al. 2007). Ecozones of the boreal and taiga forest areas in Canada experienced the greatest area burned and released most of the direct fire C emissions from 1959 to 1999 (Amiro et al. 2001). Overall, an estimated 5–15 million hectare burns annually in the boreal biome. The natural fire cycle in boreal forest stands ranges from 50 to 200 years, but the fire return intervals increased to more than 500 years because of human interference. One-third of the NPP may be consumed by fires. The estimates for boreal North America range between 35 and 85 Tg C year^{-1}, and between 58 and 273 Tg C year^{-1} for boreal Russia/Siberia (Balshi et al. 2007). Two thirds of the fuel consumed by boreal fires consists of forest floor (litter, moss, humus layer) and dead woody surface residues, and one-third is needles and fine twigs (Stocks 2004). Biomass and detritus C are partially converted by fire into gaseous forms (i.e., CO_2, smaller proportions of CO and CH_4), and between 1% and 7% of burning mass are converted to pyrogenic C (Preston and Schmidt 2006). A fraction of the C emissions may be taken up by boreal forests. In comparison to savanna fires, however, boreal fire emissions inject at much higher atmospheric heights (Stocks 2004). Wide-ranging transport of forest fire emissions can therefore remove C from the boreal ecosystem C pool.

Of potential importance to C sequestration is the observation that between 1.4% and 3.5% of the C consumed by boreal forest fires is converted to BC (charcoal) (Forbes et al. 2006). A crucial factor for sequestration of BC is permafrost (Guggenberger et al. 2008). However, whether BC is a potential C sink in boreal forests by contributing to the stable SOC fraction is debatable (Czimczik and Masiello 2007). Also, the influence of charcoal on the decomposition of SOM is poorly understood. Charcoal can cause humus loss by promoting microbial respiration and/or leaching of soluble compounds (Wardle et al. 2008). Boreal forests burn frequently but soils do not have high BC contents. Thus, if boreal forests sequester C in the long-term by charcoal formation is debatable (Schulze et al. 1999). BC plays probably a more dynamic role in the soil C budget of boreal forests because BC is also observed in soil DOC, and exported from boreal catchments in the dissolved and colloidal phase (Guggenberger et al. 2008).

Aside from wildfires, tree uprooting by strong winds is another major disturbance in boreal forests (Hytteborn et al. 2005). For example, up to 40% of even-aged boreal *Picea* forests in Russia may be disturbed by uprooting. Uprooting alters structure and dynamics of the forest community. Specifically, the gap-phase dynamics in boreal forest communities is affected by windthrow (Ulanova 2000). Major biotic disturbance factors include insect predation, pathogens, and various anthropogenic impacts (Taggart and Cross 2009). The insect-related disturbance regime controls changes in the C balance of boreal forests (Bernier and Apps 2006). In particular, photosynthetic C uptake is reduced through defoliation and R_e is enhanced by insect respiration and accelerated decomposition (Fleming et al. 2002). Further important impacts by insects originate from changes to forest succession and age-class structure (Kurz et al. 2008b). However, forest fires can have greater impact on the C storage per unit area. Yet, both insect-related and wildfire disturbances can interact, i.e., wildfire may increase several years after insect outbreaks (Fleming et al. 2002). Otherwise, the impact of pathogens such as fungal, bacterial, and viral predators is more difficult to quantify (Taggart and Cross 2009). Logging and other exploitive activities such as mining and petroleum exploration and field development are major anthropogenic impacts in the boreal forest biome.

In summary, boreal forests are net C sinks and take up as much as 0.49 Pg C year^{-1} (Robinson 2007). This is about one-third of the total northern land C uptake estimated at 1.5 Pg C year^{-1} based on the measurements of vertical atmospheric CO_2 distributions (Stephens et al. 2007). Boreal semi-arid deciduous forests, in particular, retain 15% of the photosynthetically fixed C (Luyssaert et al. 2007). Otherwise, boreal humid evergreen forests retain 13% but boreal semi-arid evergreen forests only 6% of the C fixed during photosynthesis. Boreal humid evergreen forests, however, can also be sources of atmospheric CO_2. Boreal semi-arid evergreen forests have very low C-use efficiencies as is indicated by their very high R_e:GPP ratios (Luyssaert et al. 2007). The SOC pool in the boreal forest biome is slowly cycling and exceeds the C pool in the vegetation (De Deyn et al. 2008). Most C in plant biomass is located aboveground and root biomass is often distributed shallow. Fire and insect outbreaks are the dominant natural disturbances influencing the boreal forest C dynamics and pools.

4.1.2 Effects of Climate Change

Boreal forests are likely to be especially affected by ACC because of their sensitivity to high temperatures (IPCC 2007). Effects of ACC on boreal forests are of paramount importance as large C pools are stored in their cold and frozen soils (Gower et al. 2001). The greatest temperature increase during the twenty-first century (i.e., twice the global average temperature increase) is likely to occur at high northern latitudes within the boreal forest biome (Meehl et al. 2007). Large increases in thaw depth over much of the permafrost region may then release large amounts of C (Schuur et al. 2008). Temperature increases may also release large amounts of CH_4 from deposits of hydrates covered by permafrost (Leigh Mascarelli 2009). Permafrost C reservoirs that are more than a few meters below the surface, however, can also be relatively stable (Froese et al. 2008). Higher temperatures may also increase fire danger and frequency, and geographical expanse of severe fires in the boreal biome (Stocks 2004; Nabuurs et al. 2007). Warmth appears to be permitting bark beetles to invade northern tree populations previously inaccessible to them by virtue too-short growing seasons (Solomon and Freer-Smith 2007). Otherwise, warmer temperatures are expected to stimulate plant productivity and increase plant litter inputs and SOC pools (Keeling and Phillips 2005; Reichstein 2007). Thus, ACC impacts on the boreal forest may strongly affect the global C cycle. However, the temperature threshold for boreal forest dieback due to global warming is highly uncertain because of limitations in existing models and physiological understanding (Lenton et al. 2008).

The effects of the increasing atmospheric CO_2 concentrations on forests are not studied in fumigating boreal trees by using long-running FACE experiments (Ainsworth and Long 2005; Norby et al. 2005; Hickler et al. 2008). However, after seven years of growth in elevated CO_2, the total biomass of *P. tremuloides* which also occurs in the boreal biome increased by 24.9% (King et al. 2005). Furthermore, the annual biomass production of foliage, wood, fine and coarse roots of this species increased between 30.6% and 45.1%. This FACE experiment is, however, located in the temperate region. Thus, it remains unclear how would aspen respond in the boreal forest soils as, for example, fluxes of mineral N in boreal biome are lower than those in temperate soils (Jarvis et al. 2001). The potential of boreal trees for C sequestration at elevated CO_2 levels is probably small as up to 70% of GPP are respired (Trumbore 2006). Any CO_2-induced increases in photosynthetic rates may, thus, have only small effects on NPP.

Several studies have shown that the biomass of *Sphagnum* species and vascular plants in the understory of boreal forests is not significantly affected after growing for 3 years at elevated CO_2 concentrations (Hoosbeek et al. 2001). Elevated CO_2, however, caused strong chemical changes in leaves of the widely distributed and common boreal species such as *Betula pubescens* Ehrh., *B. pendula* and *Salix myrsinifolia* (Veteli et al. 2007). The concentrations of phenolic acids, salicylates, flavonoids and condensed tannins increased which may have effects on the C cycle by altering plant–herbivore interactions (Stiling and Cornelissen 2007). Thus, leaves grown at elevated CO_2 concentrations may decompose more slowly but indirect effects of elevated CO_2 (e.g., amount and dynamics of litterfall, plant community composition, soil environment) may be more important for the boreal C balance than changes in the chemical composition (Hyvönen et al. 2007). Little is known if leaching of DOC from litter and soils in boreal forests is affected by increase in atmospheric CO_2 concentrations (Hagedorn and Machwitz 2007). Even soil with large SOC pools are not necessarily saturated at present (Jastrow et al. 2005). Boreal forest soils may, thus, store more C if elevated CO_2 concentrations result in an increase in C inputs. However, there are no experimental data to verify this effect.

Boreal forests may respond to recent warming with increases in productivity, accelerated seasonal development of some insects, changes in the distribution of some insect pests, and provenances of boreal trees from slightly warmer areas outcompeting local provenances (Boisvenue and Running 2006). In particular, the data from experimental ecosystem warming has shown increases in aboveground plant productivity, especially in colder ecosystems (Rustad et al. 2001). Boreal forests partition a smaller fraction of GPP belowground with increase in temperature because nutrient supply also increases with temperature (Litton et al. 2007; Litton and Giardina 2008). Boreal tree and shrub species responded with increased photosynthesis to increased air temperatures (Beerling 1999). Future warming, however, may likely surpass the photosynthetic temperature optimum of many boreal tree species (Way and Sage 2008). Thus, boreal forests may respond less with increase in biomass to the projected increase in temperature. In general, the C uptake may be strengthened in colder high-latitude regions (Fung et al. 2005). The responses are, however, likely to change over time (Norby et al. 2007). For example, with increasing temperature C allocation may shift to transpiring foliage but towards lower allocation to foliage with increasing soil drought (Lapenis et al. 2005).

Changes to the water balance may be more important for C uptake in high latitudes than changes in temperature (Angert et al. 2005; Zhang et al. 2008). The acclimation of trees to ongoing warming and changes in the surface water balance may, therefore, alter the boreal forest C sink. However, as warmth dries soils beyond the wilting point *P. glauca* is dying in some boreal regions (Barber et al. 2000).

In the long-term, acclimation processes of trees, soil microorganisms and fauna may result in negligible effects of increasing temperatures on litter and root decomposition (Aerts 2006). Drastic changes in vegetation abundance and composition in the Northern Hemisphere may, however, occur along with alteration in the decomposition rates (Cornelissen et al. 2007). In boreal peatlands, for example, *Picea*-dominated permafrost plateaus thaw due to climate warming, and *P. mariana* are toppling over in extensive areas as soils undergo flooding and anaerobic conditions (Solomon and Freer-Smith 2007). *Sphagnum* and *Carex*-dominated collapse scars form and result in 60–100% increases in C sequestration (Camill 2005).

A substantial release of C may occur by warming the large SOC pool in boreal uplands, and, in particular, by C release from labile OM stored in boreal forest soils (Rustad et al. 2001; Chapin et al. 2002). After labile OM is exhausted, no further increased loss from the SOC pool may occur at higher temperatures (Jandl et al. 2007). The temperature sensitivity of SOC decomposition is, however, a matter of active debate (Davidson and Janssens 2006). Permafrost in boreal forests constitutes another large C reservoir, intermediate in size between vegetation and soil (Zimov et al. 2006). Projected increases in temperatures have the potential to quickly release the OM stored in permafrost through microbial decomposition. Specifically, aerobic decomposition has a greater feedback to warming than anaerobic decomposition on a century scale as C emission rates are much higher in aerobic conditions (Schuur et al. 2008). If deep-soil respiration is triggered in response to atmospheric warming and soil microorganisms are activated to produce enough heat, the mobilization of SOC can be very strong, self-sustainable and irreversible (Khvorostyanov et al. 2008). The decomposition of previously frozen organic C is one of the most significant potential feedbacks from terrestrial ecosystems to the atmosphere in a changing climate (Schuur et al. 2008). Over the next century, the magnitude of projected loss is similar in size with the global C emissions from land-use change. Furthermore, N_2O emissions from permafrost soils may also contribute to the climatic impact of boreal forests (Repo et al. 2009).

Elevated CO_2 concentrations in combination with warming enhanced stem growth, and fine root production and mortality of two *Acer* species (Norby et al. 2007). Such a trend is likely to affect the C balance. The response of boreal forest species to the combined increases in CO_2 and warming is, however, not understood. Temperature increases may increase NPP in boreal forests but the direct CO_2 responses may be small (Hickler et al. 2008).

The increase in precipitation throughout the boreal region, particularly during the winter season, is expected with the projected ACC (Soja et al. 2007). Thus, summer dryness and winter wetness are projected to increase in the northern latitudes (Meehl et al. 2007). Precipitation trends at any given point on the Earth's surface are, however, less certain than temperature trends (Reichstein 2007). Some boreal regions may experience also warming in combination with drought, and the fraction of foliage may decrease whereas the fraction of aboveground wood and roots may increase (Lapenis et al. 2005). However, GPP and NPP in forests limited by typical climatic conditions of boreal forests (i.e., low precipitation sums (<800 mm) or low mean annual temperatures (<5°C)) may not benefit from higher mean annual temperatures or precipitation (Luyssaert et al. 2007). On the other hand, NEP in forests may be insensitive to climatic conditions, but mainly determined by successional stage, management, site history and site disturbance.

In boreal Eurasia and North America, an earlier freeze-thaw is observed with the ACC (Smith et al. 2004). During 1980–2002 the growing season length between 50°N and 90°N increased by 0.40 days year^{-1} in Eurasia and by 0.56 days year^{-1} in North America (Piao et al. 2007). The growing season length is positively correlated with the annual GPP and NPP. Thus, more C may be fixed by photosynthesis and stored in the boreal forest biomass. Boreal needleleaf summergreen forests, in particular, show large relative increase in rates of GPP associated with the growing season length (Piao et al. 2007). The poor correlation between growing season length and NEP indicates, however, that SOC decomposition is also enhanced. Furthermore, autumn warming increases respiration more than photosynthesis in northern ecosystems whereas spring warming increases photosynthesis more than respiration (Piao et al. 2008). In total, the net C uptake period is shorter and boreal forests may lose CO_2 in response to autumn warming. The CO_2 uptake enhancement by warmer springs in high latitudes may be cancelled out by the decrease in net summer uptake due to drought-induced reduction in photosynthesis (Angert et al. 2005). Thus, the increase in growing season length may result in lengthening seasonal drought conditions and not in a net increase in growing season CO_2 uptake (Schimel 2007). A warming climate may therefore not result in a net C gain in the boreal biome.

Boreal forests have a lower albedo (i.e., absorb more sunlight) than the cultivated land under snow cover, which can lead to increased solar heating regionally (Betts 2000). Boreal forests, in particular, have the greatest biogeophysical effect of all biomes on annual mean global temperature (Bonan 2008). Thus, planting forests in the boreal region helps to stabilize global atmospheric CO_2 but may accelerate climate warming (Jackson et al. 2008). However, how increase in LAI and reduction in snow cover due to ACC may affect the net climate forcing of the boreal forests is not known.

The boreal terrestrial environment is experiencing temperature increases and changes in natural disturbances like wildfires and insect outbreaks (Groisman et al. 2007; Soja et al. 2007). Thus, growth of *P. glauca* in Alaska is declining, and extreme fire seasons have been observed in Siberia, and more frequent extreme fire years in both Alaska and Canada. Boreal forests in North America and Eurasia are projected to experience relatively large changes in fire probabilities in the 21st century (Krawchuk et al. 2009). For example, direct C emissions from fires are projected to increase in western Canadian boreal forests by ACC but to decrease in parts of eastern Canada (Amiro et al. 2001). Observed increases in fire occurrence in Canadian boreal forests as result of temperature increases have important implications for CO_2 emissions and boreal forest ecosystems (Gillett et al. 2004). The greatest impact on future emissions from Canadian boreal forest fires comes from changes in area burned and less from the effect of more severe fuel consumption density (Amiro et al. 2009). Larger areas in both Canada and Alaska may change by increase in natural disturbances from coniferous species into *Betula* and *Populus* spp. (Jarvis et al. 2001). When boreal evergreen forests with low winter albedo are replaced by deciduous trees with high winter albedo, increased fire activity associated with climate warming may lead to a cooling effect (Euskirchen et al. 2007). In general, the negative forcing from postfire increase in surface albedo may exceed the small positive biogeochemical forcing from greenhouse gas emission, ozone, BC deposited on snow and ice, and aerosols (Bonan 2008). Multidecadal increases in surface albedo after fire may have larger impact on climate warming than fire-emitted greenhouse gases (Randerson et al. 2006). However, the net ACC effect of boreal forest fires is still uncertain (Amiro et al. 2009).

Extreme and geographically expansive multi-year outbreaks of the spruce beetle have been observed in Alaska. Also, previously beetle free *P. banksiana* and *P. albicaulis* populations are invaded by warmth induced northward and uphill moving mountain pine beetles (*Dendroctonus ponderosae*; Solomon and Freer-Smith 2007; Seppälä et al. 2009). The effects of these recent observations on the C sink of boreal forest biomes remains uncertain. The species composition of boreal forests may likely differ in response to ACC with possible feedbacks on the SOC pool (Menyailo et al. 2002). Whether the observed poleward shift of biomes as well as the increase in Northern Hemisphere winter storm track activity as observed over the second half of the twentieth century may result in more uprooting in boreal forests is not known (Trenberth et al. 2007).

Based on climate projection by eight general circulation models, boreal forests shift polewards in Northern Hemisphere high latitudes (Alo and Wang 2008). Vegetation models indicate, in particular, that ACC increases the evergreen vegetation along the northern edge of the boreal zone (Lucht et al. 2006). In contrast, a widespread shift toward more deciduous trees but a recession of total forest cover due to increase in drought may occur in areas of the southern boreal zone. Increase in deciduousness at the southern edge of the current boreal zone may result in a decrease in forest C storage (Weedon et al. 2009). However, the tree species migration rates are uncertain (Lucht et al. 2006). Globally, about half of the studied treeline sites have advanced to higher altitudes and/or latitudes and, in particular, those that experienced strong winter warming (Harsch et al. 2009). Migration is probably too slow as tree performance in cold environments experiencing warming is limited by branch morphological characteristics which raise leaf temperatures (Helliker and Richter 2008). Thus, lagged forest migration at the southern boundary of the boreal forests may result in increased deforestation and lead to massive loss of natural forests and a large C pulse (Kirilenko and Sedjo 2007). As climate changes more rapidly as tree population shifts, the carrying capacity of trees and biomass is likely to be greatly reduced, and tree species flora may also become increasingly dominated by weedy species (Solomon and Freer-Smith 2007). The actual rates of tree species migration may be an order of magnitude lower than necessary for species migration to track future climate warming (Petit et al. 2008). Any northward migration of boreal forest elements will be species-specific, resulting in systematic changes in forest composition (Taggart and Cross 2009). Thus, future global warming is likely to have a negative impact on the diversity and vitality of the boreal forest biome.

In summary, boreal forests store ca. 18.0% of the terrestrial biome C, and 18.2% of the annual global C sink is located in the boreal forest biome (Robinson 2007). Important to C sequestration is the observation that 22.8% of the global SOC to 1-m depth is stored in boreal forests. However, there are large data uncertainties. Effects of ACC on the boreal forest biome C pool are also highly uncertain. Vegetation growth and C input into boreal soils may increase by the projected ACC. In contrast, recent climate-driven changes indicate that summer drought led to marked NPP decreases in much of the boreal forest region but seasonal low temperatures are still a dominant limitation on regional NPP (Zhang et al. 2008). Any changes in the SOC pool associated with climate warming are, however, debatable. Thawing of permafrost will accelerate microbial decomposition and release large amounts of C to the atmosphere (Schuur et al. 2008). Disturbances by fire and insect outbreaks have increased in boreal forests, probably because of ACC. Such disturbances and, in addition, changes in tree biodiversity may alter the C balance. The net effect of growth enhancement because of ACC on the C pool in the boreal forest biome versus increased losses of SOC and C emissions by fire is also a subject of intense debate (Bond-Lamberty et al. 2007; Nabuurs et al. 2007). The boreal forest C pool may decline as a result of ACC as the enhanced growth may not offset the C losses from increases in natural disturbances (Kurz et al. 2008a).

4.2 Temperate Forests

The temperate biome extends from 25°N to 55°N, with counterparts in the Southern Hemisphere (Reich and Eswaran 2006). This biome can be sub-divided into a semiarid part covered by grassland (735 million hectare), a humid part covered by forest (1,245 million hectare), a Mediterranean warm part covered by shrubs and forest (362 million hectare), and a Mediterranean cold part covered by forbs and shrubs (79 million hectare). Temperate forests occupy about 1,420 million hectare, i.e., much of North America, Europe, Asia, Australia, and a small part of South America (Table 4.1; Reich and Bolstad 2001).

The climate of the temperate forests is characterized by marked seasonality, and alternates between warm summers and markedly cool or cold winters (Table 4.1; Reich and Bolstad 2001). The precipitation generally ranges between 500 and 1,500 mm (Martin et al. 2001). The tree species diversity in temperate forests is higher than in the boreal biome (Fig. 4.2). Dominant temperate forest types are broadleaf deciduous and needleleaf evergreen whereas needleleaf deciduous and broadleaf evergreen forests occupy smaller proportions of the temperate forest area. Species composition varies markedly with changes in topography, soil fertility and successional status. Important temperate deciduous tree genera include *Acer*, *Ailanthus*, *Albizzia*, *Betula*, *Carya*, *Castanopis*, *Fagus*, *Fraxinus*, *Juglans*,

Fig. 4.2 Temperate forest (*Fagus sylvatica*, photo credit: Malene Thyssen, http://commons.wikimedia.org/wiki/User:Malene)

Liriodendron, *Magnolia*, *Nothofagus*, *Populus*, *Quercus*, *Tilia*, and *Zelkova* (Landsberg and Gower 1997). Common temperate coniferous tree genera include *Abies*, *Picea*, *Pinus*, *Pseudotsuga*, *Thuja*, and *Tsuga*. The genera *Agathis*, *Dacrycarpus*, *Eucalyptus*, *Nothofagus*, *Podocarpus*, and *Quercus* are common to temperate broadleaf evergreen forests (Landsberg and Gower 1997). Temperate forests have the highest LAI among forest biomes (Asner et al. 2003). Values range from 5.1 for temperate deciduous broadleaf forests to 5.5 for temperate evergreen needleleaf forests and to 5.7 for temperate evergreen broadleaf forests.

The understory vegetation is well-developed and richly diversified (Martin et al. 2001). In the past, temperate forests have been generally subjected to more severe human impact than any other forest type (Reich and Bolstad 2001). Thus, just 3% of the temperate forest stands are relatively undisturbed (Bryant et al. 1997). Inconsistencies exist, however, in temperate forest area trends (Gold et al. 2006). Wildfires and insect-related disturbances have smaller impact on temperate compared to that in boreal forests. Otherwise, air pollutants such as tropospheric ozone may cause damage in trees and forests of the temperate zone (Wittig et al. 2009).

Site heterogeneity is usually large in most of the temperate zone (Reich and Bolstad 2001). Differing distributional patterns among temperate forest types appear to be related to the relative advantages among varying sites and climates of differing leaf types, leaf habits, and/or phenologies in terms of annual C gain potential. Temperate forest soils are highly variable but generally fertile (Martin et al. 2001; Gower et al. 2003). The more common soil orders include Alfisols, Entisols, Inceptisols, Mollisols and Ultisols.

4.2.1 Carbon Dynamics and Pools

Temperate forests play an important role in the global C cycle, although the biome C pool is the smallest among all forest biomes (Reich and Bolstad 2001; Robinson 2007). Various methods for quantifying C budgets indicate the temperate forests are currently net C sinks (Martin et al. 2001). Direct and indirect human impacts, however, severely affect the temperate forests and their C balance.

Among the temperate forest types, estimates of GPP are the highest for the humid evergreen forests (1,762 g C m^{-2} $year^{-1}$), followed by Mediterranean warm evergreen forests (1,478 g C m^{-2} $year^{-1}$), and humid deciduous forests (1,375 g C m^{-2} $year^{-1}$; Table 4.2; Luyssaert et al. 2007). The lowest GPP is observed for the temperate semi-arid evergreen forests (1,228 g C m^{-2} $year^{-1}$). Thus, forests of the temperate biome take up more C by photosynthesis per square meter than boreal forests. About half of the C photosynthetically fixed in temperate forests is used in metabolism, and the remaining portion is transferred to build plant tissues (Trumbore 2006). Estimates for C losses from temperate forests by R_a range between 498 (semi-arid evergreen) and 951 g C m^{-2} $year^{-1}$ (humid evergreen) (Table 4.2; Luyssaert et al. 2007).

The NPP is estimated at 783 g C m^{-2} $year^{-1}$ for temperate humid evergreen, 738 g C m^{-2} $year^{-1}$ for temperate humid deciduous, 354 g C m^{-2} $year^{-1}$ for temperate semi-arid evergreen, and 801 g C m^{-2} $year^{-1}$ for Mediterranean warm evergreen forests (Table 4.2; Luyssaert et al. 2007). Thus, temperate semi-arid evergreen forests build up less plant tissue per square meter annually than boreal semi-arid deciduous forests. Melillo et al. (1993) estimated the NPP at 465, 669, and 741 g C m^{-2} $year^{-1}$ for temperate coniferous, mixed, deciduous and broadleaf evergreen forests, respectively. In another study, the mean NPP for temperate forests was estimated at 710 g C m^{-2} $year^{-1}$, and NPP peaked in the 11–30 years age class, much earlier than that in the boreal forests (Pregitzer and Euskirchen 2004). Other studies estimated NPP values between 400 and 2,500 g C m^{-2} $year^{-1}$ (summarized in Nieder and Benbi 2008).

The NPP have not been measured for many wide-ranging temperate tree species (Reich and Bolstad 2001). Mostly available are NPP data for economically important tree species, and range between 430 g C m^{-2} $year^{-1}$ for *Pinus resinosa* Sol. ex Aiton, and 2,060 g C m^{-2} $year^{-1}$ for *Pinus taeda* L. and *Pinus contorta* Douglas (Reich and Bolstad 2001; Litton et al. 2007). The maximum NPP among temperate coniferous trees can be up to 2,185 g C m^{-2} $year^{-1}$ for *Cryptomeria japonica* (L.f.)

D.Don, 1,900 g C m^{-2} $year^{-1}$ for *Pinus radiata* D.Don and 1,810 g C m^{-2} $year^{-1}$ for *Tsuga heterophylla* (Raf.) Sarg. (calculated from Maguire et al. 2005). However, the aboveground NPP alone can also be much higher, e.g., exceeding 2,200 g C m^{-2} $year^{-1}$ in *Eucalyptus* spp. plantations (Ryan et al. 2004).

The major NPP fraction in temperate humid evergreen and deciduous, and in Mediterranean warm evergreen forests, is allocated to the production of wood (36%, 45% and 49%, respectively; Luyssaert et al. 2007). In contrast, 49% of NPP in temperate semi-arid evergreen forests are allocated to the production of roots. Among temperate trees, between 14% and 80% of NPP are allocated belowground in *P. taeda* and *Abies amabilis* Douglas ex J.Forbes, respectively (Reich and Bolstad 2001). In general, the belowground NPP is 50% of the total belowground C flux but mycorrhizal production or exudation and coarse root increment are not included in this estimate (Litton and Giardina 2008).

Similar to boreal forests, the NPP in temperate forests may be underestimated because the experimental data for many processes are lacking. In comparison with the NPP, for example, hydrologic C fluxes from the temperate forests may be significant (Fahey et al. 2005). Also, the largest CH_4 uptake rates among ecosystems are observed in coarse textured temperate forests (Dutaur and Verchot 2007). The total CH_4 flux in temperate forests is estimated at 3.3 Tg CH_4 $year^{-1}$, but variability is rather high. Furthermore, root C in temperate forests to 1-m depth may be 19% higher than previously assumed (Robinson 2007). Also, temperate coniferous and deciduous forests may have 48% and 35% of their total root biomass below 30-cm depth, respectively (Jackson et al. 1996). Maximum rooting depths have been estimated at 3.9 and 2.9 m for temperate coniferous and temperate deciduous forests, respectively (Canadell et al. 1996). The turnover of deep roots is, however, not adequately addressed in the most available NPP estimates. Also, significant BVOC emission rates by *Quercus*, *Liquidambar*, *Nyssa*, *Populus*, *Salix*, and *Robinia* species contribute to the NPP in temperate forests (Guenther et al. 2006). Specifically, high temperatures and high photosynthetically active radiation combined with soil drought in the Mediterranean forests can result in isoprene emissions which are up to twice as much as the net C uptake (Steinbrecher et al. 1997). Furthermore, frequent and recurrent wildfires occur in Mediterranean forests during hot and dry summers and affect NPP. For example, in Andalusia, Spain, up to 30,000 Mg of BC are produced each year by forest fires (González-Pérez et al. 2002).

In comparison with R_a, lower respiratory C losses occur in temperate forests by R_h, but Mediterranean warm evergreen forests lose almost the same amount of C as R_a and R_h (615 and 574 g C m^{-2} $year^{-1}$; Luyssaert et al. 2007). Another study reported the average rates of R_h in temperate forests ranging from 970 g C m^{-2} $year^{-1}$ in the youngest age class to 280 g C m^{-2} $year^{-1}$ in the oldest age class (Table 4.2; Pregitzer and Euskirchen 2004). The R_e in the temperate forest biome is higher than in the boreal forest biome but differences among forest types are smaller, and range between 1,048 and 1,336 g C m^{-2} $year^{-1}$ for temperate humid deciduous and temperate humid evergreen forests, respectively (Luyssaert et al. 2007).

The NEP in temperate humid evergreen and deciduous, and Mediterranean warm evergreen forests is higher than in forest types of the boreal biome (Table 4.2; Luyssaert et al. 2007). However, similar to the boreal forests substantial correction terms (12–58% of GPP) are needed to close the CO_2 balance in the temperate forests. The highest NEP is observed in temperate humid forests probably because of the intensive forest management. In general, the temperate semi-arid evergreen forests have lower NEP (133 g C m^{-2} $year^{-1}$) than boreal semi-arid deciduous forests. In young temperate forests, the NEP can even be negative because of the wide range of management activities such as timber harvest and site preparation (Pregitzer and Euskirchen 2004). The NEP in deciduous forests is lower than that in evergreen Mediterranean warm forests (Luyssaert et al. 2007).

The aboveground and belowground biomass in temperate forests is estimated at 6,283 and 2,238 g C m^{-2} in semi-arid evergreen, and 14,394 and 4,626 g C m^{-2} in humid evergreen forests, respectively (Luyssaert et al. 2007). Evergreen Mediterranean warm forests, on the other hand, store 5,947 g C m^{-2} in the aboveground biomass and 3,247 g C m^{-2} in the belowground biomass. However, Robinson (2007) reported higher mean values for temperate forest plant biomass. In general, temperate forests have higher above- and belowground biomass than boreal forests. Also, the pool size of CWD in temperate forests is greater than in boreal forests (4,200 g C m^{-2}; Pregitzer and Euskirchen 2004). Among temperate coniferous trees, *Sequoia sempervirens* (D. Don) Endl. may store up to 173,050 g C m^{-2} in the stem, and *Pseudotsuga menziesii* (Mirb.) Franco up to 10,450 g C m^{-2} in the root biomass (calculated from Maguire et al. 2005). Otherwise, the world's highest known total biomass C density (living plus dead) of 186,700 g C m^{-2} was recently reported for Australian temperate moist *Eucalyptus regnans* forests (Keith et al. 2009).

The biomass of understory vegetation in temperate forests can also be considerable and can be as much as 1,460 g C m^{-2} in *P. radiata* forests (calculated from Maguire et al. 2005). Otherwise, the biomass of epiphytes is only 85 g C m^{-2} in the mixed coniferous forests (calculated from Maguire et al. 2005). The forest floor and CWDs may also contain large C pools. For example, the forest floor masses range between 2.8 Mg ha^{-1} for *P. radiata* and 188.0 Mg ha^{-1} for *Pinus ponderosa* Douglas ex C. Lawson (Maguire et al. 2005). The forest floor under Norway spruce (*Picea abies* L.) stores more C than those under *Quercus robur* L. and *Fagus sylvatica* L. whereas *Acer pseudoplatanus* L., *Fraxinus excelsior* L. and *Tilia cordata* Mill. store the lowest amount of C in the forest floor (Vesterdahl et al. 2008). Between 1 Mg ha^{-1} (*P. contorta*) and 490 Mg ha^{-1} (*P. menziesii-T. heterophylla* forest) CWD are stored in temperate coniferous forests (Maguire et al. 2005). Trees in temperate coniferous forest are associated with large pools of mineral SOC, and values range between 56 Mg C ha^{-1} for *P. menziesii* and 388 Mg C ha^{-1} for *T. heterophylla-Picea sitchensis* (Bong.) Carr forests (Maguire et al. 2005). However, differences in SOC pool to 30-cm depth among temperate European tree species three decades after planting were small (Vesterdahl et al. 2008).

The total biomass C pool stored above- and belowground in temperate forests is higher than that in the boreal forests (109 and 49 Pg C; Robinson 2007). On the other hand, soils under temperate forest store less C to 1-m depth than those under boreal forest (Table 4.2; Saugier et al. 2001). Inventories of SOC pools, however, must also include the SOC pool stored below 1-m depth. It is estimated that 31% and 41% of the SOC pool to 3-m depth is stored in 1–3 m depth in temperate deciduous and temperate evergreen forests, respectively (Jobbágy and Jackson 2000). Thus, an estimated 262 Pg C is probably stored in temperate forest soils in the top 3-m depth. Long-term SOC accumulation rates in temperate forests range between 0.7 and 12.0 g C m^{-2} $year^{-1}$, and are comparable with those in boreal forest soils (Schlesinger 1990). The total C sink in the temperate forest biome is estimated to be 0.37 Pg C $year^{-1}$, and, thus, smaller than the sink in the boreal forest biome (Robinson 2007).

In summary, forests of the temperate biome are net C sinks and absorb an estimated 0.37 Pg C $year^{-1}$ (Robinson 2007). Plant biomass, stored in large woody aboveground organs and deep, coarse root systems, accounts for a large portion of the biome C pool (De Deyn et al. 2008). Due to rapid rates of C mineralization, plant biomass C exceeds SOC. Among forest types of the temperate biome, the highest percentage of C fixed through photosynthesis (ca. 26%) occurs in the Mediterranean warm evergreen forests (Luyssaert et al. 2007). In contrast, ca. 23% of C fixation occurs in temperate deciduous and temperate evergreen forests but only ca. 11% in semi-arid evergreen forests. Young temperate forests, however, can also be notable C sources. In humid evergreen forests, ca. 70% of NEP accumulates in woody biomass, suggesting that C sequestration in soils and OM pools can be important processes affecting the C budget (Luyssaert et al. 2007). In contrast, soils and OM pools are less important to C sequestration in temperate humid deciduous and temperate semi-arid evergreen forests. As is indicated by the low R_e:GPP ratios, temperate humid and Mediterranean warm forests have high C-use efficiency, probably because of an intensive management (Luyssaert et al. 2007). Thus, human activities are the most important disturbance regime in the temperate forests.

4.2.2 Effects of Climate Change

Effects of ACC on forest C pools and dynamics have been mainly studied for the temperate tree species and for the forest types in the temperate biome. The ACC-induced alterations in C input and output from temperate forest ecosystems have already been discussed in Chapter 2, and are only briefly described in the following section. Productivity increases due to recent ACC in temperate forests in North America, Northern Europe, most of Central Europe, some parts of Southern Europe, and Japan, although local conditions cause exceptions (Boisvenue and Running 2006). Otherwise, recovery from historic land use may be the dominant process leading to terrestrial C uptake in temperate forests (Caspersen et al. 2000). The consequences of ACC in the temperate biome are likely to be less severe than in other forest biomes. In particular, the net C sink in temperate forests is projected to persist under ACC scenarios due to enhanced tree growth

(Nabuurs et al. 2007). However, tree growth may be reduced in the Mediterranean-type temperate forests because of reduction in rainfall (IPCC 2007). In contrast to the boreal biome, human disturbances (i.e., land use and forest management practices) are of paramount importance to the future C sink strengths of the temperate forest biome.

The main response of young temperate trees to increases in CO_2 concentrations is the increase in photosynthetic rates (Hyvönen et al. 2007). In FACE experiments with young temperate trees, Norby et al. (2005) observed a 23% increase in NPP at 550 ppm CO_2. *Liquidambar styraciflua* L. partitioned this increase to the production of fine roots (Norby et al. 2004). Thus, at elevated CO_2 more C may be sequestered in the SOC pool. Based on increases in NEP of young, productive forests at elevated CO_2, Hamilton et al. (2002) estimated that the temperate forest could sequester an extra 1.47 Pg C $year^{-1}$. The net response of photosynthesis, forest growth and C sequestration in the temperate biome, however, depends on the effects of elevated CO_2 during the entire cycle of forest stand development. Indeed, some studies indicate that mature temperate trees may acclimate to the projected higher atmospheric CO_2 concentrations (Körner 2006). Whether the net growth is enhanced by increases in CO_2 and the C cycle strongly altered in temperate forests is a topic of priority research. Specifically, growth enhancement due to CO_2 fertilization is probably far less important whereas recovery from historic land use may be the dominant process leading to C uptake in temperate forests (Caspersen et al. 2000).

In contrast to boreal forests, the fraction of GPP partitioned belowground increases with increase in temperature in temperate forests (Litton and Giardina 2008). However, the main direct response of temperate forests to increases in temperature may be increase in the length of the growing season (Hyvönen et al. 2007). Between 1980 and 2002, the growing season length has increased by 0.43 days $year^{-1}$ between 25°N and 50°N in Eurasia, and by 0.13 days $year^{-1}$ for the same latitudinal range in North America (Piao et al. 2007). Because of relatively high GPP, large increases in the ratio of annual GPP to the growing season length (9.8 g C m^{-2} $year^{-1}$ day^{-1}) have been observed, in particular, for temperate broadleaf summergreen forests. Advances in the onset of the photosynthetically active period during a warm spring 2007 after a record warm autumn 2006–winter 2007 period in European temperate forests resulted in increases in GPP that exceeded increases in total ecosystem respiration (Delpierre et al. 2009). Thus, for 5 months the net C uptake was higher than the long-term mean.

Similar to the boreal forests growing season length is not a good predictor for the net C balance (i.e., NEP) in temperate forests because the enhanced SOC decomposition may also occur in combination with the GPP increase (Piao et al. 2007). Temperate forests may lose C in response to autumn warming (Piao et al. 2008). Also similar to the boreal forests, drier summer may cancel out the CO_2 uptake enhancement induced by warmer springs (Angert et al. 2005). The extension of the growing season length in middle latitudes may, thus, not cause a net increase in growing season CO_2 uptake because seasonal drought conditions may also extend in length (Schimel 2007). Furthermore, severe and recurrent drought may lead to accelerated rates of tree decline and mortality in temperate forests and alter the C cycle (Bréda et al. 2006). Temperate forests in the western United States, for example, are affected by increases in tree mortality rates most likely from regional warming and consequent increases in water deficits (van Mantgem et al. 2009). Also, GPP and net C uptake of temperate forests were reduced during the European heatwave in 2003, and some forests temporarily become a net CO_2 source to the atmosphere (Ciais et al. 2005b). Specifically, pronounced soil water deficits compensated for the effect of warmer temperatures in reducing soil respiration. The potential effects of such extreme events on the long-term C balance of temperate forests are, however, uncertain.

How ACC may alter wildfire activity is still largely unknown (Krawchuk et al. 2009). However, during the twenty-first century temperate forests in North America and Eurasia are projected to experience relatively large changes in fire probabilities. For example, the extent and frequency of western U.S. forest fires has increased due to changing climate although a century of fire suppression contributed to this increase (Wiedinmyer and Neff 2007). Needleleaf forests, in particular, are the dominant source of U.S. fire CO_2 emissions.

Similar to boreal forests, climate projections by general circulation models project a poleward shift of temperate forests in Northern Hemisphere high latitudes (Alo and Wang 2008). Future extinctions of temperate tree species in response to ACC are, thus, possible as the migration capacity of trees may be too low to keep up with the rate of future warming (Petit et al. 2008). For example, tree migrations since the last glaciation in northeastern North America were much slower than needed to keep up with current and future climate warming (Mohan et al. 2009). Otherwise, in areas where dispersal is not a major constraint the elevation distribution of temperate trees may shift rapidly upwards as consequence of changes in regional climate (Kelly and Goulden 2008; Harsch et al. 2009).

Indirect effects of ACC on the C pool in unmanaged temperate forests may also involve a shift in tree species composition. For example, rising CO_2 concentrations have the potential to change the future composition and productivity of the northern temperate forests (Zak et al. 2007). Elevated CO_2 may alter competitive interactions among temperate tree species (Bradley and Pregitzer 2007). This may potentially alter the forest C cycle by affecting root production and decomposition, and plant-soil feedbacks. Inter- and intraspecific differences in biochemistry among temperate species may change forest ecosystem C storage (Bradley and Pregitzer 2007). Evergreen plant functional types in the southeastern US, Europe and in parts of eastern China are projected to increase at the expense of the deciduous vegetation by the projected ACC (Lucht et al. 2006).

As was shown in the previous section, temperate tree species differ in their biomass C pool. More important to C sequestration, however, is the observation that various temperate trees growing at the same site also differ in the amount of the forest floor C and mineral SOC pools and their dynamics (Hobbie et al. 2007; Vesterdahl et al. 2008). While *Abies alba* Mill., *P. abies*, *Pinus nigra* J.F.Arnold and *P. sylvestris* substantially accumulate C in the forest floor, *T. cordata* and *Acer platanoides* L. and *A. pseudoplatanus* rapidly loose it (Hobbie et al. 2006). After 67 years of growing at the same site, Norway spruce reportedly stored higher amounts of SOC in the O-horizon compared to that stored by elm (*Ulmus glabra* Huds.) (Oostra et al. 2006). The SOC pool in the O-horizon, however, is susceptible to disturbances by fire and erosion. Otherwise, elm trees can store more SOC pool in 0–20 cm depth than do spruce trees. The ACC-induced shifts in vegetation can result in higher proportions of *U. glabra* which can strengthen the rate and magnitude of SOC sequestration. A shift to higher proportions of tree species that increase the stabilized SOC pool in the mineral soil is highly relevant to the permanence of C sequestration (Jandl et al. 2007). Thus, C sequestration in temperate forest can be strengthened by higher proportions of trees characterized by a high turnover rate of the mycorrhizal external mycelium. This turnover is most probably a very fundamental mechanism for the transfer of root-derived C into the SOC pool (Godbold et al. 2006). In general, temperate forests dominated by coniferous species may sequester C more effectively and store it for longer periods than temperate ecosystems dominated by deciduous species (Hyvönen et al. 2007). Effects of temperate tree species on mineral soil C pools are, however, a subject of strong debate (Jandl et al. 2007). How temperate forest SOC dynamic and pools are affected by the projected ACC is difficult to predict. As was indicated in Chapter 2, quality and quantity of C inputs into temperate forests may change. In addition, the temperature response of SOM decomposition to temperature increase is a matter of intense debate.

Aside causing a warming of the atmosphere, current tropospheric ozone concentrations have the potential to reduce the C sink in Northern Hemisphere temperate forests (Wittig et al. 2009). In particular, the total tree biomass and above- and belowground productivity may be reduced at current ozone levels. Furthermore, large increases in tropospheric ozone are projected for the twenty-first century and may enhance damaging effects of ozone on the C sink in temperate forests (Meehl et al. 2007). For example, indirect radiative forcing by ozone effects on plants may contribute more to ACC than the direct radiative forcing due to tropospheric ozone increases (Sitch et al. 2007). However, our current understanding about effects of elevated ozone on trees is based on studies using small trees in chambers. Thus, the knowledge about elevated ozone impacts on large forest trees and C sequestration in forest ecosystems is limited. Also, it is not known how impacts of elevated ozone will interact with projected changes in other atmospheric and climatic variables. Thus, accurate projections of impacts of rising ozone levels on forest ecosystems must be based on long-term studies with mature trees grown in open-air conditions with changing ozone concentrations and changes in other variables such as CO_2 concentrations, water and nutrient supply, and temperature (Karnosky et al. 2005; Wittig et al. 2009).

In summary, temperate forests store 13.9% of the terrestrial biome C pool, and ca. 13.8% of the land C sink is located in temperate forests (Robinson 2007). However, only 10.3% of the global SOC pool to 1-m depth is stored in temperate forests. Thus, temperate forests are less important to the global C cycle than boreal forests. The net effect of ACC is, however, projected to increase the C pool in the temperate forest biome through an enhanced forest growth. Disturbance by land use and forest management have a large potential to affect the future C sink strength of temperate forests. Otherwise, the ozone responses of temperate forest C dynamics and pools in a changing climate are under discussion. Also, biogeophysical feedbacks on climate forcing through the evaporative cooling effect are unclear, and the net climate forcing of temperate forests highly uncertain (Bonan 2008).

4.3 Tropical Forests

The Tropical biome occupies about 3,480 million hectare of which the humid tropics cover 1,451 million hectare (Reich and Eswaran 2006). Tropical forests are located between the tropics of Cancer and Capricorn although a consistent, precise and universal definition of tropical forest does not exist (Table 4.1; Lewis 2006). Before the 1990s, 460 million hectare of land area were covered by tropical deciduous forests and 1,740 million hectare by the evergreen forests (Melillo et al. 1993). Thus, tropical evergreen forests (generally called rainforests) comprise the largest single forest biome in the world, and the Amazon basin is the largest land area covered by this forest type (Landsberg and Gower 1997). Furthermore, in 2005 mangrove forests covered 15.2 million hectare in sheltered coastlines, deltas and along river banks in the tropics and subtropics (FAO 2007). The long-term global trend in topical forest area is, however, difficult to track (Grainger 2008).

The climate of tropical evergreen forests is characterized by high temperatures, with little seasonal or diurnal fluctuations (20–25°C). The mean annual precipitation exceeds 2,000 mm year^{-1}, and the relatively high humidity is observed uniformly throughout the year (Table 4.1; Landsberg and Gower 1997). Due to its immense size, however, generalizations about the climatic conditions in the tropical forest biome are difficult to make. Tropical deciduous forests, in particular, differ on the basis of the water balance. It is the drought that mainly controls the leaf shedding in the tropical deciduous forests. Tropical forests cool their climate through strong evaporative cooling (Bonan 2008).

Tropical evergreen forests have the highest tree diversity among all forest types (Fig. 4.3; Table 4.1). For example, the Amazonian forest alone contains more than 2,500 tree species (Landsberg and Gower 1997). The most common species are *Brosimum guianense* (Aubl.) Huber, *Casearia commersoniana* Cambess., *Rhamnus sphaerosperma* Sw., *Guarea guidonia* (L.) Sleumer, *Hymenaea courbarii* L., and *Trichilia quadrijuga* Kunth. The tropical rainforests in the eastern regions contain more conifers. Otherwise, the African rainforests are relatively poor in species composition. Undisturbed tropical forests can have a complex and species-rich mycorrhizal fungal community but the importance of this complexity to tropical forest diversity is not known (Alexander and Selosse 2009).

Tropical forests have a strong vertical structure, with much of the leaf biomass and fruits in the brightly lit canopy, and seed germination, seedling growth, and juvenile recruitment in the dark understory (Gilbert and Strong 2007). Tropical forest canopies have a layered structure. Tall trees comprise the upper layer, followed by a main canopy layer, and a subcanopy of smaller trees and shrubs near the ground level (Landsberg and Gower 1997). Trees in tropical forests have relatively large leaves and are often characterized by buttresses, and palms, climbing plants, epiphytes and hemi-epiphytes (Lewis 2006). In contrast to tropical evergreen forests, tree species diversity is lower in tropical deciduous forests. Furthermore, canopies are shorter and the structure is more open compared to the tropical rainforests.

Fig. 4.3 Tropical forest (photo credit: H.-D. Viktor Boehm)

Epiphytes, ferns and herbs contribute much less to total LAI of tropical rainforests compared to trees, palms and lianas (Clark et al. 2008). Trees are often the most important functional group followed by palms and lianas. The average LAI for tropical deciduous broadleaf forests is 3.9 whereas it is 4.8 for tropical evergreen broadleaf forests (Asner et al. 2003).

Tropical forests contain more than half of the Earth's terrestrial species (Myers et al. 2000). Furthermore, tropical forests predominantly contribute to global biodiversity 'hotspots' or areas featuring exceptional concentrations of endemic species and experiencing exceptional loss of habitat. Biodiversity is generally high but little is known as tropical forests are extensive, highly variable and generally more difficult to study than any other vegetation type (Grace et al. 2001). About 44% of the world's relatively undisturbed forest lies in the tropics (Bryant et al. 1997). The most dramatic changes in tropical forest ecosystems involve rapid conversions to other land uses (Lewis 2006). Half of the earth's closed-canopy tropical forest, in particular, has been converted to other uses. Thus, the tropical landscape increasingly comprises a fragmented network of relatively intact patches, separated by secondary forests and cultivated land (Lewis 2009). A widespread trend towards 'biotic homogenisation' is observed in tropical forests but the consequences of this development for important ecosystem functions such as decomposition are not known.

Similar to soils under temperate forests, those under tropical forest are highly variable (Table 4.1; Landsberg and Gower 1997). Oxisols, in particular, cover extensive areas in the humid tropics (620 million hectare Reich and Eswaran 2006). These highly weathered soils contain much of the plant nutrients in the top 5 cm, and nutrients are recycled through plant uptake and litter fall. Further, predominant soil orders in the humid tropics include Ultisols, Inceptisols and Entisols covering ca. 324, 250 and 109 million hectare respectively. Infertile soils, from an agricultural point of view, pre-dominate in the tropical forest biome (Lewis 2006).

4.3.1 Carbon Dynamics and Pools

Old growth tropical forests contain large pools of C, and account for a major fraction of the global NPP (Denman et al. 2007). Changes in tropical forests may, thus, have significant effects on the global C balance but the importance of tropical forests for the global C cycle is not well understood (Grace et al. 2001). For example, a reevaluation of the terrestrial productivity gradient indicated that annual NPP in tropical forests is not different than annual NPP in temperate forests (Huston and Wolverton 2009). Also, whether tropical rainforests are sinks or sources of C is matter of debate (Levy 2007; Sierra et al. 2007; Malhi et al. 2008). Otherwise, combining all standardized inventory data from tropical Africa, America and Asia indicates a tree C sink of 1.3 Pg C $year^{-1}$ across all tropical forests during recent decades (Lewis et al. 2009). Large-scale biomass inventories may, however, not adequately survey tropical forests, and not adequately consider tree mortality and dead wood decomposition (Denman et al. 2007; Fisher et al. 2008). Sampling biases are, however, too small to explain currently observed biomass gains for intact forests across the Amazon (Gloor et al. 2009). Large uncertainties still exist for the C budget of mangrove forests as >50% of NPP is unaccounted for (Bouillon et al. 2008). Less well known is also the role of mycorrhiza in maintaining tropical forest productivity (Alexander and Selosse 2009).

The forest turnover rates, i.e., tree mortality and recruitment rates, are higher in tropical than in temperate forests (Stephenson and van Mantgem 2005). In contrast to temperate forests, climate is the primary driver of root and leaf litter decomposition, especially during early stages of decomposition (Cusack et al. 2009). The C balance of a tropical forest based on atmospheric, eddy covariance or ground-based studies may differ among each other (Clark 2007). The global CO_2 flux caused by the land use changes, however, is dominated by tropical deforestation as about 13 million hectare tropical forest are felled or grazed each year (FAO 2006; Denman et al. 2007). Relatively well studied is the largest tropical rainforest in the Amazon which is intimately connected to the global climate but lost 85% of the original area by 2003 (Soares-Filho et al. 2006; Malhi et al. 2008). In contrast, Africa has the second largest block of rainforest in the world, but is the least known in terms of C stocks and rates of conversion (Baccini et al. 2008). The second largest rainforest in the Congo River Basin, in particular, is the least exploited yet most scantily studied of the world's humid forest regions (Koenig 2008). Furthermore, swamp forests in the Congo Basin have also received little attention (Keddy et al. 2009).The Congo River Basin forest, however, has among the highest C contents per hectare of any rainforest.

The database compiled by Luyssaert et al. (2007) shows that tropical humid evergreen forests have the highest observed GPP values among forest types (3,551 g C m^{-2} $year^{-1}$; Table 4.2). For the Central African tropical forests a GPP of 2,558 g C m^{-2} $year^{-1}$ has been reported (Bombelli et al. 2009). The GPP in tropical forests does not correlate with the mean annual precipitation. Other environmental variables such as soil fertility seem to be more important to GPP (Malhi et al. 2004). Important for C uptake in rainforests are trees, palms and lianas as indicated by their large contribution to the total LAI whereas epiphytes, ferns and herbs are of minor importance (Clark et al. 2008). Tropical semi-arid forests although covering 30% of the area of forested biomes, seem to be less studied. In comparison to temperate forests, a much higher fraction of C fixed in tropical forests is respired quickly (Trumbore 2006). Thus, large amounts of C are lost from tropical humid evergreen forests by R_a (2,323 g C m^{-2} $year^{-1}$; Luyssaert et al. 2007).

The NPP in tropical humid evergreen forests is estimated at 864 g C m^{-2} $year^{-1}$, and is thus lower than that in temperate forests (Luyssaert et al. 2007). Specifically, losses by respiration and/or non-CO_2 emissions may be responsible for the observation that both the GPP and NPP are not high in the tropical humid forests. A comparable mean value of NPP of 830 g C m^{-2} $year^{-1}$ has been estimated for tropical forests by Pregitzer and Euskirchen (2004). For 39 tropical forest sites, the estimated NPP values spanned a very wide range between 170 and 2,170 g C m^{-2} $year^{-1}$ (Table 4.2; Clark et al. 2001). The available data for this study, however, were extremely limited. Thus, the data on NPP values are rough approximations. In comparison, the NPP values are estimated at 1,098 g C m^{-2} $year^{-1}$ for tropical deciduous and at 418 g C m^{-2} $year^{-1}$ for tropical evergreen forests (Melillo et al. 1993). Bombelli et al. (2009) estimated a mean NPP of 1,151 g C m^{-2} $year^{-1}$ for tropical forests of Central Africa. Other studies reported mean NPP values between 1,000 and 3,500 g C m^{-2} $year^{-1}$ (summarized in Nieder and Benbi 2008). Across the humid Amazonian forest, Malhi et al. (2004) estimated a mean of 880 g C m^{-2} $year^{-1}$ for the aboveground NPP. However, the quantification of belowground NPP in tropical forests is still lacking.

Among tropical humid evergreen forests, the NPP is allocated in comparable amounts to the production of foliage and roots (316 and 324 g C m^{-2} $year^{-1}$, respectively; Luyssaert et al. 2007). Only 25% of the NPP is allocated to the production of wood. Thus, among the forests studied by Luyssaert et al. (2007), the tropical humid evergreen forests allocate the least percentage of NPP to the production of wood. In the Neotropics, which include South and Central America, between 150 and 550 g C m^{-2} $year^{-1}$ of NPP are allocated to the production of aboveground coarse wood (Malhi et al. 2004). Allocation of NPP to live wood, however, is probably not a fixed proportion of the total GPP as is often assumed in most model studies of NPP of tropical forest regions.

The estimates of NPP for forests of the tropical biome are probably less certain than those for the boreal and temperate biomes, probably because most of the production studies in tropical forests have been conducted on relatively few sites and direct measurements address only a few aspects of NPP, i.e., the aboveground biomass

(Clark et al. 2001; Huston and Wolverton 2009). Specifically, robust measurements of the belowground NPP in tropical forests are scanty (Raich et al. 2006). For example, estimates for fine root production range from 75 to 1,380 g m^{-2} $year^{-1}$, estimates for fine root turnover from 0.3 to 2.5 $year^{-1}$, and from 114 to 361 g m^{-2} $year^{-1}$ for the fine root C flux in the upper 20 cm of the soil (summarized in Hertel et al. 2009).

Emissions of BVOCs constitute an important aspect of the NPP in tropical forests. Tropical broadleaf trees are responsible for half of the annual global isoprene emission, and the magnitude can exceed 150 mg isoprene m^{-1} day^{-1} in the Amazon (Guenther et al. 2006). Aside attracting pollinators and repelling herbivores, BVOCs may contribute to the recycling of hydroxyl radicals to sustain the atmospheric oxidation capacity above tropical forests (Lelieveld et al. 2008). The loss of leaf C by leaching is another aspect of NPP that has not been amply quantified in tropical forests (Clark et al. 2001). Furthermore, tropical forests may oxidize up to 3.3 kg CH_4 ha^{-1} $year^{-1}$ amounting to a total CH_4 flux of 6.4 Tg $year^{-1}$ (Dutaur and Verchot 2007). In particular, tropical upland forest soils are CH_4 sinks (Potter et al. 1996). Otherwise, tropical rainforest soils are one of the major global sources of atmospheric N_2O, and emissions may range between 0.88 and 2.37 Tg N $year^{-1}$ (Werner et al. 2007).

The estimated R_h value (877 g C m^{-2} $year^{-1}$) for tropical humid evergreen forests is much smaller than that for R_a (Table 4.2). Thus, tropical humid evergreen forests have the largest total, autotrophic and heterotrophic fluxes among global forest types (Luyssaert et al. 2007). For tropical forests of more than 120 years of age, Pregitzer and Euskirchen (2004) estimated the R_h values of 460 g C m^{-2} $year^{-1}$. Soil respiration in tropical forests ranges between 359 and 1,700 g C m^{-2} $year^{-1}$ (Raich et al. 2006; Metcalfe et al. 2007).

The estimated NEP in tropical humid evergreen forests (408 g C m^{-2} $year^{-1}$) is comparable to that in temperate humid forests (Table 4.2; Luyssaert et al. 2007). Pregitzer and Euskirchen (2004) estimated the mean NEP of 360 g C m^{-2} $year^{-1}$ across a range of tropical forests. A similar value of 345 g C m^{-2} $year^{-1}$ was reported for tropical forests of Central Africa (Bombelli et al. 2009). It is important to consider all uncertainties in correct interpretations of CO_2 fluxes in tropical rain forests measured on calm nights (Luyssaert et al. 2009). Thus, the NEP values for tropical forests based on CO_2 balances are probably systematically overestimated. Luyssaert et al. (2007) estimated a correction term of 22% of GPP to close the CO_2 balance for tropical humid evergreen forests.

Jackson et al. (1996) observed that tropical evergreen forests have the highest estimated root biomass (5 kg m^{-2}) among global terrestrial biomes. The observed maximum rooting depths are 3.7 and 7.3 m for tropical deciduous and tropical evergreen forests, respectively (Canadell et al. 1996). Data on complete tree root profiles of diverse forest types are, however, scanty (Schenk and Jackson 2005). That is why Robinson (2007) estimated that the root C pool to 1-m depth in tropical forests may be 49% larger than hitherto assumed.

Estimates of the aboveground and belowground biomass of tropical humid forests included in the database by Luyssaert et al. (2007) are 11,389 and 2,925 g C m^{-2}, respectively. Clark et al. (2001) reported the aboveground biomass values for tropical forests between 2,250 and 32,450 g C m^{-2}. Estimates for the aboveground C range between 2,250 and 20,300 g C m^{-2} for mature tropical forests, and between 2,300 and 27,300 g C m^{-2} for mature seasonally dry tropical forests (Vargas et al. 2008). In general, values for aboveground live biomass range between 1,800 and 26,600 g C m^{-2}, and those for the root biomass between 400 and 5,700 g C m^{-2} for moist tropical forests (Raich et al. 2006; Bombelli et al. 2009). The average biomass of natural tropical forests is estimated at 9,400 g C m^{-2} (Houghton 2005). Average above-ground biomass estimates for tropical forests of Africa range from 85 Mg ha^{-1} for closed deciduous forests to 251 Mg ha^{-1} for swamp forests (Baccini et al. 2008). Bamboo stands are important forest types in China, and contain between 95 and 160 Mg ha^{-1} in living biomass (Chen et al. 2009). Lianas are other important components of the aboveground biomass in tropical forests as they possess up to five times the leaf mass of trees of the same diameter at breast height (Vargas et al. 2008). However, the basis of variations in aboveground C pools for all but the driest Amazon forests are not understood (Lloyd and Farquhar 2008).

The amount of C stored in the surface litter in moist broad-leaved evergreen tropical forests is small and ranges between 100 and 500 g C m^{-2} (Raich et al. 2006). In contrast, much more C in these forests is stored in CWD (200–2,700 g C m^{-2}). Pregitzer and Euskirchen (2004) reported a mean pool size of 1,750 g C m^{-2} for CWD. Chao et al. (2009) reported 2,730 g C m^{-2} for Amazonian forests with high stem mortality rates, and 5,850 g C m^{-2} for forests with low stem mortality rates. The total Amazonian CWD stock was estimated to be 9.6 Pg C and is, thus, a significant component of the C budget relative to aboveground biomass.

The C pool stored in organic horizons is highly variable among tropical forests, and ranges between 0 and 16,200 g C m^{-2} (Pregitzer and Euskirchen 2004). Between 11,000 and 600,000 g C m^{-2} is stored in tropical forest soils to 1-m soil depth (Pregitzer and Euskirchen 2004; Raich et al. 2006; calculated from Henry et al. 2009). Carbon in the sub-soil, however, must be included to obtain credible estimates of the SOC pools, because as much as 84% and 50% of the SOC pool to 3-m depth is stored between 1 and 3 m depth in tropical deciduous and tropical evergreen forests, respectively (Jobbágy and Jackson 2000). In total, ca. 435 Pg C is stored in tropical soils to 1-m depth, but 692 Pg C in the top 3 m (Table 4.2). Long-term SOC accumulations rates in tropical forests seem to be low, and estimated at 2.3–2.5 g C m^{-2} $year^{-1}$. These rates are approximations at best and are based on only a few field studies (Schlesinger 1990).

The tropical live biomass sink has been estimated at 1.2–1.3 Pg C $year^{-1}$ (Denman et al. 2007; Lewis et al. 2009). For tropical forests, in particular, the C sink is estimated to range between 0.72 and 1.3 Pg C $year^{-1}$, the largest C sink among global forest biomes (Table 4.2; Robinson 2007; Lewis et al. 2009). Otherwise, for mangrove forests ca. 0.1 Pg C $year^{-1}$ are unaccounted for by current budgets (Bouillon et al. 2008). Tidal export of DIC and CO_2 fluxes from intertidal sediments are probably higher than currently assumed. It is, however, highly uncertain if a C sink exists in the tropics (Houghton 2005). For example, when river transport is taken into account the tropics are a net C source (1.5 Pg C $year^{-1}$). Also, the high Amazon C sink suggested by eddy covariance studies may be caused by systematic underestimation of the night time respiratory fluxes (Lloyd et al. 2007;

Malhi et al. 2008). The major disturbance of the C balance of tropical forests is deforestation which dominates the global CO_2 flux from land use change (Denman et al. 2007). During the 1990s, 1.58 Pg C were released annually. However, estimates of the rates of tropical deforestation are extremely uncertain (Houghton 2005; Nabuurs et al. 2007). The data on the rates of tropical deforestation for the 1990s, in particular, are extremely unreliable (Laurance et al. 2006).

The occurrence of wildfires is another major disturbance to tropical forests although a rare occurrence without anthropogenic ignition sources (Malhi et al. 2008). Thus, the majority of closed tropical evergreen forests of the central Amazon and the Congo are relatively fire-free (Krawchuk et al. 2009). Production rates of BC as charcoal are relatively high during tropical forest fires (Forbes et al. 2006). Per kilogram dry matter burned, 1,580 g CO_2, 104 g CO, 6.8 g CH_4 and 8.1 g non-methane hydrocarbons are emitted from biomass in tropical forests, and a total of 1,330 Tg dry matter burned in the late 1990s (Andreae 2004). In the majority of most moist tropical forests, however, neither fuel loading nor fuel condition are conductive to large or intense fires (Binkley et al. 1997). Otherwise, during unusually dry weather patterns or after large-scale tree mortality caused by hurricanes or typhoons forest fires may play a significant role. Deforestation increases the fire damage as burning for deforestation in the tropics can lead to large uncontrolled fires in dry conditions and increase the fire danger indirectly through forest fragmentation (Golding and Betts 2008). For example, surface fires during severe drought by El Niño Southern Oscillation (ENSO) may be responsible for 10% of the annual anthropogenic C emissions in the Amazonian forests by burning leaf litter and debris but also by killing large trees (Barlow et al. 2003). In 1998 forest fires following severe drought in Amazonia burnt 40,000 km^2 and released about 0.4 Pg C by killing trees (Nepstad et al. 2004; de Mendoça et al. 2004). In Indonesia, 0.81 Pg C were released from peat and 2.57 Pg C from vegetation by burning of forested tropical peatlands in 1997 (Page et al. 2002). Average fire emissions over 2000–2006 from equatorial Asia were 0.128 Pg C $year^{-1}$, but rates were increasing due to increase in peatland area becoming vulnerable to fire in drought years and increasing rates of forest loss (van der Werf et al. 2008). Deforestation and fires in the tropics are associated with slash-and-burn agriculture (Laurance et al. 2006).

In summary, undisturbed tropical forests are responsible for ca. 27% of the global terrestrial C sink (Robinson 2007). Wood growth, in particular, accounts for 50% of the sink (Luyssaert et al. 2007). Carbon pools are large and diverse and plant biomass is the major C pool in tropical forests (De Deyn et al. 2008). Rates of C cycling are high. Tropical humid evergreen forests retain, for example, only ca. 12% of the C fixed by photosynthesis (Luyssaert et al. 2007). Large uncertainties, however, exist for the belowground C pools and dynamics. The C-use efficiency is intermediate compared to other forest types. ENSO is the major natural and both deforestation and fire the major anthropogenic disturbances of the tropical forest C pool. Whether or not a C sink exists in the tropics is, however, highly uncertain (Houghton 2005). Also, the importance of a long-term C sink in wood is a matter of debate. However, C sequestration in soils and OM is an important process in tropical humid forests (Luyssaert et al. 2007).

4.3.2 *Effects of Climate Change*

The modeled and observed temperature trends in the tropical troposphere are consistent with the amplification of tropical surface warming (Santer et al. 2008). However, observation networks to monitor ACC impacts on physical and biological systems in tropical regions are rare (Rosenzweig et al. 2008). Thus, the response of the C dynamics and pools of tropical forests to the projected ACC is highly uncertain (Clark 2004, 2007; Laurance et al. 2006; Lewis 2006; Chave et al. 2008). The ACC may result in significant changes in root and leaf litter decomposition rates (Cusack et al. 2009). The fate of the Amazon rainforest, in particular, depends on a complex interplay between direct land-use change and the response of regional precipitation and ENSO to global forcing (Lenton et al. 2008). Currently observed biomass gains for intact forests across the Amazon are presumably a response to ACC (Gloor et al. 2009). However, Amazon forests are vulnerable to increasing moisture stress, with the potential for large C losses that might accelerate ACC (Phillips et al. 2009). The rapid transformation of the Amazon by fire is sharply elevated by logging and forest fragmentation (Cochrane and Laurance 2008). Future global warming is projected to increase temperature, and synergisms with ENSO may greatly increase drought making the Amazon increasingly vulnerable to fire.

Aside from the uncertain data on C pools in the tropical biome, FACE experiments are not conducted fumigating a tropical forest with elevated CO_2 concentrations (Houghton 2005; Wright 2005; Hickler et al. 2008). Only one open-top chamber experiment to study effects of elevated CO_2 is located in a tropical climate (Nowak et al. 2004). Furthermore, only a few long-term experiments have been conducted in secondary tropical forests. Conclusions about the productivity of tropical forests over the recent decades are uncertain (Boisvenue and Running 2006). If the magnitude and pattern of increases in forest dynamics across Amazonia observed over the last few decades is consistent with a CO_2-induced stimulation of tree growth is under discussion (Lloyd and Farquhar 2008). Otherwise, wide-spread changes in resource availability such as increasing atmospheric CO_2 concentrations may have contributed to an increase in tree C stocks in African tropical forests (Lewis et al. 2009). On the other hand, growth trends in Amazonian forests may be explained by recovery from past disturbances and changes in resource availability such as an increase in atmospheric CO_2 concentrations (Chave et al. 2008). Thus, the trends in the C balance of tropical forests under future climate are highly uncertain. However, it is likely that tropical forests mitigate global warming through evaporative cooling and C sequestration (Bonan 2008).

The effects of increase in CO_2 concentrations on tropical tree C have only been studied for individual branches of canopy trees and for understory seedlings (Wright 2005). These studies have not documented any enhancement in either biomass production or LAI at elevated CO_2 levels. The NSC concentrations in leaves or twigs, however, have increased. Whether the associated decreased leaf or twig tissue quality has consequences to nutrient cycling, herbivory and other forest processes is unclear (Wright 2005). Early successional tree species with high relative growth rates may probably have growth advantages at higher CO_2 concentrations relative to late successional species (Körner 2004). Furthermore, tropical lianas may also have significant advantages from atmospheric CO_2 enrichment, and attendant changes in composition may also occur (Laurance et al. 2006). As early successional forests store much less C than late successional forests, a biodiversity shift by increasing CO_2 concentrations may lead to the reduction in tropical forest C storage (Körner 2004). In undisturbed Amazonian forests, for example, compositional changes in tree communities have been observed and these may be explained by rising atmospheric CO_2 concentrations (Laurance et al. 2004).

The response of mature tropical trees to increasing CO_2 concentrations is uncertain (Lewis et al. 2004; Hickler et al. 2008). The observed increase in growth of tropical trees is probably not the result of elevated resource availability caused by the CO_2 fertilization effect (Feeley et al. 2007). Thus, a direct CO_2 fertilization effect on tree growth in tropical forests probably does not exist (Würth et al. 2005). In particular, physiological benefits to plants from atmospheric CO_2 may have reached already a plateau (Körner 2003). Otherwise, the growth of tropical epiphytes is stimulated by 3–4% at current CO_2 concentrations (Fernandez Monteiro et al. 2009). However, the CO_2 effect observed in tropical epiphytes is significantly smaller than that reported for lianas and other tropical species, and may be negligible compared to effects of ACC and alterations of the water regime. Thus, the existence of a strong tropical CO_2 fertilization sink is probably unlikely (Jacobsen et al. 2007; Schimel 2007). The potential role of tropical forests to C sequestration depending on increase in photosynthetic rates in the future is also uncertain (Trumbore 2006). Otherwise, indirect effects of increasing CO_2 concentrations on water-use and nutrient efficiencies of tropical trees have not been studied by long-term experiments (Lewis et al. 2004). For example, the ability of some tropical forests growing on highly weathered acidic soils to acquire new N in response to CO_2 fertilization may be constrained by molybdenum limitation (Barron et al. 2008).

Long-term warming experiments have not been conducted in tropical ecosystems (Rustad et al. 2001). Thus, the effects of the observed 0.26°C increases in air temperatures per decade on the C dynamics and pools in the tropical forest biome are highly uncertain (Lewis et al. 2004). Similar to temperate forests, however, the fraction of GPP partitioned belowground is projected to increase (Litton and Giardina 2008). As respiratory C losses also increase with increase in temperature, forest growth may be reduced (Laurance et al. 2006). Thus, warming may result in a decline the GPP:R_a ratio, and the attendant C emissions may increase (Clark 2004). On the other hand, increased soil temperatures may enhance nutrient availability and the NPP of tropical forests. Based on current understanding, models project a decline in the productivity of tropical forests as warming proceeds (Clark 2007).

Lowland species are unlikely to be threatened directly by the rising temperatures (Laurance et al. 2006). In contrast, cool-adapted montane species are vulnerable to the rising air temperatures but direct effects of global warming on C dynamics and pools of montane forests have not been experimentally studied (Laurance et al. 2006). Species near their thermal limit may respond with sharp decline in photosynthetic rates and decreased daytime C uptake with temperature increase (Clark 2004).

Tropical forest trees, however, may shift upward their photosynthetic temperature optimum. On the other hand, tropical plants have a relatively constricted ambient temperature range and may be ill-equipped to adjust to any increase in temperature (Clark 2004). Tropical lowland plants, for example, may face net lowland biotic attrition without parallel shifts at higher latitudes due to global warming, and a high proportion of tropical species may soon face gaps between current and projected elevation ranges (Colwell et al. 2008; Harsch et al. 2009). The photosynthetic optimum temperature for tropical forests is, however, poorly understood as is the acclimation capacity of photosynthesis and respiration (Clark 2007). For example, as consequence of higher tissue temperatures, increases in photosynthetic rates associated with increases in ambient CO_2 should more than offset any decline in photosynthetic productivity due to higher leaf temperatures (Lloyd and Farquhar 2008).

Against damages by short-term episodes of high temperatures, tropical forest trees emit isoprene, one of the major BVOCs (Section 2.3.1.3; Lerdau 2007; Lelieveld et al. 2008). Thus, at higher temperatures more assimilated C is lost as BVOCs but how much photosynthetically fixed C is currently lost is not known (Clark 2004; 2007). It remains to be tested, however, whether tropical trees also increase isoprene emissions during the prolonged periods of drought which accompany the warming despite reductions in photosynthesis and the NPP, as is observed for the temperate forest trees (Monson et al. 2007).

In tropical regions, atmospheric warming will increase heavy rain events as a distinct link between rainfall extremes and temperature was recently observed (Allan and Soden 2008). This may cause a decrease in biomass C sequestration in wet tropical forests as NPP declines at high precipitation (Schuur 2003). Otherwise, in regions where precipitation may decline with ACC, tropical rainforests are likely to be especially affected by the future ACC (IPCC 2007). For example, more intense and longer droughts as a result of decreased precipitation and increasing temperatures have become more common in the tropics (Trenberth et al. 2007).

The altered large-scale patterns of precipitation, in particular, may have larger effects on tropical forests than increasing temperatures (Laurance et al. 2006). For eastern Amazonia, vegetation-climate models predict increasing seasonal water deficit (Malhi et al. 2008). Short-term drought in the Amazon forests, however, may cause increased greenness of trees when more sunlight is available due to decreased cloudiness, and especially when water is not limiting (Saleska et al. 2007). In contrast, the response to long-term drought may be different when plant-available soil water declines below the critical minimum thresholds (Nepstad et al. 2007). For example, during a 7-year experimental severe drought in an Amazon forest, lianas were found to be more vulnerable to drought-induced mortality than trees or palms (Nepstad et al. 2007). The mortality of large trees, however, generated 3.4 times more dead biomass compared to the control. Furthermore, ANPP declined, soil N_2O emissions were reduced, NO production rates and CH_4 consumption rates increased whereas litterfall was less strongly affected, belowground C cycling processes remained almost unaffected and soil CO_2 efflux was not affected by the experimental drought (Brando et al. 2008; Davidson et al. 2008). More profound changes would probably require significant changes in vegetation cover.

The intense 2005 drought in the Amazon, for example, killed selectively trees and may alter species composition (Phillips et al. 2009). Most importantly, this drought reversed a large long-term C sink and caused the loss of 1.2–1.6 Pg of biomass C. Therefore, if climatic changes follow current trends, the global C cycle may be adversely affected by long-term drought episodes in the humid tropics.

The likelihood of forest burning and tree mortality may be increased in the future (Alo and Wang 2008). Fire increases the forest vulnerability to future burning and C release (Nepstad et al. 1999). Thus, for most of Amazonia climate models project increased fire danger for the twenty-first century, and as early as 2020 significant portions of Amazonian forests may be lost by ACC-induced fire and interactions with deforestation (Golding and Betts 2008). Human activities can amplify drought-induced biomass burning. For example, patterns of change in land use and population density contributed to drought-induced burning of peat fires and fires associated with deforestation in Indonesia since 1960 (Field et al. 2009). Increases in aerosol concentrations from tropical forest burning may have contributed to the observed dimming in the tropics between 1979 and 2006 (Wang et al. 2009). The potential impact of the reduced solar radiation on forest productivity is, however, not known (Wild 2009). Otherwise, severe drought events in moist tropical forests are often associated with ENSO episodes which may become more frequent or severe in the future (Laurance et al. 2006; Nepstad et al. 2007). In summary, tropical forests are vulnerable to a warmer, drier climate, and substantial vegetation changes are projected for the tropics in the future (Bonan 2008; Alo and Wang 2008).

Tropical storms may increase in intensity and duration by the ACC (Trenberth et al. 2007). In particular, island, coastal and sub-coastal but also fragmented tropical forests may be damaged by high winds (Laurance et al. 2006). Major changes in C dynamics and pools may, thus, occur. In recent decades, the tropical belt and probably also the tropical forest biome has expanded polewards (Seidel et al. 2008). This may result in shifts in storm tracks, precipitation patterns and ecosystems and also affect the C sequestration in tropical forests.

In summary, how C dynamics and pools in tropical forest ecosystems are altered by ACC is difficult to predict (Lloyd and Farquhar 2008). The importance of history for today's tropical forests is not adequately addressed (Clark 2007). Diverse types of natural disturbances interfere with the identification of effects global change. Specifically, data from long-term experiments manipulating single factors associated with ACC are lacking. Whether the accelerating dynamism in tropical forests and widespread biodiversity changes affect C sequestration in the tropical biome is a matter of debate (Stokstad 2009).

4.4 Conclusions

Forest ecosystems are of paramount importance to terrestrial C sequestration and the global C cycle because of their area extent, biomass C and soil C pool sizes and likely vulnerability to the projected ACC. Specifically, forest biomes cover ca. one third of the land surface area, of which tropical forests cover 15%, boreal forests 11% and temperate forests 5% of the global land surface. These estimates, however, are tentative and vary widely. Among the C pool of global biomes, forests store large amounts of C in plant biomass comprising 389 Pg C in tropical forests, 159 Pg C in temperate forests and 67 Pg C in boreal forests. In total, 81% of the global plant C pool is stored in forest ecosystems. The SOC pool to 1-m depth is the highest in boreal forests (338 Pg C) followed by tropical (214 Pg C) and temperate forests (153 Pg C). However, tropical forests store more SOC in 1–3 m than in 0–1 m depth. Furthermore, boreal and temperate forests also store large amounts of C below 1-m soil depth. Credible estimates of the SOC pool of forest biomes are, however, not available because the data on the SOC pool to 30 cm-depth for 51% of the total forest area are lacking and non existent. Also, the data on SOC storage below 30 cm depth in forests are scanty, despite the widespread recognition of the importance of the subsoil to the long-term stabilization of atmospheric CO_2. Between 11% and 26% of photosynthetically fixed C remains in temperate forests, compared to between 6% and 15% is retained in boreal forests and only 12% of the C fixed by photosynthesis is retained in tropical forests. Tropical forests, however, contribute 27% of the terrestrial C sink and 0.72–1.3 Pg C is stored annually in tropical forests. The C sink in boreal and temperate forests, on the other hand, is much smaller (0.49 and 0.37 Pg C $year^{-1}$, respectively). However, the existence of a significant C sink in tropical forest ecosystems is a matter of debate. The major C sinks of mangrove are not well characterized.

The magnitude of the direct effects of ACC on C sequestration differs among global forest biomes. Boreal forest C pools are increasingly disturbed by increased frequency and severity of wildfires and insect outbreaks due to projected changes in temperatures and precipitation patterns. Whether warming induced SOC losses in boreal forests are balanced by increase in NPP is, however, a matter of debate. On the other hand, the temperate forest biome C pool may increase due to ACC-induced growth enhancement although recovery from historic land use may be more important. The direct responses of tropical forests to ACC are not known as manipulative experiments are mainly performed in temperate and boreal forests. However, in boreal, temperate and tropical forests ACC is indirectly changing the plant biodiversity and this may affect C sequestration.

4.5 Review Questions

1. Photosynthesis is the main C input to forests. What factors determine plant photosynthetic activity in boreal, temperate and tropical forests?
2. Are there any differences in the main processes responsible for C losses from boreal, temperate and tropical forests?
3. Describe the pathways of C through boreal, temperate and tropical forest biomes.
4. What do we know about direct and indirect ACC effects on C sequestration in boreal, temperate and tropical forests?

5. Describe the consequences for C sequestration that may occur by tree species extinction in boreal, temperate and tropical forest biomes.
6. Discuss research priorities to achieve credible C budgets in boreal, temperate and tropical forest biomes.
7. Do forest biomes possess a maximum C storage capacity? Discuss!
8. What would happen to the C dynamics and pools in forest ecosystems if mankind disappears?

References

Aber JD, Melillo JM (2001) Terrestrial ecosystems. Academic, San Diego, CA

Aerts R (2006) The freezer defrosting: global warming and litter decomposition rates in cold biomes. J Ecol 94:713–724

Ainsworth EA, Long SP (2005) What have we learned from 15 years of free-air CO_2 enrichment (FACE)? A meta-analytical review of responses to rising CO_2 in photosynthesis, canopy properties and plant production. New Phytol 165:351–371

Alexander I, Selosse M-A (2009) Mycorrhizas in tropical forests: a neglected research imperative. New Phytol 182:14–16

Allan RP, Soden BJ (2008) Atmospheric warming and the amplification of precipitation extremes. Science 321:1481–1484

Alo CA, Wang G (2008) Potential future changes of the terrestrial ecosystem based on climate projections by eight general circulation models. J Geophys Res 113, G01004, doi:10.1029/2007JG000528

Amiro BD, Cantin A, Flannigan MD, de Groot WJ (2009) Future emissions from Canadian boreal forest fires. Can J For Res 39:383–395

Amiro BD, Todd JB, Wotton BM, Logan KM, Flannigan MD, Stocks BJ, Mason JA, Martell DL, Hirsch KG (2001) Direct carbon emissions from Canadian forest fires, 1959–1999. Can J For Res 31:512–525

Andreae MO (2004) Assessment of global emissions from vegetation fires. Int For Fire News 31:112–121

Angert A, Biraud S, Bonfils C, Henning CC, Buermann W, Pinzon J, Tucker CJ, Fung I (2005) Drier summer cancel out the CO_2 uptake enhancement induced by warmer springs. Proc Natl Acad Sci USA 102:10823–10827

Apps MJ, Kurz WA, Luxmoore RJ, Nilsson LO, Sedjo RA, Schmidt R, Simpson LG, Vinson TS (1993) Boreal forests and tundra. Water Air Soil Pollut 70:39–53

Asner GP, Scurlock JMO, Hicke JA (2003) Global synthesis of leaf area index observations: implications for ecological and remote sensing studies. Global Ecol Biogeogr 12:191–205

Baccini A, Laporte N, Goetz SJ, Sun M, Dong H (2008) A first map of tropical Africa's above-ground biomass derived from satellite imagery. Environ Res Lett 3, doi:10.1088/1748-9326/3/4/045011

Baldocchi DD (2003) Assessing the eddy covariance technique for evaluating carbon dioxide exchange rates of ecosystems: past, present and future. Glob Change Biol 9:479–492

Balshi MS, McGuire AD, Zhuang Q, Melillo J, Kicklighter DW, Kasischke E, Wirth C, Flannigan M, Harden J, Clein JS, Burnside TJ, McAllister J, Kurz WA, Apps M, Shvidenko A (2007) The role of historical fire disturbance in the carbon dynamics of the pan-boreal region: a process-based analysis. J Geophys Res 112:G02029. doi:10.1029/2006JG000380

Barber VA, Juday GP, Finne BP (2000) Reduced growth of Alaskan white spruce in the twentieth century from temperature-induced drought stress. Nature 405:668–673

Barlow J, Peres CA, Lagan B, Haugaasen T (2003) Forest biomass collapse following Amazonian wildfires. Ecol Lett 6:6–8

Barron AR, Wurzburger N, Bellenger JP, Wright SJ, Kraepiel AML, Hedin LO (2008) Molybdenum limitation of asymbiotic nitrogen fixation in tropical forest soils. Nat Geosci 2:42–45

Beerling DJ (1999) Long-term responses of boreal vegetation to global change: an experimental and modeling investigation. Glob Change Biol 5:55–74

Bernier PY, Apps MJ (2006) Knowledge gaps and challenges in forest ecosystems under climate change: a look at the temperate and boreal forests of North America. In: Bhatti JS, Lal R, Apps MJ, Price MA (eds) Climate change and managed ecosystems. Taylor & Francis, Boca Raton, FL. pp 333–353

Betts RA (2000) Offset of the potential carbon sink from boreal forestation by decreases in surface albedo. Nature 408:187–190

Binkley CS, Apps MJ, Dixon RK, Kauppi PE, Nilsson LO (1997) Sequestering carbon in natural forests. Crit Rev Env Sci Technol 27:S23–S45

Black TA, Gaumont-Guay D, Jassal RS, Amiro BD, Jarvis PG, Gower ST, Kelliher FM, Dunn A, Wofsy SC (2005) Measurement of CO_2 exchange between Boreal forest and the atmosphere. In: Griffiths H, Jarvis PG (eds) The carbon balance of forest biomes. Taylor & Francis, Boca Raton, FL, pp 151–185

Boisvenue C, Running SW (2006) Impacts of climate change on natural forest productivity – evidence since the middle of the 20th century. Global Change Biol 12:862–882

Bombelli A, Henry M, Castaldi S, Adu-Bredu S, Arneth A, de Grandcourt A, Grieco E, Kutsch WL, Lehsten V, Rasile A, Reichstein M, Tansey K, Weber U, Valentini R (2009) The sub-saharan Africa carbon balance, an overview. Biogeosci Discuss 6:2085–2123

Bonan GB (2008) Forests and climate change: forcings, feedbacks, and the climate benefits of forests. Science 320:1444–1449

Bond-Lamberty B, Peckham SD, Ahl DE, Gower ST (2007) Fire as dominant driver of central Canadian boreal forest carbon balance. Nature 450:89–93

Bouillon S, Borges AV, Castañeda-Moya E, Diele K, Dittmar T, Duke NC, Kristensen E, Lee SY, Marchand C, Middelburg JJ, Rivera-Monroy VH, Smith III TJ, Twilley RR (2008) Mangrove production and carbon sinks: a revision of global estimates. Global Biogeochem Cy 22, GB2013, doi:10.1029/2007GB003052

Bradley KL, Pregitzer KS (2007) Ecosystem assembly and terrestrial carbon balance under elevated CO_2. Trends Ecol Evol 22:538–547

Brando PM, Nepstad DC, Davidson EA, Trumbore SE, Ray D, Camargo P (2008) Drought effects on litterfall, wood production and belowground carbon cycling in an Amazon forest: results of a throughfall reduction experiment. Philos Trans R Soc B 363:1839–1848

Bréda N, Huc R, Granier A, Dreyer E (2006) Temperate forest trees and stands under severe drought: a review of ecophysiological responses, adaptation processes and long-term consequences. Ann For Sci 63:625–644

Bryant D, Nielsen D, Tangley L (1997) The last frontier forests: ecosystems and economics on the edge. World Resources Institute, Washington, DC

Camill P (2005) Permafrost thaw accelerates in boreal peatlands during late-20th century climate warming. Clim Change 68:135–152

Canadell J, Jackson RB, Ehleringer JR, Mooney HA, Sala OE, Schulze E-D (1996) Maximum rooting depth of vegetation types at the global scale. Oecologia 108:583–595

Caspersen JP, Pacala SW, Jenkins JC, Hurtt GC, Moorcroft PR, Birdsey RA (2000) Contributions of land-use history to carbon accumulation in U.S. forests. Science 290:1148–1151

Chao K-J, Phillips OL, Baker TR, Peacock J, Lopez-Gonzalez G, Martínez RV, Monteagudo A, Torres-Lezama A (2009) After trees die: quantities and determinants of necromass across Amazonia. Biogeosci Discuss 6:1979–2006

Chapin FS III, Matson PA, Mooney HA (2002) Principles of terrestrial ecosystem ecology. Springer, New York

Chave J, Condit R, Muller-Landau HC, Thomas SC, Ashton PS, Bunyavejchewin S, Co LL, Dattaraja HS, Davies SJ, Esufali S, Ewango CEN, Feeley KJ, Foster RB, Gunatilleke N, Gunatilleke S, Hall P, Hart TB, Hernández C, Hubbell SP, Itoh A, Kiratiprayoon S, LaFrankie JV, de Lao SL, Makana J-R, Noor MNS, Kassim AR, Samper C, Sukumar R, Suresh HS, Tan S,

Thompson J, Tongco MDC, Valencia R, Vallejo M, Villa G, Yamakura T, Zimmerman JK, Losos EC (2008) Assessing evidence for a pervasive alteration in tropical tree communities. PLoS Biol 6:455–462
Chen X, Zhang X, Zhang Y, Booth T, He X (2009) Changes of carbon stocks in bamboo stands in China during 100 years. For Ecol Manage 258:1489–1496
Ciais P, Janssens I, Shvidenko A, Wirth C, Malhi Y, Grace J, Schulze E-D, Heimann M, Phillips O, Dolman AJ (2005a) The potential for rising CO_2 to account for the observed uptake of carbon by tropical, temperate, and Boreal forest biomes. In: Griffiths H, Jarvis PG (eds) The carbon balance of forest biomes. Taylor & Francis, Boca Raton, FL, pp 109–149
Ciais P, Reichstein M, Viovy N, Granier A, Ogée J, Allard V, Aubinet M, Buchmann N, Bernhofer C, Carrara A, Chevallier F, De Noblet N, Friend AD, Friedlingstein P, Grünwald T, Heinesch B, Keronen P, Knohl A, Krinner G, Loustau D, Manca G, Matteucci G, Miglietta F, Ourcival JM, Papale D, Pilegaard K, Rambal S, Seufert G, Soussana J-F, Sanz M-J, Schulze E-D, Vesala T, Valentini R (2005b) Europe-wide reduction in primary productivity caused by the heat and drought in 2003. Nature 437:529–533
Clark DA (2004) Sources or sinks? The responses of tropical forests to current and future climate and atmospheric composition. Philos Trans R Soc Lond B 359:477–491
Clark DA (2007) Detecting tropical forests' response to global climatic and atmospheric change: current challenges and a way forward. Biotropica 39:4–19
Clark DA, Brown S, Kicklighter DW, Chambers JQ, Thomlinson JR, Ni J, Holland EA (2001) Net primary production in tropical forests: an evaluation and synthesis of existing field data. Ecol Appl 11:371–384
Clark DB, Olivas PC, Oberbauer SF, Clark DA, Ryan MG (2008) First direct landscape-scale measurement of tropical rain forest Leaf Area Index, a key driver of global primary productivity. Ecol Lett 11:163–172
Cochrane MA, Laurance WF (2008) Synergisms among fire, land use, and climate change in the Amazon. Ambio 37:522–527
Colwell RK, Brehm G, Cardelús CL, Gilman AC, Longino JT (2008) Global warming, elevational range shifts, and lowland biotic attrition in the wet tropics. Science 322:258–261
Cornelissen JHC, van Bodegom PM, Aerts R, Callaghan TV, Van Logtesijn RSP, Alatalo J, Chapin FS, Gerdol R, Guðmundsson J, Gwynn-Jones D, Hartley AE, Hik DS, Hofgaard A, Jónsdóttir IS, Karlsson S, Klein JA, Laundre J, Magnusson B, Michelsen A, Molau U, Onipchenko VG, Quested HM, Sandvik SM, Schmidt IK, Shaver GR, Solheim B, Soudzilovskaia NA, Stenström A, Tolvanen A, Totland Ø, Wada N, Welker JM, Zhao X, Team MOL (2007) Global negative vegetation feedback to climate warming responses of leaf litter decomposition rates in cold biomes. Ecol Lett 10:619–627
Cusack DF, Chou WW, Yang WH, Harmon ME, Silver WL, Team TLIDET (2009) Controls on long-term root and leaf litter decomposition in neotropical forests. Global Change Biol 15:1339–1355
Czimczik CI, Masiello CA (2007) Controls on black carbon storage in soils. Global Biogeochem Cy 21, GB3005, doi:10.1029/2006GB002798
Davidson EA, Janssens IA (2006) Temperature sensitivity of soil carbon decomposition and feedbacks to climate change. Nature 440:165–173
Davidson EA, Nepstad DC, Ishida FY, Brando PM (2008) Effects of an experimental drought and recovery on soil emissions of carbon dioxide, methane, nitrous oxide, and nitric oxide in a moist tropical forest. Glob Change Biol 14:2582–2590
De Deyn GB, Cornelissen JHC, Bardgett RD (2008) Plant functional traits and soil carbon sequestration in contrasting biomes. Ecol Lett 11:516–531
Delpierre N, Soudani K, François C, Köstner B, Pontailler J-Y, Nikinmaa E, Misson L, Aubinet M, Bernhofer C, Granier A, Grünewald G, Heinesch B, Longdoz B, Ourcival JM, Rambal S, Vesala T, Dufrêne E (2009) Exceptional carbon uptake in European forests during the warm spring of 2007: a data-model analysis. Glob Change Biol 15:1455–1474
De Mendoça MJC, MdCV D, Nepstad D, da Motta RS, Alencar A, Gomes JC, Ortiz RA (2004) The economic cost of the use of fire in the Amazon. Ecol Econ 49:89–105

Denman KL, Brasseur G, Chidthaisong A, Ciais P, Cox PM, Dickinson RE, Hauglustaine D, Heinze C, Holland E, Jacob D, Lohmann U, Ramachandran S, da Silva Dias PL, Wofsy SC, Zhang X (2007) Couplings between changes in the climate system and biogeochemistry. In: Intergovernmental panel on climate change (ed). Climate change 2007: the physical science basis, Chapter 7. Cambridge University Press, Cambridge

Dutaur L, Verchot LV (2007) A global inventory of the soil CH_4 sink. Global Biogeochem Cy 21, GB4013, doi:10.1029/2006GB002734

Euskirchen ES, McGuire AD, Chapin FSIII (2007) Energy feedbacks of northern high-latitude ecosystems to the climate system due to reduced snow cover during 20th century warming. Glob Change Biol 13:2425–2438

Fahey TJ, Siccama TG, Driscoll CT, Likens GE, Campbell J, Johnson CE, Battles JJ, Aber JD, Cole JJ, Fisk MC, Groffman PM, Hamburg SP, Holmes RT, Schwarz PA, Yanai RD (2005) The biogeochemistry of carbon at Hubbard Brook. Biogeochemistry 75:109–176

FAO (Food and Agricultural Organization of the United Nations) (2006) Global forest resources assessment 2005. Progress towards sustainable forest management. FAO Forestry paper 147. FAO, Rome

FAO (Food and Agricultural Organization of the United Nations) (2007) The world's mangroves 1980–2005. FAO Forestry paper 153. FAO, Rome

Feeley KJ, Wright SJ, Nur Supardi MN, Kassim AR, Davies SJ (2007) Decelerating growth in tropical forest trees. Ecol Lett 10:461–469

Field RD, Van Der Werf GR, Shen SSP (2009) Human amplification of drought-induced biomass burning in Indonesia since 1960. Nat Geosci 2:185–188

Fernandez Monteiro JA, Zotz G, Körner C (2009) Tropical epiphytes in a CO_2-rich atmosphere. Acta Oecol 35:60–68

Fisher JI, Hurtt GC, Thomas RQ, Chambers JQ (2008) Clustered disturbances lead to bias in large-scale estimates based on forest sample plots. Ecol Lett 11:554–563

Fleming RA, Candau JN, McAlpine RS (2002) Landscape-scale analysis of interactions between insect defoliation and forest fire in central Canada. Clim Change 55:251–272

Forbes MS, Raison RJ, Skjemstad JO (2006) Formation, transformation and transport of black carbon (charcoal) in terrestrial and aquatic ecosystems. Sci Total Environ 370:190–206

Froese DG, Westgate JA, Reyes AV, Enkin RJ, Preece SJ (2008) Ancient permafrost and a future, warmer arctic. Science 321:1648

Fung IY, Doney SC, Lindsay K, John J (2005) Evolution of carbon sinks in a changing climate. Proc Natl Acad Sci USA 102:11201–11206

Gilbert GS, Strong DR (2007) Fungal symbionts of tropical trees. Ecology 88:539–540

Gillett NP, Weaver AJ, Zwiers FW, Flannigan MD (2004) Detecting the effect of climate change on Canadian forest fires. Geophys Res Lett 31:L18211. doi:10.1029/2004GL020876

Gloor M, Phillips OL, Lloyd JJ, Lewis SL, Malhi Y, Baker TR, López-Gonzalez G, Peacock J, Almeida S, Alves de Oliveira AC, Alvarez E, Amaral I, Arroyo L, Aymard G, Banki O, Blanc L, Bonal D, Brando P, Chao K-J, Chave J, Dávila N, Erwin T, Silva J, Di Fiore A, Feldpausch T, Freitas A, Herrera R, Higuchi N, Honorio E, Jiménez E, Killeen T, Laurance W, Mendoza C, Monteagudo A, Andrade A, Neill D, Nepstad D, Núñez Vargas P, Peñuela MC, Peña Cruz A, Prieto A, Pitman N, Quesada C, Salomáo R, Marcos Silveira M, Schwarz J, Stropp F, Ramírez H, Ramírez A, Rudas H, ter Steege N, Silva N, Torres A, Terborgh J, Vásquez R, van der Heijden G (2009) Does disturbance hypothesis explain the biomass increase in basin-wide Amazon forest plot data? Glob Change Biol doi: 10.1111/j.1365-2486.2009.01891.x

Godbold DL, Hoosbeek MR, Lukac M, Cotrufo MF, Janssens IA, Ceulemans R, Polle A, Velthorst EJ, Scarascia-Mugnozza G, De Angelis P, Miglietta F, Peressotti A (2006) Mycorrhizal hyphal turnover as a dominant process for carbon input into soil organic matter. Plant Soil 281:15–24

Gold S, Korotkov AV, Sasse V (2006) The development of European forest resources, 1950–2000. For Pol Econ 8:183–192

Golding N, Betts R (2008) Fire risk in Amazonia due to climate change in the HadCM3 climate model: potential interactions with deforestation. Global Biogeochem Cy 22, GB4007, doi:10.1029/2007GB003166

González-Pérez JA, González-Vila FJ, Polvillo O, Almendros G, Knicker H, Salas F, Costa JC (2002) Wildfire and black carbon in Andalusian Mediterranean forest. In: Viegas DX (ed) Forest fire research and wildland fire safety [CD-ROM]. Millpress, Rotterdam, The Netherlands, pp 1–7

Gower ST, Krankina O, Olson RJ, Apps M, Linder S, Wang C (2001) Net primary production and carbon allocation patterns of boreal forest ecosystems. Ecol Appl 11:1395–1411

Gower ST, Landsberg JJ, Bisbee KE (2003) Forest biomes of the world. In: Young RA, Giese RL (eds) Forest ecosystem science and management. Wiley, Hoboken, NJ, pp 57–74

Grace J, Malhi Y, Higuchi N, Meir P (2001) Productivity of tropical rain forests. In: Roy J, Saugier B, Mooney HA (eds) Terrestrial global productivity. Academic, San Diego, CA, pp 401–426

Grainger A (2008) Difficulties in tracking the long-term global trend in tropical forest area. Proc Natl Acad Sci USA 105:818–823

Groisman PY, Sherstyukov BG, Razuvaev VN, Knight RW, Enloe JG, Stroumentova NS, Whitfield PH, Førland E, Hannsen-Bauer I, Tuomenvirta H, Aleksandersson H, Mescherskaya AV, Karl TR (2007) Potential forest fire danger over Northern Eurasia: changes during the 20th century. Glob Planet Change 56:371–386

Guenther A, Karl T, Harley P, Wiedinmyer C, Palmer PI, Geron C (2006) Estimates of global terrestrial isoprene emissions using MEGAN (model of emissions of gases and aerosols from nature). Atmos Chem Phys 6:3181–3210

Guggenberger G, Rodionov A, Shibistova O, Grabe M, Kasansky OA, Fuchs H, Mikheyeva N, Zrazhevskaya G, Flessa H (2008) Storage and mobility of black carbon in permafrost soils of the forest-tundra ecotone in northern Siberia. Glob Change Biol 14:1367–1381

Hagedorn F, Machwitz M (2007) Controls on dissolved organic matter leaching from forest litter grown under elevated atmospheric CO_2. Soil Biol Biochem 39:1759–1769

Hamilton JG, DeLucia EH, George K, Naidu SL, Finzi AC, Schlesinger WH (2002) Forest carbon balance under elevated CO_2. Oecologia 131:250–260

Harsch MA, Hulme PE, McGlone MS, Duncan RP (2009) Are treelines advancing? A global meta-analysis of treeline response to climate warming. Ecol Lett (in press) doi: 10.1111/j.1461-0248.2009.01355.x

Helliker BR, Richter SL (2008) Subtropical to boreal convergence of tree-leaf temperatures. Nature 454:511–515

Henry M, Valentini R, Bernoux M (2009) Soil carbon stocks in ecoregions of Africa. Biogeosci Discuss 6:797–823

Hertel D, Harteveld MA, Leuschner C (2009) Conversion of a tropical forest into agroforest alters the fine root-related carbon flux to the soil. Soil Biol Biochem 41:481–490

Hickler T, Smith B, Prentice IC, Mjöfors K, Miller P, Arneth A, Sykes MT (2008) CO_2 fertilization in temperate FACE experiments not representative of boreal and tropical forests. Glob Change Biol 14:1531–1542

Hobbie SE, Ogdahl M, Chorover J, Chadwick OA, Oleksyn J, Zytkowiak R, Reich PB (2007) Tree species effects on soil organic matter dynamics: the role of soil cation composition. Ecosystems 10:999–1018

Hobbie SE, Reich PB, Oleksyn J, Ogdahl M, Zytkowiak R, Hale C, Karolewski P (2006) Tree species effects on decomposition and forest floor dynamics in a common garden. Ecology 87:2288–2297

Hoosbeek MR, van Breemen N, Berendse F, Grosvernier P, Vasander H, Wallén B (2001) Limited effect of increased atmospheric CO_2 concentration on ombrotrophic bog vegetation. New Phytol 150:459–463

Houghton RA (2005) Aboveground forest biomass and the global carbon balance. Glob Change Biol 11:945–958

Houghton RA, Butman D, Bunn AG, Krankina ON, Schlesinger P, Stone TA (2007) Mapping Russian forest biomass with data from satellites and forest inventories. Environ Res Lett 2, doi:10.1088/1748-9326/2/4/045032

Huston MA, Wolverton S (2009) The global distribution of net primary production: resolving the paradox. Ecol Monogr 79:343–377

Hytteborn H, Maslov AA, Nazimova DI, Rysin LP (2005) Boreal forests of Eurasia. In: Andersson F (ed) Ecosystems of the world 6. Coniferous forests. Elsevier, Amsterdam, The Netherlands, pp 23–99

Hyvönen R, Ågren GI, Linder S, Persson T, Cotrufo MF, Ekblad A, Freeman M, Grelle A, Janssens IA, Jarvis PG, Kellomäki S, Lindroth A, Loustau D, Lundmark T, Norby RJ, Oren R, Pilegaard K, Ryan MG, Sigurdsson BD, Strömgren M, Oijen M, Wallin G (2007) The likely impact of elevated [CO_2], nitrogen deposition, increased temperature and management on carbon sequestration in temperate and boreal forest ecosystems: a literature review. New Phytol 173:463–480

IPCC (Intergovernmental Panel on Climate Change) (2007) Summary for policymakers of the synthesis report of the IPCC fourth assessment report

Jackson RB, Canadell J, Ehleringer JR, Mooney HA, Sala OE, Schulze E-D (1996) A global analysis of root distributions for terrestrial biomes. Oecologia 108:389–411

Jackson RB, Randerson JT, Canadell JG, Anderson RG, Avissar R, Baldocchi DD, Bonan GB, Caldeira K, Diffenbaugh NS, Field CB, Hungate BA, Jobbágy EG, Kueppers LM, Nosetto MD, Pataki DE (2008) Protecting climate with forests. Environ Res Lett 3, doi:10.1088/1748-9326/3/4/044006

Jacobsen AR, Fletcher SEM, Gruber N, Sarmiento JL, Gloor M (2007) A joint atmosphere-ocean inversion for surface fluxes of carbon dioxide: 1. methods and global-scale fluxes. Global Biogeochem Cy 21, GB1019, doi:10.1029/2005GB002556

Jandl R, Lindner M, Vesterdahl L, Bauwens B, Baritz R, Hagedorn F, Johnson DW, Minkkinen K, Byrne KA (2007) How strongly can forest management influence soil carbon sequestration? Geoderma 137:253–268

Jarvis PG, Saugier B, Schulze E-D (2001) Productivity of boreal forests. In: Roy J, Saugier B, Mooney HA (eds) Terrestrial global productivity. Academic, San Diego, CA, pp 211–244

Jastrow JD, Miller RM, Matamala R, Norby RJ, Boutton TW, Rice CW, Owensby CE (2005) Elevated atmospheric carbon dioxide increases soil carbon. Glob Change Biol 11:2057–2064

Jobbágy EG, Jackson RB (2000) The vertical distribution of soil organic carbon and its relation to climate and vegetation. Ecol Appl 10:423–436

Karnosky DF, Pregitzer KS, Zak DR, Kubiske ME, Hendrey GR, Weinstein D, Nosal M, Percy KE (2005) Scaling ozone responses of forest trees to the ecosystem level in a changing climate. Plant Cell Environ 28:965–981

Kasischke ES (2000) Boreal ecosystems in the global carbon cycle. In: Kasischke ES, Stocks BJ (eds) Fire, climate change, and carbon cycling in the boreal forest. Springer, New York, pp 19–30

Keddy PA, Fraser LH, Solomeshch AI, Junk WJ, Campbell DR, Arroyo MTK, Alho CJR (2009) Wet and wonderful: the world's largest wetlands are conservation priorities. Bioscience 59:39–51

Keeling HC, Phillips OL (2005) The global relationship between forest productivity and biomass. Global Ecol Biogeogr 16:618–631

Keith H, Mackey BG, Lindenmayer DB (2009) Re-evaluation of forest biomass carbon stocks and lessons from the world's most carbon-dense forests. P Natl Acad Sci USA 106:11635–11640

Kelly AE, Goulden ML (2008) Rapid shifts in plant distribution with recent climate change. Proc Natl Acad Sci USA 105:11823–11826

Khvorostyanov DV, Ciais P, Krinner G, Zimov SA (2008) Vulnerability of east Siberia's frozen carbon stores to future warming. Geophys Res Lett 35:L10703. doi:10.1029/2008GL033639

King JS, Kubiske ME, Pregitzer KS, Hendrey GR, McDonald EP, Giardina CP, Quinn VS, Karnosky DF (2005) Tropospheric O_3 compromises net primary production in young stands of trembling aspen, paper birch and sugar maple in response to elevated atmospheric CO_2. New Phytol 168:623–636

Kirilenko AP, Sedjo RA (2007) Climate change impacts on forestry. Proc Natl Acad Sci USA 104:19697–19702

Koenig R (2008) Critical time for African rainforests. Science 320:1439–1441

Körner C (2003) Carbon limitation in trees. J Ecol 91:4–17

Körner C (2004) Through enhanced tree dynamics carbon dioxide enrichment may cause tropical forests to lose carbon. Philos Trans R Soc Lond B 359:493–498

Körner C (2006) Plant CO_2 responses: an issue of definition, time and resource supply. New Phytol 172:393–411

Krawchuk MA, Moritz MA, Parisien M-A, Van Dorn J, Hayhoe K (2009) Global pyrogeography: the current and future distribution of wildfire. PloS ONE 4(4):e5102. doi:10.1371/journal.pone.0005102

Kurz WA, Stinson G, Rampley G (2008a) Could increased boreal forest ecosystem productivity offset carbon losses from increased disturbances? Philos Trans R Soc B 363:2261–2269

Kurz WA, Stinson G, Rampley GJ, Dymond CC, Neilson ET (2008b) Risk of natural disturbances makes future contribution of Canada's forest to the global carbon cycle highly uncertain. Proc Natl Acad Sci USA 105:1551–1555

Landsberg JJ, Gower ST (1997) Applications of physiological ecology to forest management. Academic, San Diego, CA

Lapenis A, Shvidenko A, Shepaschenko D, Nilsson S, Aiyyer A (2005) Acclimation of Russian forests to recent changes in climate. Glob Change Biol 11:2090–2102

Laurance WF, Oliveira AA, Laurance SG, Condit R, Nascimento HEM, Sanchez-Thorin AC, Lovejoy TE, Andrade A, D'Angelo S, Ribeiro JE, Dick CW (2004) Pervasive alteration of tree communities in undisturbed Amazonian forests. Nature 428:171–175

Laurance WF, Peres CA, Jansen PA, D'Croz L (2006) Emerging threats to tropical forests: what we know and what we don't know. In: Laurance WF, Peres CA (eds) Emerging threats to tropical forests. The University of Chicago press, Chicago, IL/London, pp 437–462

Leigh Mascarelli A (2009) A sleeping giant? Nat Rep Clim Change. doi:10.1038/climate.2009.24

Lelieveld J, Butler TM, Crowley JN, Dillon TJ, Fischer H, Ganzeveld L, Harder H, Lawrence MG, Martinez M, Taraborrelli D, Williams J (2008) Atmospheric oxidation capacity sustained by a tropical forest. Nature 452:737–740

Lenton TM, Held H, Kriegler E, Hall JW, Lucht W, Rahmstorf S, Schellnhuber HJ (2008) Tipping element's in the Earth's climate system. Proc Natl Acad Sci USA 105:1786–1793

Lerdau M (2007) A positive feedback with negative consequences. Science 316:212–213

Lewis OT (2009) Biodiversity change and ecosystem function in tropical forests. Basic Appl Ecol 10:97–102

Lewis SL (2006) Tropical forests and the changing earth system. Philos Trans R Soc Lond B 361:195–210

Lewis SL, Lopez-Gonzalez G, Sonké B, Affum-Baffoe K, Baker TR, Ojo LO, Phillips OL, Reitsma JM, White L, Comiskey JA, M-N DK, Ewango CEN, Feldpausch TR, Hamilton AC, Gloor M, Hart T, Hladik A, Lloyd J, Lovett JC, Makana J-R, Malhi Y, Mbago FM, Ndangalasi HJ, Peacock J, S-H PK, Sheil D, Sunderland T, Swaine MD, Taplin J, Taylor D, Thomas SC, Votere R, Wöll H (2009) Increasing carbon storage in intact African tropical forests. Nature 457:1003–1007

Lewis SL, Malhi Y, Phillips OL (2004) Fingerprinting the impacts of global change on tropical forests. Philos Trans R Soc Lond B 359:437–462

Levy S (2007) Running hot and cold: Are rainforests sinks or taps for carbon? BioScience 57:552–557

Limpens J, Berendse F, Blodau C, Canadell JG, Freeman C, Holden J, Roulet N, Rydin H, Schaepman-Strub G (2008) Peatlands and the carbon cycle: from local processes to global implications – a synthesis. Biogeosciences 5:1475–1491

Litton CM, Giardina CP (2008) Below-ground carbon flux and partitioning: global patterns and response to temperature. Funct Ecol 22:941–954

Litton CM, Raich JW, Ryan MG (2007) Review: carbon allocation in forest ecosystems. Glob Change Biol 13:2089–2109

Lloyd J, Farquhar GD (2008) Review. Effects of rising temperatures and [CO_2] on the physiology of tropical forest trees. Philos Trans R Soc B 363:1811–1817

Lloyd J, Kolle O, Fritsch H, de Freitas SR, Silva Dias MAF, Artaxo P, Nobre AD, de Araújo AC, Kruijt B, Sogacheva L, Fisch G, Thielmann A, Kuhn U, Andreae MO (2007) An airborne regional carbon balance for Central America. Biogeosciences 4:759–768

Lucht W, Schaphoff S, Erbrecht T, Heyder U, Cramer W (2006) Terrestrial vegetation redistribution and carbon balance under climate change. Carbon Bal Manag 1:6

Luyssaert S, Inglima I, Jung M, Richardson AD, Reichstein M, Papale D, Piao SL, Schulze E-D, Wingate L, Matteucci G, Aragao L, Aubinet M, Beer C, Bernhofer C, Black KG, Bonal D, Bonnefond J-M, Chambers J, Ciais P, Cook B, Davis KJ, Dolman AJ, Gielen B, Goulden M, Grace J, Granier A, Grelle A, Griffis T, Grünwald T, Guidolotti G, Hanson PJ, Harding R, Hollinger DY, Hutyra LR, Kolari P, Kruijt B, Kutsch W, Lagergren F, Laurila T, Law BE, Le Maire G, Lindroth A, Loustau D, Malhi Y, Mateus J, Migliavacca M, Misson L, Montagnani L, Moncrieff J, Moors E, Munger JW, Nikinmaa E, Ollinger SV, Pita G, Rebmann C, Roupsard O, Saigusa N, Sanz MJ, Seufert G, Sierra C, Smith M-L, Tang J, Valentini R, Vesala T, Janssens IA (2007) The CO_2-balance of boreal, temperate and tropical forests derived from a global database Glob Change Biol 13:2509–2537

Luyssaert S, Reichstein M, Schulze E-D, Janssens IA, Law BE, Papale D, Dragoni D, Goulden ML, Granier A, Kutsch WL, Linder S, Matteucci G, Moors E, Munger JW, Pilegaard K, Saunders M, Falge EM (2009) Toward a consistency cross-check of eddy covariance flux-based and biometric estimates of ecosystem carbon balance. Global Biogeochem Cy 23, GB3009, doi:10.1029/2008GB003377

Maguire DA, Osawa A, Batista JLF (2005) Primary production, yield and carbon dynamics. In: Andersson F (ed) Ecosystems of the world 6. Coniferous forests. Elsevier, Amsterdam, The Netherlands, pp 339–383

Malhi Y, Baker TR, Phillips OL, Almeida S, Alvarez E, Arroyo L, Chave J, Czimczik CI, Di Fiore A, Higuchi N, Killeen TJ, Laurance SG, Laurance WF, Lewis SL, Montoya LMM, Monteagudo A, Neill DA, Vargas PN, Patiño S, Pitman NCA, Quesada CA, Salomão R, Silva JNM, Lezama AT, Martínez RV, Terborgh J, Vinceti B, Lloyd J (2004) The above-ground coarse wood productivity of 104 Neotropical forest plots. Glob Change Biol 10:563–591

Malhi Y, Roberts JT, Betts RA, Killeen TJ, Li W, Nobre CA (2008) Climate change, deforestation, and the fate of the Amazon. Science 319:169–172

van Mantgem PJ, Stephenson NL, Byrne JC, Daniels LD, Franklin JF, Fulé PZ, Harmon ME, Larson AJ, Smith JM, Taylor AH, Veblen TT (2009) Widespread increase of tree mortality rates in the western United States. Science 323:521–524

Martin PH, Nabuurs G-J, Aubinet M, Karjalainen T, Vine EL, Kinsman J, Heath LS (2001) Carbon sinks in temperate forests. Annu Rev Energy Environ 26:435–465

Mather AS (2005) Assessing the world's forests. Global Environ Chang A 15:267–280

Meehl GA, Stocker TF, Collins WD, Friedlingstein P, Gaye AT, Gregory JM, Kitoh A, Knutti R, Murphy JM, Noda A, Raper SCB, Watterson IG, Weaver AJ, Zhao Z-C (2007) Global climate projections. In: Solomon S, Qin D, Manning M, Chen Z, Marquis M, Averyt KB, Tignor M, Miller HL (eds) Climate change 2007: The physical science basis. Contribution of working group I to the fourth assessment report of the intergovernmental panel on climate change. Cambridge University Press, Cambridge, United Kingdom/New York, NY, pp 747–845

Melillo JM, McGuire AD, Kicklighter DW, Moore B III, Vorosmarty CJ, Schloss AL (1993) Global climate change and terrestrial net primary production. Nature 363:234–240

Menyailo OV, Hungate BA, Zech W (2002) Tree species mediated soil chemical changes in a Siberian artificial afforestation experiment. Plant Soil 242:171–182

Metcalfe DB, Meir P, Aragão LEOC, Malhi Y, da Costa ACL, Braga A, Gonçalves PHL, de Athaydes J, de Almeida SS, Williams M (2007) Factors controlling spatio-temporal variation in carbon dioxide efflux from surface litter, roots, and soil organic matter at four rain forest sites in the eastern Amazon. J Geophys Res 112:G04001. doi:10.1029/2007JG000443

Mohan JE, Cox RM, Iverson LR (2009) Composition and carbon dynamics of forests in northeastern North America in a future, warmer world. Can J For Res 39:213–230

Monson RK, Trahan N, Rosenstiel TN, Veres P, Moore D, Wilkinson M, Norby RJ, Volder A, Tjoelker MG, Briske DD, Karnosky DF, Fall R (2007) Isoprene emission from terrestrial ecosystems in response to global change: minding the gap between models and observations. Philos Trans R Soc A 365:1677–1695

Myers N, Mittermeier RA, Mittermeier CG, da Fonseca GAB, Kent J (2000) Biodiversity hotspots for conservation priorities. Nature 403:853–858

Nabuurs GJ, Masera O, Andrasko K, Benitez-Ponce P, Boer R, Dutschke M, Elsiddig E, Ford-Robertson J, Frumhoff P, Karjalainen T, Krankina O, Kurz WA, Matsumoto M, Oyhantcabal W,

Ravindranath NH, Sanz Sanchez MJ, Zhang X (2007): Forestry. In: Metz B, Davidson OR, Bosch PR, Dave R, Meyer LA (eds) Climate Change 2007: Mitigation. Contribution of working group III to the fourth assessment report of the intergovernmental panel on climate change. Cambridge University Press, Cambridge, United Kingdom/New York, NY, pp 541–584

Nepstad D, Lefebvre P, da Silva UL, Tomasella J, Schlesinger P, Solórzano L, Moutinho P, Ray D, Benito JG (2004) Amazon drought and its implications for forest flammability and tree growth: a basin-wide analysis. Glob Change Biol 10:704–717

Nepstad DC, Tohver IM, Ray D, Moutinho P, Cardinot G (2007) Mortality of large trees and lianas following experimental drought in an amazon forest. Ecology 88:2259–2269

Nepstad DC, Veríssima A, Alencar A, Nobre C, Lima E, Lefebvre P, Schlesinger P, Potter C, Moutinho P, Mendoza E, Cochrane M, Brooks V (1999) Large-scale impoverishment of Amazonian forests by logging and fire. Nature 398:505–508

Nieder R, Benbi DK (2008) Carbon and nitrogen in the terrestrial environment. Springer, Berlin

Norby RJ, Ledford J, Reilly CD, Miller NE, O'Neill EG (2004) Fine-root production dominates response of a deciduous forest to atmospheric CO_2 enrichment. Proc Natl Acad Sci USA 101:9689–9693

Norby RJ, DeLucia EH, Gielen B, Calfapietra C, Giardina CP, King JS, Ledford J, McCarthy HR, Moore DJP, Ceulemans R, De Angelis P, Finzi AC, Karnosky DF, Kubiske ME, Lukac M, Pregitzer KS, Scarascia-Mugnozza G, Schlesinger WH, Oren R (2005) Forest response to elevated CO_2 is conserved across a broad range of productivity. Proc Natl Acad Sci USA 102:18052–18056

Norby RJ, Rustad LE, Dukes JS, Ojima DS, Parton WJ, Del Grosso SJ, McMurtie RE, Pepper DA (2007) Ecosystem responses to warming and interacting global change factors. In: Canadell JG, Pataki DE, Pitelka LF (eds) Terrestrial ecosystems in a changing world. Springer, Berlin, pp 23–36

Nowak SR, Ellsworth DS, Smith SD (2004) Functional responses of plants to elevated atmospheric CO_2 - do photosynthetic and productivity data from FACE experiments support early predictions? New Phytol 162:253–280

Oostra S, Majdi H, Olsson M (2006) Impact of tree species on soil carbon stocks and soil acidity in southern Sweden. Scan J For Res 21:364–371

Page SE, Siegert F, Rieley JO, Boehm H-D V, Jaya A, Limin S (2002) The amount of carbon released from peat and forest fires in Indonesia during 1997. Nature 420:61–65

Petit RJ, Sheng Hu F, Dick CW (2008) Forests of the past: a window to future changes. Science 320:1450–1452

Phillips OL, Aragão LEOC, Lewis SL, Fisher JB, Lloyd J, López-González G, Malhi Y, Monteagudo A, Peacock J, Quesada CA, van der Heijden G, Almeida S, Amaral I, Arroyo L, Aymard G, Baker TR, Bánki O, Blanc L, Bonal D, Brando P, Chave J, de Oliveira ÁCA, Cardozo ND, Czimczik CI, Feldpausch TR, Freitas MA, Gloor E, Higuchi N, Jiménez E, Lloyd G, Meir P, Mendoza C, Morel A, Neill DA, Nepstad D, Patiño S, Peñuela MC, Prieto A, Ramírez F, Schwarz M, Silva J, Silveira M, Thomas AS, ter Steege H, Stropp J, Vásquez R, Zelazowski P, Dávila EA, Andelman S, Andrade A, Chao K-J, Erwin T, Di Fiore A, Honorio E, Keeling H, Killeen TJ, Laurance WF, Cruz AP, Pitman NCA, Vargas PN, Ramírez-Angulo H, Rudas A, Salamão R, Silva N, Terborgh J, Torres-Lezama A (2009) Drought sensitivity of the Amazon rainforest. Science 323:1344–1347

Piao S, Ciais P, Friedlingstein P, Peylin P, Reichstein M, Luyssaert S, Margolis H, Fang J, Barr A, Chen A, Grelle A, Hollinger DY, Laurila T, Lindroth A, Richardson AD, Vesala T (2008) Net carbon dioxide losses of northern ecosystems in response to autumn warming. Nature 451:49–53

Piao S, Friedlingstein P, Ciais P, Viovy N, Demarty J (2007) Growing season extension and its impact on terrestrial carbon cycle in the Northern Hemisphere over the past 2 decades. Global Biogeochem Cy 21, GB3018, doi:10.1029/2006GB002888

Potter CS, Davidson EA, Verchot LV (1996) Estimation of global biogeochemical controls and seasonality on soil methane consumption. Chemosphere 32:2219–2246

Pregitzer KS, Euskirchen ES (2004) Carbon cycling and storage in world forests: biome patterns related to forest age. Glob Change Biol 10:2052–2077

Preston CM, Schmidt MWI (2006) Black(pyrogenic) carbon: a synthesis of current knowledge and uncertainties with special consideration of boreal regions. Biogeosciences 3:397–420

Raich JW, Russell AE, Kitayama K, Parton WJ, Vitousek PM (2006) Temperature influences carbon accumulation in moist tropical forests. Ecology 87:76–87

Randerson JT, Chapin FSIII, Harden JW, Neff JC, Harmon ME (2002) Net ecosystem production: a comprehensive measure of net carbon accumulation by ecosystems. Ecol Appl 12:937–947

Randerson JT, Liu H, Flanner MG, Chambers SD, Jin Y, Hess PG, Pfister G, Mack MC, Treseder KK, Welp LR, Chapin FS, Harden JW, Goulden ML, Lyons E, Neff JC, Schuur EAG, Zender CS (2006) The impact of boreal forest fire on climate warming. Science 314:1130–1132

Reich PB, Bolstad P (2001) Productivity of evergreen and deciduous temperate forests. In: Roy J, Saugier B, Mooney HA (eds) Terrestrial global productivity. Academic, San Diego, CA, pp 245–283

Reich P, Eswaran H (2006) Global resources. In: Lal R (ed) Encyclopedia of soil science. Taylor & Francis, Boca Raton, FL, pp 765–768

Reichstein M (2007) Impacts of climate change on forest soil carbon: principles, factors, models, uncertainties. In: Freer-Smith PH, Broadmeadow MSJ, Lynch JM (eds) Forestry and climate change. CAB International, Wallingford, U.K., pp 127–135

Repo ME, Susiluoto S, Lind SE, Jokinen S, Elsakov V, Biasi C, Virtanen T, Martikainen PJ (2009) Large N_2O emissions from cryoturbated peat soil in tundra. Nat Geosci 2:189–192

Robinson D (2007) Implications of a large global root biomass for carbon sink estimates and for soil carbon dynamics. Proc R Soc B 274:2753–2759

Rosenzweig C, Karoly D, Vicarelli M, Neofotis P, Wu Q, Casassa G, Menzel A, Root TL, Estrella N, Seguin B, Tryjanowski P, Liu C, Rawlins S, Imeson A (2008) Attributing physical and biological impacts to anthropogenic climate change. Nature 453:353–358

Ruckstuhl KE, Johnson EA, Miyanishi K (2008) Introduction. The boreal forest and global change. Philos Trans R Soc B 363, doi:10.1098/rstb.2007.2196

Rustad LE, Campbell JL, Marion GM, Norby RJ, Mitchell MJ, Hartley AE, Cornelissen JHC, Gurevitch J, GCTE-NEWS (2001) A meta-analysis of the response of soil respiration, net nitrogen mineralization, and aboveground plant growth to experimental ecosystem warming. Oecologia 126:543–562

Ryan MG, Binkley D, Fownes JH, Giardina CP, Senock RS (2004) An experimental test of the causes of forest growth decline with stand age. Ecol Monogr 74:393–414

Saleska SR, Didan K, Huete AR, da Rocha HR (2007) Amazon forests green-up during 2005 drought. Science 318:612

Santer BD, Thorne PW, Haimberger L, Taylor KE, Wigley TML, Lanzante JR, Solomon F, Free M, Gleckler PJ, Jones PD, Karl TR, Klein SA, Mears C, Nychka D, Schmidt GA, Sherwood SC, Wentz FJ (2008) Consistency of modeled and observed temperature trends in the tropical troposphere. Int J Climatol 28:1703–1722

Saugier B, Roy J, Mooney HA (2001) Estimations of global terrestrial productivity: converging toward a single number? In: Roy J, Saugier B, Mooney HA (eds) Terrestrial global productivity. Academic, San Diego, CA, pp 543–557

Schenk HJ, Jackson RB (2005) Mapping the global distribution of deep roots in relation to climate and soil characteristics. Geoderma 126:129–140

Schimel D (2007) Carbon cycle conundrums. Proc Natl Acad Sci USA 104:18353–18354

Schlesinger WH (1990) Evidence from chronosequence studies for a low carbon-storage potential of soils. Nature 348:232–234

Schlesinger WH (1997) Biogeochemistry – an analysis of global change. Academic, San Diego, CA

Schulze E-D, Lloyd J, Kelliher FM, Wirth C, Rebmann C, Lühker B, Mund M, Knohl A, Milyukova IM, Schulze W, Ziegler W, Varlagin AB, Sogachev AF, Valentini R, Dore S, Grigoriev S, Kolle O, Panfyorov MI, Tchebakova N, Vygodskaya NN (1999) Productivity of forests in the Eurosiberian boreal region and their potential to act as a carbon sink – a synthesis. Glob Change Biol 5:703–722

Schuur EAG (2003) Productivity and global climate revisited: the sensitivity of tropical forest growth to precipitation. Ecology 84:1165–1170

Schuur EAG, Bockheim J, Canadell JG, Euskirchen E, Field CB, Goryachkin SV, Hagemann S, Kuhry P, Lafleur PM, Lee H, Mazhitova G, Nelson FE, Rinke A, Romanovsky VE, Shiklomanov N, Tarnocai C, Venevsky S, Vogel JG, Zimov SA (2008) Vulnerability of permafrost carbon to climate change: implications for the global carbon cycle. BioScience 58:701–714

Seidel DJ, Fu Q, Randel WJ, Reichler TJ (2008) Widening of the tropical belt in a changing climate. Nat Geosci 1:21–24

Seppälä R, Buck A, Katila P (eds) (2009) Adaptation of forests and people to climate change. A global assessment report. International Union of Forest Research Organizations (IUFRO), Helsinki, Finnland

Shorohova E, Kuuluvainen T, Kangur A, Jõgiste K (2009) Natural stand structures, disturbance regimes and successional dynamics in the Eurasian boreal forests: a review with special reference to Russian studies. Ann For Sci 66:200–220

Sierra CA, Harmon ME, Moreno FH, Orrego SA, Del Valle JI (2007) Spatial and temporal variability of net ecosystem production in a tropical forest: testing the hypothesis of a significant carbon sink. Glob Change Biol 13:838–853

Sitch S, Cox PM, Collins WJ, Huntingford C (2007) Indirect radiative forcing of climate change through ozone effects on the land-carbon sink. Nature 448:791–795

Smith NV, Saatchi S, Randerson JT (2004) Trends in high northern latitude soil freeze and thaw cy from 1998 to 2002. J Geophys Res 109:D12101. doi:10.1029/2003JD004472

Smith P, Fang C, Dawson JJC, Moncrieff JB (2008) Impact of global warming on soil organic carbon. Adv Agron 97:1–43

Soares-Filho BS, Nepstad DC, Curran LM, Cerqueira GC, Garcia RA, Ramos CA, Voll E, McDonald A, Lefebvre P, Schlesinger P (2006) Modelling conservation in the Amazon basin. Nature 440:520–523

Soja AJ, Tchebakova NM, French NHF, Flannigan MD, Shugart HH, Stocks BJ, Sukhinin AI, Parfenova EI, Chapin FSIII, Stackhouse PW Jr (2007) Climate-induced boreal forest change: predictions versus current observations. Glob Planet Change 56:274–296

Solomon AM, Freer-Smith PH (2007) Forest responses to global change in North America: interacting forces define a research agenda. In: Freer-Smith PH, Broadmeadow MSJ, Lynch JM (eds) Forestry and climate change. CAB International, Wallingford, U.K., pp 151–159

Steinbrecher R, Hauff K, Rabong R, Steinbrecher J (1997) Isoprenoid emission of oak species typical for the Mediterranean area: source strength and controlling variables. Atmos Environ 31:79–88

Stephens BB, Gurney KR, Tans PP, Sweeney C, Peters W, Bruhwiler L, Ciais P, Ramonet M, Bousquet P, Nakazawa T, Aoki S, Machida T, Inoue G, Vinnichenko N, Lloyd J, Jordan A, Heimann M, Shibistova O, Langenfelds RL, Steele LP, Francey RJ, Denning AS (2007) Weak northern and strong tropical land carbon uptake from vertical profiles of atmospheric CO_2. Science 316:1732–1735

Stephenson NL, van Mantgem PJ (2005) Forest turnover rates follow global and regional patterns of productivity. Ecol Lett 8:524–531

Stiling P, Cornelissen T (2007) How does elevated carbon dioxide (CO_2) affect plant-herbivory interactions? A field experiment and meta-analysis of CO_2-mediated changes on plant chemistry and herbivore performance. Glob Change Biol 13:1823–1842

Stocks BJ (2004) Forest fires in the boreal zone: climate change and carbon implications. Int For Fire News 31:122–131

Stokstad E (2009) Debate continues over rainforest fate – with a climate twist. Science 323:448

Taggart RE, Cross AT (2009) Global greenhouse to icehouse and back again: the origin and future of the boreal forest biome. Glob Planet Change 65:115–121

Trenberth KE, Jones PD, Ambenje P, Bojariu R, Easterling D, Klein Tank A, Parker D, Rahimzadeh F, Renwick JA, Rusticucci M, Soden B, Zhai P (2007) Observations: Surface and Atmospheric Climate Change. In: Solomon S, Qin D, Manning M, Chen Z, Marquis M, Averyt KB, Tignor M, Miller HL (eds) Climate change 2007: The physical science basis. Contribution of working group I to the fourth assessment report of the intergovernmental panel on climate change. Cambridge University Press, Cambridge, United Kingdom/New York, NY, pp 235–336

Trumbore S (2006) Carbon respired by terrestrial ecosystems – recent progress and challenges. Glob Change Biol 12:141–153

Tunved P, Hansson H-C, Kerminen V-M, Ström J, Dal Maso M, Lihavainen H, Viisanen Y, Aalto PP, Komppula M, Kulmala M (2006) High natural aerosol loading over boreal forests. Science 312:261–263

Ulanova NG (2000) The effects of windthrow on forests at different spatial scales: a review. For Ecol Manag 135:155–167

van der Werf GR, Dempewolf J, Trigg SN, Randerson JT, Kasibhatla PS, Giglio L, Murdiyarso D, Peters W, Morton DC, Collatz GJ, Dolman AJ, DeFries RS (2008) Climate regulation of fire emissions and deforestation in equatorial Asia. Proc Natl Acad Sci USA 105:20350–20355

Vargas R, Allen MF, Allen EB (2008) Biomass and carbon accumulation in a fire chronosequence of a seasonally dry tropical forest. Glob Change Biol 14:109–124

Vesterdahl L, Schmidt IK, Callesen I, Nilsson LO, Gundersen P (2008) Carbon and nitrogen in forest floor and mineral soil under six common European tree species. For Ecol Manag 255:35–48

Veteli TO, Mattson WJ, Niemelä P, Julkunen-Tiitto R, Kellomäki S, Kuokkanen K, Lavola A (2007) Do elevated temperature and CO_2 generally have counteracting effects on phenolic phytochemistry of boreal trees? J Chem Ecol 33:287–296

Wang K, Dickinson RE, Liang S (2009) Clear sky visibility has decreased over land globally from 1973 to 2007. Science 323:1468–1470

Wardle DA, Nilsson M-C, Zackrisson O (2008) Fire-derived charcoal causes loss of forest humus. Science 320:629

Waring RW, Running SW (2007) Forest ecosystems – analysis at multiple scales. Elsevier Academic, Burlington, MA

Way DA, Sage RF (2008) Elevated growth temperatures reduce the carbon gain of black spruce. Glob Change Biol 14:624–636

Weedon JT, Cornwell WK, Cornelissen JHC, Zanne AE, Wirth C, Coomes DA (2009) Global meta-analysis of wood decomposition rates: a role for trait variation among tree species? Ecol Lett 12:45–56

Werner C, Butterbach-Bahl K, Haas E, Hickler T, Kiese R (2007) A global inventory of N2O emissions from tropical rainforest soils using a detailed biogeochemical model. Global Biogeochem Cy 21, GB3010, doi:10.1029/2006GB002909

Wiedinmyer C, Neff JC (2007) Estimates of CO_2 from fires in the United States: implications for carbon management. Carbon Bal Manag 2:10

Wild M (2009) Global dimming and brightening: a review. J Geophys Res 114, D00D16, doi:10.1029/2008JD011470

Wittig VE, Ainsworth EA, Naidu SL, Karnosky DF, Long SP (2009) Quantifying the impact of current and future tropospheric ozone on tree biomass, growth, physiology and biochemistry: a quantitative meta-analysis. Glob Change Biol 15:396–424

Woodward FI, Kelly CK (2008) Responses of global plant diversity capacity to changes in carbon dioxide concentration and climate. Ecol Lett 11:1–9

Woodward FI, Lomas MR, Kelly CK (2004) Global climate and the distribution of plant biomes. Philos Trans R Soc Lond B 359:1465–1476

Wright SJ (2005) Tropical forests in a changing environment. Trends Ecol Evol 20:553–560

Würth MKR, Peláez-Riedl S, Wright SJ, Körner C (2005) Non-structural carbohydrate pools in a tropical forest. Oecologia 143:11–24

Zak DR, Holmes WE, Pregitzer KS, King JS, Ellsworth DS, Kubiske ME (2007) Belowground competition and the response of developing forest communities to atmospheric CO_2 and O_3. Glob Change Biol 13:2230–2238

Zhang K, Kimball JS, Hogg EH, Zhao M, Oechel WC, Cassano JJ, Running SW (2008) Satellite-based model detection of recent climate-driven changes in northern high-latitude vegetation productivity. J Geophys Res 113:G03033. doi:10.1029/2007JG000621

Zimov SA, Schuur EAG, Chapin FSIII (2006) Permafrost and the global carbon budget. Science 312:1612–1613

Chapter 5
Nutrient and Water Limitations on Carbon Sequestration in Forests

In addition to CO_2, light, warmth, water, nutrients and growing media are also required for plant growth and net primary production (NPP). Nutrient supply drives C allocation and biomass partitioning into leaves, stems, roots, and in storage and reproductive organs (Ericsson et al. 1996). Thus, C sequestration in forest ecosystems depends on nutrient inputs and their availability (Hessen et al. 2004). Lack of adequate nutrient supply constrains, in particular, the productivity of boreal and temperate forests but nutrient constraints on tropical forest productivity are less well studied (Chapin et al. 2002; Clark 2007). Because the C:nutrient stoichiometry of vegetation and soils differ greatly, C sequestration depends on the distribution of C and nutrients between vegetation and soil (Hessen et al. 2004). Interactions of nutrients affect organic matter (OM) production and decomposition (Melillo et al. 2003). Tree species share the same basic nutrient requirements (Ericsson 1994). However, the quantity of nutrients taken up and returned annually to the forest floor is lower for evergreen compared to other tree species. Important macronutrients are nitrogen (N), phosphorus (P), sulfur (S), potassium (K), calcium (Ca) and magnesium (Mg). Examples of important micronutrients are boron (B), iron (Fe), copper (Cu), zinc (Zn), manganese (Mn) and molybdenum (Mo). In most cases, N is the major limiting nutrient explaining production increases in fertilization experiments (Binkley et al. 1997). In some forests, however, low availability of P, K, Mg, B, Cu, Mn or Zn also limits NPP.

The cycling of nutrients through forest ecosystems is closely linked with cycling of C (Waring and Running 2007). To meet growth requirements, nutrients must be available in appropriate forms and sufficient amounts, especially at critical growth stages. Internal cycling of nutrients supplies a large fraction of nutrients required for growth of new tissues (Ericsson 1994). N and P withdrawal from senescent leaves to woody tissues in stems and roots, and from older needle age-classes to younger temporarily store nutrients. The major fraction of the aboveground nutrient pool is, however, confined to the tree parts most frequently harvested (i.e., bole or total aboveground woody biomass).

A decrease in tissue nutrient concentration may alter C allocation patterns to leaves, stems and roots (Waring and Running 2007). Photosynthetically fixed C supplied from roots is the energy source for the activity of heterotrophic soil organisms. Heterotrophic activity releases nutrients which are readily adsorbed by plants whereas the heterotrophic soil organisms contribute to the soil organic carbon (SOC) pool. Photosynthates are also used to recycle nutrients within plants from older to newer

K. Lorenz and R. Lal, *Carbon Sequestration in Forest Ecosystems*,
DOI 10.1007/978-90-481-3266-9_5, © Springer Science+Business Media B.V. 2010

tissues (Waring and Running 2007). Nutrient and, in particular, N or P deficiency enhances biomass partitioning to roots whereas increased nutrient supply induces decreased root:shoot ratio (Ericsson 1994). Long-term effects of CO_2 on forest growth are probably intimately linked with the availability of N (Oren et al. 2001; Jarvis and Linder 2007). Furthermore, the N–P interactions may have a strong influence on forest ecosystem functioning (Wardle et al. 2004; Wang et al. 2007; Houlton et al. 2008). Data regarding nutrient effects on C cycling by elements other than N are, however, too limited to support broad based generalizations (Hessen et al. 2004). Thus, a better understanding is needed about how nutrient cycles interact with C cycling, especially in tropical forest ecosystems. Also, more knowledge is needed how human perturbations of Earth's natural biogeochemical cycles and of the cycles of metals affect the C cycle (Falkowski et al. 2000; Rauch and Pacyna 2009).

The C cycle in forests is closely linked with the water cycle (Waring and Running 2007). Water is essential for metabolism and growth of plants (Ericsson et al. 1996). Photosynthetic capture of CO_2, C partitioning within plants (i.e., root:shoot allometry), C losses by decomposition, and dissolved and particulate C losses from forest ecosystems depend on water availability and fluxes. Thus, water influences gross primary production (GPP), NPP and the net ecosystem carbon balance (NECB), and C sequestration in forest ecosystems.

In the following section the better studied effects of N and P on C sequestration in forest ecosystems are discussed. Readers are referred to other reports for discussions about the effects of the availability of other mineral nutrients on C sequestration (e.g., Hüttl and Schaaf (1997) for Mg). Some major effects of water on C sequestration in forests are also discussed in this Chapter.

5.1 Nitrogen

N is a major constituent of amino acids, enzymes, proteins, nucleic acids and chlorophyll (Waring and Running 2007; Davidson 2008). Most of the N in the plant cells in leaves, stems and roots is associated with proteins (Reich et al. 2008). N controls the primary production in the biosphere and is the principal limiting nutrient in terrestrial ecosystems (Elser et al. 2007; Gruber and Galloway 2008). Thus, global perturbation of the terrestrial landscape may be the reason for the widespread occurrence of N limitation to achieve high NPP in terrestrial ecosystems (Hungate et al. 2003).

5.1.1 Nitrogen Dynamics in Forest Ecosystems

Biological N fixation (BNF), involving the conversion of atmospheric N_2 into assimilable forms of N by bacteria, is the main process by which N enters forest ecosystems (Fig. 5.1). Microbiotic crusts on forest plants and soil are able to fix N from the atmosphere, and are likely to play major roles in the global biogeochemical

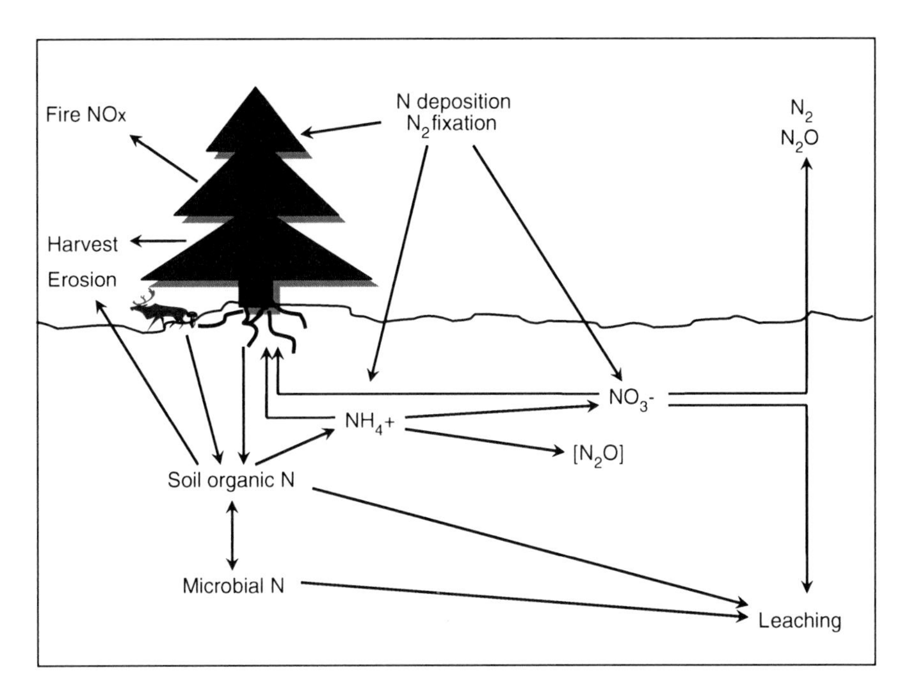

Fig. 5.1 Schematic representation of the major elements of the forest nitrogen cycle (N_2O = nitrous oxide; NO_x = nitrogen oxides; NO_3^- = nitrate; NH_4^+ = ammonium; major sources and losses underlined) (modified from Robertson and Groffman 2007)

cycle of N (Elbert et al. 2009). In addition, the atmospheric deposition may also contribute to N inputs (Table 5.1; Chapin et al. 2002; Davidson 2008; Sparks 2009). Other inputs may originate from N_2O as forest soils can be temporal N_2O sinks (Goldberg and Gebauer 2009). The largest inputs of organic N to soil are amino acids (Jones et al. 2009). The reactions in the microbial N cycle include N_2 fixation, aerobic ammonium (NH_4^+) oxidation, aerobic nitrite (NO_2^-) oxidation, denitrification, anaerobic ammonium oxidation, and dissimilatory nitrate (NO_3^-) and nitrite reduction to ammonium (Jetten 2008). Specifically, ammonia (NH_3) is oxidized to NO_3^- during nitrification. Ammonia oxidation, the first step in this process is carried out by bacteria and by soil Crenarchaeota, microorganisms that belong to the domain Archaea (Leininger et al. 2006). Whether NO_3^- is then incorporated into SOM by abiotic processes is discussed controversially (Colman et al. 2008).

Losses of N from forests occur in gaseous forms as a result of denitrification or conversion of NO_3^- to N_2, and during forest fires (Austin et al. 2003; Kimmins 2004). Some N_2O, an intermediate by-product of denitrification which is also produced by other groups of microorganisms, may escape from the soil (Chapuis-Lardy et al. 2007). The process of denitrification is more widespread and of greater significance than realized just a few years ago (Schlesinger 2009). Aside gaseous, other processes governing N losses occur with water fluxes as NH_4^+, NO_3^- and dissolved organic N (DON). Fluxes of DON in throughfall, stemflow,

Table 5.1 Ecosystem balance for N, P and water, and effects of their increased availability on C sequestration in forest biomass and soil

	Nitrogen	Phosphorus	Water
Major source	Atmospheric N_2	Soil phosphates	Atmospheric moisture
Major losses	Gaseous and dissolved N forms Harvested biomass	Particulate P flux in runoff and erosion Harvested biomass	Evapotranspiration
Biomass C	+ Boreal and temperate forests ? Tropical forests	+ Tropical forests in particular	+
Soil organic C	+ Boreal and temperate forests ? Tropical forests	+	+ In particular accumulation in wetland forests

forest floor leachates and soil solutions in temperate forests are, for example, highly correlated with those of dissolved organic carbon (DOC) (Michalzik et al. 2001). Forest disturbances may, thus, increase N losses in conjunction with fluxes of water (Chapin et al. 2002). The removal of forest biomass during harvest also contributes to the loss of N (Kimmins 2004). Silvicultural activities such as prescribed burning, herbicide application, soil plowing or ripping may also contribute to N losses (Attiwill 1981).

Most N acquired by plants becomes available through the activity of microbial decomposers on OM (Schimel and Bennett 2004; van der Heijden et al. 2008). Similar to the BNF, however, decomposition depends on C supply from plant photosynthesis (Aber et al. 1998). Thus, plant versus microbial competition determines the residence time and ability of N deposition to stimulate C uptake and storage (Holland and Carroll 2003). Internal cycling of N is an important process for the N-economy of trees (Ericsson 1994). Thus, the N pool in forests is biologically renewable (Wardle et al. 2004).

The initial N concentration of leaf litter is the dominant driver of net N release and immobilization during long-term litter decomposition (Parton et al. 2007). Climate, other litter chemical properties, edaphic conditions, or soil microbial communities are less important drivers. Roots release N linearly with mass loss during decomposition and exhibit little N immobilization. However, plants do not necessarily rely on microbes and soil fauna for the breakdown of OM as they can also use protein as N source (Paungfoo-Lonhienne et al. 2008). Specifically, amino acids are released in forest soils through hydrolysis of proteins and the amino acids are taken up by trees (Gallet-Budynek et al. 2009). Thus, organic N can be assimilated directly by tree roots and indirectly via ECM or AM fungi or other microbial symbionts (Whiteside et al. 2009). However, the quantitative importance of the uptake of organic N by plants is a matter of discussion (Näsholm et al. 2009). A clear test of direct root uptake of amino acids would be compound specific ^{13}C or ^{15}N analysis of the added amino acid as recovered from roots and/or shoots if ^{13}C or ^{15}N ratios of added and recovered amino acids are equal (Rasmussen and Kuzyakov 2009). Most of the studies on the possibility of direct uptake of organic N sources (mainly amino acids) by plants use labeled amino acids but do not test this prerequisite.

The interactions of the N cycle with the C cycle and the climate system are poorly understood (Falkowski et al. 2000; Reich et al. 2006; Gruber and Galloway 2008). Forests retain greater proportions of N deposition compared to other biomes (Holland and Carroll 2003). The effects of N deposition on forests are, however, much more complex than simple stimulation of C accumulation by N fertilization (Austin et al. 2003). In particular, N plays a crucial role in controlling key aspects of the C cycle. The effects of N inputs on C sequestration in forests are relatively large if primary recipients are trees with woody tissues, high C:N mass ratios (>200) and long turnover times (Nadelhoffer et al. 1999). The resultant effects of N inputs on C sequestration are, however, lower if N is sequestered mainly in soils (C:N between 10 and 30), and lost as NO_3^- with percolating water or as NO_x to the atmosphere (Holland and Carroll 2003). The microbial biomass has C:N ratios between 4 and 15, and microbial N residence time is less than a year. In comparison, the wood-N residence times range between decades and centuries (Holland and Carroll 2003). In the soil, the partitioning of amino-acid C into catabolic and anabolic processes in microbial communities is conservative and independent of global position (Jones et al. 2009). Thus, there may be commonalities within C and N cycling in the terrestrial biosphere.

5.1.2 Nitrogen Impacts on Biomass Carbon Sequestration

Nitrogen is generally the most limiting nutrient for tree growth and C sequestration (Oliver and Larson 1996; Oren et al. 2001). The most characteristic symptom of N deficiency is general chlorosis of leaves and young shoots (Perry 1994). New growth is retarded and slow, and trees have an unthrifty, spindly appearance. The more mature tree components are first affected by N deficiency as N is translocated from older to actively growing regions. A strong respiration-N scaling relationship for roots, stems and leaves in woody angiosperms and woody gymnosperms emphasize the importance of N for the C balance in trees (Reich et al. 2008). N availability affects C sequestration in trees through its effects on photosynthesis, foliar biomass, canopy characteristics and woody tissue growth. N availability, in particular, impacts the genes that regulate tree growth and wood chemistry (Novaes et al. 2009).

Conifers have a higher N-use efficiency (NUE) as they produce more dry matter per unit amount of N in biomass compared to deciduous broadleaved trees (Ericsson 1994). In general, high photosynthetic capacity in leaves is linked to high leaf N concentration (Wright et al. 2004). For example, the light-use efficiency of canopy in boreal and temperate ecosystems is controlled by leaf N concentrations (Kergoat et al. 2008). Thus, in N-limited forests, N deposition likely increases foliar N concentrations with a positive effect on photosynthetic rates, the net ecosystem carbon balance (NECB) and C sequestration rate (Aber et al. 1998; Hyvönen et al. 2007). Also, the amount of N sequestered in tree biomass affects forest productivity and C sequestration in trees (Waring and Running 2007). For example, aboveground C accumulation in European forests is 15–40 kg C kg^{-1} deposited N (de Vries et al. 2009). On the other hand, temperate forest trees are not the primary sink for increased N deposition (Nadelhoffer et al. 1999). Less well studied are, however, boreal and tropical forests. For example, published literature shows no empirical relationship between N deposition and annual CO_2 net flux in N-limited forests. Thus, forest growth responses to additional N are not always predictable and all stand factors need to be included to account for effects of N deposition (Hyvönen et al. 2007; Davis et al. 2008). For example, the stand C sequestration or annual NEP increase in a N-fertilized Douglas fir stand growing on N-deficient soil remote from N pollution sources (Jassal et al. 2008). Although N_2O losses occur in the fertilized stand, the net change in GHG global warming potential is generally negative, indicating favorable effect of N fertilization.

At the leaf scale, N may affect NPP by its impacts on chlorophyll synthesis, level and activity of photosynthetic enzymes, and stomatal conductance to CO_2 exchange (Pallardy 2008). Thus, N is an important regulator of C assimilation through the relationship between leaf-level photosynthetic capacity and foliar N concentrations (Wright et al. 2004). This is why the photosynthesis rate often increases after application of N fertilizers. However, environmental conditions such as light intensity, temperature and water availability modify the effect of N on photosynthesis (Pallardy 2008). In comparison, N effects on the activity of the photosynthesizing enzyme RUBISCO, an N-rich enzyme that plays the key role in C fixation, are very well documented (Chapter 2; Vrede et al. 2004; Millard et al. 2007). For example, it is very well known that increase in leaf N supply increases the pool of the inactive RUBISCO protein but causes a decline in the pool of RUBISCO in the activation stage. In addition, N is also temporarily stored in RUBISCO, and used for subsequent leaf growth (Millard et al. 2007). Under elevated CO_2, RUBISCO may be down regulated by a general decrease in leaf N (Reich et al. 2006). This process of enhancing NUE may cause negligible changes in leaf C assimilation rates (Millard et al. 2007). Whether N availability limits the response to elevated CO_2 ('CO_2 fertilization effect') is, however, a debatable issue (Long et al. 2004). In particular, whether increase in N deposition at elevated CO_2 levels results in increase in foliar N concentration with a positive effect on photosynthetic rates and C sequestration needs to be studied (Reich et al. 2006).

A likely response to increase in N deposition in forests with a low N supply is increase in production of foliar biomass (Hyvönen et al. 2007). In some forests, N fertilization increases C allocation to foliage (Litton et al. 2007). This may lead to greater photosynthesis per unit area of forest (Högberg 2007). Also, canopy characteristics may change although forest canopies have considerable acclimation capability. Thus, only transient increases in foliar biomass production may occur as the distribution of N in the canopy is adjusted to make effective use of both light and N in photosynthesis (Kull 2002).

Increased N deposition in boreal and temperate forests may result in increased production of wood, including coarse roots (Hyvönen et al. 2007). Increase in soil N availability, for example, results in increased biomass accumulation in branches and wood but less increases in leaf and root biomass (Xia and Wan 2008). Tree species with fast root turnover are probably more prone to progressive N limitation to C sequestration in woody biomass than species with slow root turnover such as evergreens (Franklin et al. 2009). Among tree species, broadleaved tree biomass responds more positively to N addition than the biomass of coniferous trees whereas no differences are observed between evergreen and deciduous trees (Xia and Wan 2008). Broadleaved trees also have a greater N response in aboveground than in the belowground biomass (Pregitzer et al. 2008). Thus, broadleaved trees have a higher response to N input than coniferous trees.

The net effects of N deposition on NPP depend also on the form of N deposited and its interaction with other pollutants, the role that N oxides play in generating ozone, and the cation loss from soil by leaching of NO_3^--N (Austin et al. 2003). The loss of the basic cations Ca^{2+}, Mg^{2+}, K^+, and Na^+ in combination with toxicity of Al^{3+}, Fe^{3+}, and Mn^{3+} may contribute to a decrease in plant biomass with increase in N inputs (Aber et al. 1998; Bowman et al. 2008). The net effect of N deposition on NPP, however, appears to be positive in Europe but often negative in North America, a discrepancy that requires additional research.

Up to 30% of the applied N in forest fertilization trials is retained in biomass (Nadelhoffer et al. 1999). Fertilization with N usually increases C partitioning into woody biomass in a range of forests (Litton et al. 2007). Sequestration of N in woody biomass has large effects on C sequestration as wood C:N ratios are high and turnover times are long (Nadelhoffer et al. 1999). However, only relatively little N added into forests becomes immobilized in woody biomass. In contrast to boreal and temperate forests, however, N supply does not limit plant production in the majority of tropical forests (Matson et al. 1999). Thus, increase in N deposition may have little direct effect on C sequestration in biomass in most tropical forests. Yet, N limitation has important impact on the aboveground NPP in montane and secondary lowland tropical forests (LeBauer and Treseder 2008). Indirectly, higher N inputs may lead to a lower productivity and reduced C storage in moist tropical forests because of the losses of basic cations, decrease in P availability and inhibition of plant growth due to increases in concentration of soluble Al (Matson et al. 1999). On the other hand, in boreal and temperate forests losses of basic cations such as Mg^{2+}, Ca^{2+} and K^+ may occur from plant canopies and soils through increase in NO_3^- leaching (Reichstein 2007). These nutrient losses can reduce NPP and partially offset gains resulting from increase in N availability.

The effects of increase in N supply on C sequestration in trees by direct absorption of N deposition and gaseous N forms in the canopy are not very well known (Sparks 2009). This direct pathway of N uptake short-circuits plant microbial competition for N uptake and increases the likelihood of N incorporation into the plant biomass (Holland and Carroll 2003). Specifically, NH_3 derived from foliar uptake interacts directly with photorespiration processes and, thus, the loss of assimilated CO_2 (Section 2.1.1; Sparks 2009). In many trees, especially conifers, foliar uptake of atmospheric N may be the only or most significant source of NO_3^-/NO_2^- to the leaves. Canopy uptake of N is variably reported as 0–50% of plant N demand (Harrison et al. 2000). However, large-scale experiments involving direct application of N to the canopy are scanty (Sparks 2009). The canopy of a mature coniferous forest, for example, may retain as much as 70% of applied N. However, the fraction assimilated by trees and its effect on growth are not known (Gaige et al. 2007). On the other hand, additional CO_2 uptake may likely result from enhanced NEE that is driven by canopy N uptake in a mixed coniferous forest (Sievering et al. 2007). Additional C gain at elevated CO_2 in otherwise N-limited forests may be driven by increases in available N at the site of carboxylation in the leaf after foliar N uptake (Sparks 2009). Thus, there is some indication that increase in canopy-N uptake may enhance C sequestration in forests.

In summary, the N impacts on C sequestration in forest biomass remain a matter of debate and in need of additional research. In particular, whether N deposition enhances C sequestration in boreal and temperate forests by promoting the production of wood needs further research (Magnani et al. 2007; De Schrijver et al. 2008; de Vries et al. 2008). Modeling estimates of enhancement of forest growth due to atmospheric N deposition vary widely between 0.1 and 2 Pg C $year^{-1}$ (Sparks 2009). N deposition probably enhances C uptake only in growing forests but this effect may decrease as forests reach maturity (Churkina et al. 2007). Aboveground growth is generally a linear function of the total N concentration in the foliage (Waring and Running 2007). Increase in N deposition probably stimulates aboveground NPP in temperate, montane and secondary lowland tropical forests (LeBauer and Treseder 2008). However, the NPP stimulation and tree growth may decrease when critical thresholds in the N saturation process are exceeded (Aber et al. 1998; Galloway et al. 2003).

The interactions between N and high CO_2 levels need additional research since only three long-term treatments have been conducted which simultaneously manipulated both N and CO_2 levels (Reich et al. 2006; Reichstein 2007; McMurtrie et al. 2008). However, these experiments were restricted to juvenile trees and the response of mature trees and forest stands is not known (Wang 2007). The available experimental data reveal strong CO_2–N interactions (Dewar et al. 2009). Specifically, on short timescales both the stomatal conductance and leaf N concentrations are reduced under elevated CO_2. However, the increased plant N uptake at elevated CO_2 has negligible effects on the fraction of photosynthates partitioned to foliage production at intermediate timescales but the fraction partitioned to wood production increases whereas that to belowground pools decreases (Litton et al. 2007). On longer timescales, an increased N immobilization in plant litter, biomass and soil may cause a decrease soil N availability under elevated CO_2 (Luo et al. 2004). Also, molybdenium, a cofactor in the N-fixing enzyme nitrogenase, may constrain the ability of some forests on highly weathered acidic soils to acquire new N in response to CO_2 fertilization (Barron et al. 2009). A recent meta-analysis has indicated that the biomass of broadleaved trees shows a greater response to N addition under CO_2 enrichment than coniferous tree biomass (Xia and Wan 2008). Biomass enhancement of deciduous trees is often higher than that of evergreen trees. Partitioning of deposited

N among foliar vs. soil pathways of incorporation is the key in understanding the influence of N deposition on forest ecosystem productivity (Sparks 2009).

5.1.3 Nitrogen Impacts on Soil Organic Carbon Sequestration

Increase in N inputs may affect SOC through changes in litter quality and quantity, and through changes in the microbial decomposer community. The known responses of DOC to N addition are, however, inconsistent (Evans et al. 2008). Specifically, DOC increases in European and North American forests are largely consistent with an acidity-change mechanism attributable to decreases in acidifying deposition, primarily sulfur, and less to increases in N deposition. In a range of forests, partitioning of C into the total belowground C flux decreases with N fertilization (Litton et al. 2007). However, N addition usually increases SOC accumulation in forest ecosystems (Johnson and Curtis 2001). As low soil CO_2 emissions are observed in forest areas with high N-deposition rates, it is hypothesized that N cycling drives a within-soil CO_2-sink but the mechanisms remain speculative (Fleischer and Bouse 2008).

Forest soils retain about 30% of the N applied in fertilization trials (Schlesinger 2009). The forest floor and mineral soil, for example, is the primary sink for N inputs in temperate forests (Nadelhoffer et al. 1999). C inputs by litterfall and rhizodepostion generally increases at elevated N and also results in increases in the SOC pool (Jandl et al. 2007). For example, between 5 and 35 kg C are sequestered in European forest soils per kg deposited N (de Vries et al. 2009). The decomposition of high-lignin woody litter, in particular, is inhibited by N additions (Knorr et al. 2005). Such inhibition by N addition may increase SOC pools in surface layers in Northern temperate forests over decadal time scale (Pregitzer et al. 2008). Decomposition of angiosperm wood is faster at higher wood N concentrations (Weedon et al. 2009). Available data on effects of N on litter decomposition rates are, however, inconsistent and contradictory (Hyvönen et al. 2007). For example, N fertilization increases decomposition in pine forests, has no effects on decomposition in jack pine and tropical forests, but decreases decomposition in temperate forests (Austin et al. 2003). Otherwise, increase in nutrient concentration in litter at higher levels of N availability may stimulate SOM decomposition but the fraction of old and stable SOM may increase and enhance forest C sequestration (Jandl et al. 2007). Anthropogenic N application, in particular, may affect terrestrial C storage by increasing NPP more than the decomposition (Vitousek et al. 1997). Thus, the effects of increase in N availability on litter decomposition are highly variable (Austin et al. 2003). Otherwise, N deposition effects on SOM chemistry vary also across forest ecosystems and soil size fractions, and may be linked to long-term changes in enzyme activities (Grandy et al. 2008).

The majority of N added to forests becomes incorporated into SOM (Aber et al. 1998). The dominant processes involved are likely to be mycorrhizal assimilation in fungal tissues and mycorrhizal production of recalcitrant compounds (Treseder et al. 2007). Increase in rhizodeposition at elevated N may also increase plant carbohydrate supply for mycorrhizal growth. However, slightly negative effects of N additions on standing stocks of mycorrhizal fungi in forests indicate that plants allocate carbohydrates elsewhere and mycorrhizal fungi become C-limited (Treseder 2004). Also, in the moist tropics, increase in N deposition can lead to increase in soil acidity and the inhibition of microbial activity due to increases in concentrations of soluble Al (Matson et al. 1999). Effects of N on the microbial community in general are, however, unclear and less studied. For example, significant declines in microbial biomass and, in particular, bacterial biomass are observed by N additions in boreal forests (Treseder 2008). This reduction may be accompanied by decline in soil CO_2 emissions. The sparse knowledge of microbial diversity and ecological dynamics, however, may cause error in estimates of fluxes in the N cycle (Claire Horner-Devine and Martiny 2008). Also, the interactions between microbial activity and C:N ratios in plant and soils and increase in soil fertility are poorly understood (Gruber and Galloway 2008).

Little information is available on the net radiative forcing from N_2O emissions as a result of forest N fertilization. N_2O emissions may occur after N fertilization, thereby altering the GHG global warming potential of a forest stand. Gobally, N_2O emissions are the single most important ozone depleting substance emissions (Ravishankara et al. 2009). Studies on denitrification losses in forest soils following fertilizer N application are, however, scanty. For example, higher N_2O emissions have been observed after N additions to temperate and tropical forest soils (Koehler et al. 2009; Jassal et al. 2008). More data of this type are needed for a wide range of forest ecosystems. More research is also needed on the influence of soil moisture on N_2O dynamics as forest soils can serve as a small but persistent sink for this GHG, given the predicted increase in drought frequency and rainfall variability in many regions (Billings 2008).

N availability in soils and N uptake by trees are major challenges to modeling forest growth and biomass productivity at a range of scales (Landsberg 2003). To project future changes of forest ecosystems based on climate projections, dynamic global vegetation models (DGVMs; Ostle et al. 2009) are coupled to general circulation models (GCMs; Alo and Wang 2008). The predictions about plant responses to changes in atmospheric CO_2, N, and water availability, however, differ among complex vegetation models (Dewar et al. 2009). Thus, optimization models are discussed as alternative approaches to explain forest responses to global change. While GCMs are major tools for ACC assessments, the GCMs used in the latest IPCC report of 2007 did not incorporate vegetation dynamics (Meehl et al. 2007). Components of current climate models, for example, represent the atmospheric general circulation, ocean general circulation, sea-ice dynamics and thermodynamics, and relevant land processes (Donner and Large 2008). The current generation of global climate systems models, however, does not consider linkages between the C and N cycles (Chapin et al. 2009). Emerging earth systems models (ESM) may also contain component models representing vegetation and its dynamic evolution, cycles of C and N in terrestrial ecosystems, ocean biochemistry, atmospheric chemistry, continental ice sheets, and human activities (U.S. DOE 2008).

State-of-the-art DGVMs are currently the best choice to predict the effects of future changes in climate elements on the SOC balance (Reichstein 2007). Aside N deposition, climatic elements (e.g., water balance, atmospheric CO_2 concentration and temperature) control NPP whereas temperature and water balance control decomposition, and vertical and lateral C transport. Decomposition, vertical and lateral C transport, and primary production are coupled through feedback mechanisms, and, thus, determine the SOC pool by affecting the balance between C input to the soil and C loss by decomposition and erosion. However, the fate of C from decomposition of dead wood is inadequately treated by several leading global C models despite the importance of this C pool for predicting C cycle responses to global change (Cornwell et al. 2009). DGVM could be strengthened, in particular, by integrating biodiversity and height-structured competition for light (Purves and Pacala 2008). DGVM do not contain information about wood although one of the key model predictions is the C storage per unit area at each time-step, essentially an estimate of the mass of wood (Chave et al. 2009). Current DGVM do also not adequately address interactions and limitations of NPP by N and other nutrients such as P and, thus, likely overestimate the CO_2 fertilization effect (Reichstein 2007; Wang et al. 2007). In particular, none of eleven coupled climate-carbon cycle models to study the coupling between ACC and the C cycle account for an explicit treatment of the N cycle (Friedlingstein et al. 2006). Thus, less SOC storage is likely at elevated CO_2 than predicted (Reichstein 2007).

5.1.4 Conclusions

Trees and soils appear to be modest sinks for N but large enough to support a significant global sink for C (Schlesinger 2009). The response of trees and soil to increase in N availability may differ among boreal, temperate and tropical forests (Nadelhoffer et al. 1999). Additional N deposition has been responsible for about 10% of the total C sequestration in European forests during the second half of the 20th century (De Vries et al. 2006). Addition of N may increase tree C and SOC pools in temperate and boreal forests provided that other nutrients do not limit tree growth (Hyvönen et al. 2008). In boreal forests, increase in N throughfall is closely related to decrease in BNF as associative N-fixing cyanobacteria that colonize feather mosses can be the main N source (DeLuca et al. 2008).

Global cycles of N have been amplified by ca. 100% through post-industrial anthropogenic activities since about 1750 (Falkowski et al. 2000). However, the magnitude of the impacts of N deposition on the C cycle through N fertilization and saturation is difficult to quantify (Holland and Carroll 2003). For example, model analyses indicate that stand-lifetime NEP of European boreal and temperate forest stands in response to N deposition is much smaller than previously estimated (i.e., 50–75 kg C kg^{-1} deposited N) (Sutton et al. 2008). The upper limit of the response of C sequestration in northern forests is probably 200 kg C kg^{-1} N input, and the lower limit is about 40 kg C kg^{-1} N (Magnani et al. 2007; Reay et al. 2008). Some forest soils show increases in SOC pool between 7 and 23 kg C kg^{-1} N (Reay et al. 2008). The total C sequestration in European forests range between 20 and 40 kg C kg^{-1} deposited N (de Vries et al. 2009). Because of continued perturbation of the global N cycle at a record pace, the question arises about how N availability interacts with the C cycle and what consequences such an interaction may have to C sequestration in forests, in particular, in response to elevated atmospheric CO_2 levels (Reich et al. 2006; Högberg 2007; Jarvis and Linder 2007; Galloway et al. 2008).

Prior to human intervention, the Earth's N cycle was almost entirely controlled by microbes (Falkowski et al. 2008). Specifically, the creation of reactive N (i.e., all N forms except non-reactive N_2) by humans as fertilizers is increasing every year, with drastic transformation of the global N cycle (Galloway et al. 2008). However, the fate of the human-enhanced N inputs to the land surface is little understood (Schlesinger 2009). Furthermore, definite conclusions about the net effects of the attendant increase in N deposition on NPP, decomposition and the NECB are difficult to make (Austin et al. 2003). Changes in CH_4, N_2O and NO fluxes may occur through N deposition, which have strong feedbacks to climate warming. Specifically, N addition may increase CH_4 emission, reduce CH_4 uptake and increase N_2O emission (Liu and Greaver 2009). Thus, multiple element interactions cause N fertilization to be a counterproductive measure for reducing radiative forcing when N_2O fluxes from forest soils are progressively increased (Austin et al. 2003).

The potential interactions of higher N availability with C sequestration in forests are poorly understood (Reich et al. 2006). Increase in N supply results in higher C partitioning into aboveground NPP in foliage and wood biomass (Litton et al. 2007; LeBauer and Treseder 2008). In boreal and temperate forests, both CO_2 uptake capacity and canopy N concentration are strongly and positively correlated with shortwave surface albedo (Ollinger et al. 2008). Thus, higher canopy N may increase gross C assimilation and surface albedo, both representing negative feedbacks to warming. In general, N effects on the belowground NPP of forests are not widely studied. Furthermore, the available data depicting the effects of increase in N availability on the large C pool in forest soils are highly variable (Jandl et al. 2007). The magnitude of available N often determines the magnitude of the CO_2 fertilization effect, but the ability of forest ecosystems to be a net CO_2 sink in the future at elevated levels of both N and CO_2 is unclear (de Graaff et al. 2006; Gruber and Galloway 2008; Heimann and Reichstein 2008). Specifically, the current generation of coupled carbon-climate models do not account for the limitation of N availability (Doney and Schimel 2007). Otherwise, the future persistence of an N-stimulated C sink in forests is highly debatable because of the progressive alleviation of N limitation (Galloway et al. 2008). Temperature and CO_2 increases in temperate forests could lead to reduced canopy albedo and a positive feedback to warming (Ollinger et al. 2008). It is thus possible that any future forest C sequestration may be most affected in Central and South America, Africa and part of Asia as substantial increases in N deposition are projected for these regions whereas air pollutant emission controls over most of Europe and North America may decrease the N-response of forest C sequestration (Reay et al. 2008). For example, a substantial decrease in C sequestration in both coniferous and deciduous forests in large parts of Europe may occur by the gradual decrease in N deposition (Wamelink et al. 2009).

5.2 Phosphorus

Macronutrient P is a constituent of nucleic acids, nucleoproteins and phospholipids (Waring and Running 2007; Pallardy 2008). Phosphate esters and energy-rich phosphates represent the metabolic machinery of the plant cell (Marschner 1995). As most plants exhibit a limited range of element ratios, P limits NPP when it is most strongly limiting plant growth (Chapin et al. 2002). In undisturbed soil or regolith, P is extracted at soil depth by organisms, pumped upward, stored in biota and minerals, and recycled (Brantley 2008). However, P losses to groundwater and occlusion in strongly weathered soils cause ecosystem degradation over millenia (Wardle et al. 2004). Post-industrial human activities profoundly alter the global P cycle, and increase P inputs to the biosphere by ca. 400% (Falkowski et al. 2000). However, whether the anthropogenic amplification of the P cycle enhances NPP and terrestrial C sequestration is not clearly understood.

5.2.1 *Phosphorus Dynamics in Forest Ecosystems*

In contrast to atmospheric inputs which sustain N availability in forests, cycling processes sustain the P availability in forests (Fig. 5.2; Attiwill and Adams 1993).

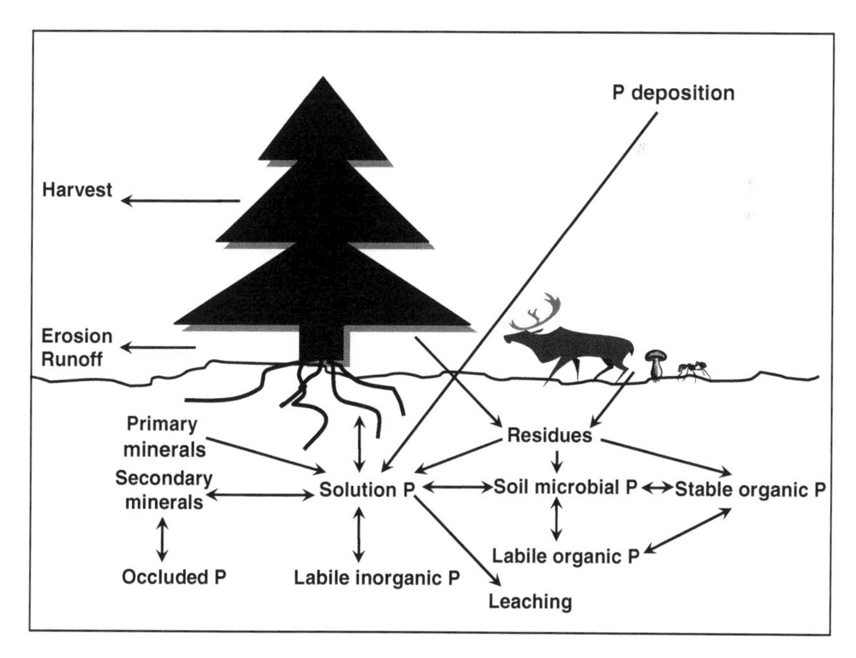

Fig. 5.2 Schematic representation of the major elements of the forest phosphorus cycle (major sources and losses underlined) (modified from Walbridge 1991)

Plant roots, microorganisms, soil mineral and organic components are the P sinks in soil. More than 50% of total P in forests soils, in particular, is in organic forms (Attiwill and Adams 1993). The size of the organic P pool generally decreases in the order inositol phosphate > polymer organic phosphate > nucleic acid P > phospholipid P (Plante 2007). In forest soils up to 20% of P is bound in the microbial biomass. Through geologic time, occluded and organic P forms become dominant in soil profiles (Walker and Syers 1976).

The availability of soil P to plants is determined by the competition between biological and geochemical sinks. Soluble inorganic P forms are released by the activity of phosphatase enzymes and plant roots. Specifically, plants acquire P predominantly through roots and the decomposer activity of mycorrhizal fungi (van der Heijden et al. 2008). Carbohydrates provided by plants support the activity of mycorrhizal fungi on organic forms of soil P (Waring and Running 2007). Phosphates are released enzymatically from organic P forms during OM decomposition and are later absorbed by ECM or AM roots before they interact with soil minerals. Otherwise, insoluble compounds are formed by binding of P to Ca^{2+}, Al^{3+} or Fe^{3+} (Thomas and Packham 2007). The organically bound and Al/Fe oxide bound components of P appear to play a key role in preventing its loss (McGroddy et al. 2008).

P is scavenged and recycled from decomposing litter in a well-developed root mat in tropical forests established on old, highly weathered and infertile soils. In contrast to boreal and temperate forests, N-fixing trees are highly abundant in tropical forests because the synthesis of phosphatase enzymes requires large amounts of N (Houlton et al. 2008). Otherwise, soil N can limit production of phosphatase enzymes but less information exists on the strength of N and P interactions in forest ecosystems (Wang et al. 2007). Aside from SOM, the microbial biomass itself is an important reservoir of potentially available P. Once P is absorbed by plant roots, it is either incorporated into organic molecules or remains in an inorganic form. The turnover of P, however, is less tightly linked to OM decomposition than is the turnover of N (Chapin et al. 2002). Consequently, organic P is lost from forest soils in diester forms following repeated forest disturbances on geological time scales.

The major source of new P in forests is the soil through the weathering of primary minerals (Chapin et al. 2002). Apatite is the dominant form of P in many parent materials (Walker and Syers 1976). Dust deposition can also be a major P source in highly weathered soils of the tropics (Vitousek 2004). Other sources are primary biogenic aerosols and combustion of biomass and biofuels (Mahowald et al. 2009). Transport of parent material by erosion or tectonic uplift can also result in higher quantities of P in some soils of the tropics than is predictable from soil age and the climatic factor (Townsend et al. 2008). Thus, geochemical processes such as absorption and precipitation of phosphates are most important at high P levels whereas biological processes often dominate P budgets at low P levels (Vitousek et al. 1988).

The major avenues of P loss are surface runoff and erosion of particulate P, because it is tightly bound to OM or soil minerals. Some P losses in dissolved organic form have been observed in highly weathered soils (Vitousek 2004). Losses of P can occur also through biomass extraction during forest harvesting and by silvicultural activities (Attiwill 1981; Waring and Running 2007). Otherwise, P is largely conserved in soils during forest fires, but may be converted into less soluble or unavailable forms (Austin et al. 2003). Biomass burning, however, can contribute to P loss from forested regions by atmospheric transport (Mahowald et al. 2009).

5.2.2 *Phosphorus Impacts on Carbon Sequestration in Forest Ecosystems*

One of the first symptoms of P deficiency in many species is the appearance of dark green or blue-green foliage (Perry 1994). Most importantly, growth rate is reduced and severe P deficiency leads to stunting. Thus, C sequestration by trees may be limited by P deficiency (Kimmins 2004). The leaf P concentration, in particular, is an important determinant of photosynthetic capacity in some tree species (Schlesinger 1997). The use efficiency of P for conifers is lower than those for deciduous trees (3.0 Mg aboveground biomass kg^{-1} P vs. 3.7 Mg aboveground biomass kg^{-1} P; Ericsson 1994). Also, the lack of P can limit the utilization of N and, thus, reduce C accumulation as biomass. Severe limitation of P can also reduce water uptake (i.e., in mangrove trees) and, thus, reduce CO_2 assimilation rates (Lovelock et al. 2006).

The response of forest biomass production to P is particularly high compared to the biomass production in other ecosystems (Elser et al. 2007). Tropical forests, for example, show a stronger response to added P than to added N fertilization because most tropical forests are high in N, and P is increasingly sequestered in old and highly weathered soils of the tropics through mineralogical transformations (Lewis et al. 2004; Vitousek 2004; Elser et al. 2007). Thus, P additions to a highly weathered tropical montane forest soil increased tree diameter, LAI, ANPP and root turnover, but BNPP did not change (Herbert and Fownes 1995; Harrington et al. 2001; Ostertag 2001). P additions, in particular, increased NPP per unit of light absorbed or the radiation conversion efficiency (Harrington et al. 2001). P may limit photosynthetic CO_2 uptake as the relationship between photosynthetic capacity and leaf N in P-limited ecosystems may be constrained by low P (Reich et al. 2009). A likely mechanism is a limited ribulose 1,5-bisphosphate regeneration in P-deficient plants.

Any generalizations about effects of P on NPP are difficult because P fertilization experiments are scanty especially in old-growth, lowland, tropical forest ecosystems (McGroddy et al. 2008). Thus, the available data on extent and severity of P limitation on NPP in tropical forests are highly variable (Townsend et al. 2008). Soil P, on the other hand, may also be linked to soil C sequestration in highly weathered soils (Giardina et al. 2004). Decomposition of angiosperm wood is faster at higher P concentrations, which may cause a decrease in the accumulation of woody debris (Weedon et al. 2009). Furthermore, positive effects of P on leaf litter decomposition may contribute to the observed positive relationship between the rate of decomposition and concentration of N in leaf and litter biomass (Cornwell et al. 2008).

The C:P ratio of OM controls the balance between mineralization and immobilization in forests with low P availability and, therefore, the P supply to plants (Chapin et al. 2002). The NPP of the terrestrial biosphere appears to be P limited over geologic time scale (Schlesinger 1997). Thus, growth of forests on highly weathered and old soils is more likely limited by P or some other element than N (Attiwill and Adams 1993). Over thousands of years, there is a consistent pattern of decline in plant biomass in those forest ecosystems which are not subject to catastrophic disturbances (Wardle et al. 2004). Such a decline in biomass probably occurs in boreal, temperate and tropical forest ecosystems. However, a biomass decline due to P deficiency occurs in forests growing on highly weathered tropical soils only when atmospheric inputs of Ca, Mg and K are adequate (Townsend et al. 2008).

5.2.3 Conclusions

The interactions among P availability, ACC and C sequestration in forest ecosystems are not thoroughly understood. Some conclusions are, however, possible based on the scientific knowledge about P cycle processes in forests. Whether or not ACC enhances C sequestration in forest biomass may in part depend on the ACC effects on plant roots and microbial processes. Enhanced enzymatic release of P may result in increased C sequestration in biomass through growth stimulation. Increase in plant growth may further stimulate P release through its effects on root growth and microbial activity. Otherwise, intensified weathering processes in a warmer climate may contribute to increases in P availability. However, runoff and erosion processes may also be intensified, which may reduce the P supply needed for the vegetative growth. In tropical forests, soil P limitations and reductions in C:P ratios of litterfall may reduce the decomposition rates and soil CO_2 efflux caused by increase in temperatures and CO_2 concentrations (Silver 1998). Increase in frequency of fires in dry tropical forests and hurricanes in humid tropical forests may reduce the P storage over longer time periods, and aggravate P constraints to enhanced forest growth. Also, in old tropical forest soils the pace of mineralogical transformations may be accelerated under ACC and, thus, P may be increasingly sequestered thereby reducing NPP.

5.3 Water

Water is essential to numerous functions in plants (Kramer and Boyer 1995). Specifically, it is an important quantitative and qualitative plant constituent; a solvent for gases, minerals and other solutes; a reactant or substrate in many important processes; and is important to maintaining the turgor pressure in plant cells (Chang 2006). Water is also essential to metabolism (i.e., C turnover) and growth (i.e., C accumulation) from the cellular scale to the whole plant level (Ericsson et al. 1996). Thus, the C cycle in forest ecosystems is closely linked with the water cycle (Waring and Running 2007). Furthermore, water availability strongly modulates the influence of N deposition and foliar uptake of N on growth in arid and semi-arid regions (Sparks 2009).

The most sustainable and best quality fresh water sources in the world originate in forest ecosystems (Neary et al. 2009). Forest ecosystems are also a major user of water and, thus, play an important role in the terrestrial water cycle (FAO (2008). Most importantly, the long-term stability of the water cycle is probably unachievable without the recovery of natural, self-sustaining forests on continental scale (Makarieva and Gorshkov 2007). Specifically, the high LAI of forest canopies causes evapotranspiration (sum of transpiration flux from within plants and evaporation flux from wet surfaces, soils, and open water) exceeding evaporation flux from the open water surface of the ocean (Makarieva et al. 2009). Thus, evaporation and condensation generate atmospheric pressure differences. Horizontal fluxes of air and water vapor are directed from the ocean or areas with weak evaporation and small evaporative force to forests or areas with intensive evaporation and large evaporative force. According to this hypothesis undisturbed natural forests are therefore 'biotic pumps' of atmospheric water vapor by 'transporting' moist air from the ocean. However, this hypothesis needs to be scrutinized and evaluated (Sheil and Murdiyarso 2009).

Trees use about 15% more precipitation than grasses or shrubs (Farley et al. 2005). This is why reducing tree cover by about 20% of the canopy causes temporary increases in water yield (FAO 2008). However, the basin scale water consumption in forests is higher than that for any other vegetation type (Andréassian 2004). Otherwise, the importance of forest cover in regulating hydrological flows has often been overestimated, and may be restricted to the micro-level whereas natural processes are responsible for flooding at the macro-scale (FAO 2008). Forests, on the other hand, affect water yield, flood peaks and flows on the watershed scale. Forest cover is essential to managing watersheds for drinking water-supply because soil erosion in forests is minimal, sediments in water bodies are reduced, water pollutants are trapped/filtered and forestry activities are not associated with the use of possible pollutants such as fertilizers and pesticides. Therefore, cloud forests, swamp forests, forests with saline subsoils or groundwater, and forested riparian buffer zones are extremely important for water resources and their management (FAO 2008).

Large-scale hydrological processes are influenced by climate warming, and many forest regions may experience either severe drought or moisture surplus in the future with potential feedbacks to C sequestration in forest ecosystems (Luo 2007). In particular, a broad range of forests around the world is susceptible to ACC-induced drought with potential impacts on the forest C balance (Allen 2009). Otherwise, more extreme rainfall regimes may decrease soil water in mesic forests by increasing runoff and deep drainage, but increase it in xeric woodlands and hydric forests by increasing percolation depth and decreasing evaporative losses (Knapp et al. 2008). Furthermore, the effects of catastrophic events such as floods, drought and landslides may be influenced by the forest cover (FAO 2008). Atmospheric water vapor being an important greenhouse gas has a large and positive feedback effect on climate warming (Dessler et al. 2008). Thus, forests also influence the radiation balance by altering atmospheric water contents (van Dijk and Keenan 2007). For example, CO_2 induced stomatal closure increases radiative forcing via rapid reduction in low cloud over the Amazon and mid and high latitude forests (Doutriaux-Boucher et al. 2009). The extra radiative forcing associated with the CO_2 physiological forcing by doubling CO_2 is 0.5 W m^{-2} and causes increases in the equilibrium warming by 0.4 K. Otherwise, global warming leads to an increase in the humidity of the atmosphere and this causes additional warming. This water vapor effect is strongly positive with a magnitude of 1.5–2.0 W m^{-2} K^{-1}, sufficient to roughly double the warming that would otherwise occur (Dessler and Sherwood 2009). Furthermore, the hydrological cycle in forests is potentially sensitive to changes in the surface radiative forcings induced by surface solar radiation dimming and brightening (Wild 2009).

5.3.1 Water Cycle in Forest Ecosystems

Water enters forest ecosystems through precipitation and other inputs of moisture such as dew and fog (Fig. 5.3; Kimmins 2004). A fraction of the gross precipitation (*PP*) is intercepted and evaporated from the surface of vegetation, from the litter layer and soil (*I*), while another fraction reaches the soil as stemflow (*SF*) and throughfall (*TF*) (Crockford and Richardson 2000). Also, foliar water uptake can occur (Limm et al. 2009). Forests generally use more water than crops, grass, or other natural short vegetation (Bates et al. 2008). Evaporation from forests is limited by radiation, advection, tree physiology, soil water, tree size and raindrop size (Calder 1998). The mass balance for *I* can be expressed by Eq. *(5.1)*:

$$I = PP - TF - SF \tag{5.1}$$

The factors affecting *I* include ground cover, geography and landscape location and climate. Features of forest types affecting *I* include the canopy storage capacity, LAI, the shrub and litter layer storage capacity, hydrophobicity of leaf and wood, and projecting tree crowns. Climatic features that affect *I* are amount, intensity and duration of rainfall, wind speeds and directions, and air temperature and humidity (Crockford and Richardson 2000). *I* and evaporation from the canopy are greater for needle-leaved than for broad-leaved trees (Cannell 1999). An estimated 10–20% of *PP* in hardwood forests and 20–40% of *PP* in conifer plantations is intercepted by the canopy (Levia and Frost 2003). Also, the percentage of rainfall intercepted is higher in wet temperate than in wet tropical forests but evaporation is higher in wet tropical compared to wet temperate forests (Calder 1998). However, *I* is extremely variable and difficult to measure. For example, *I* ranged from 3.6% to 49% of *PP* for conifers, and from 0.15% to 100% of *PP* for tropical rainforests (Crockford and Richardson 2000).

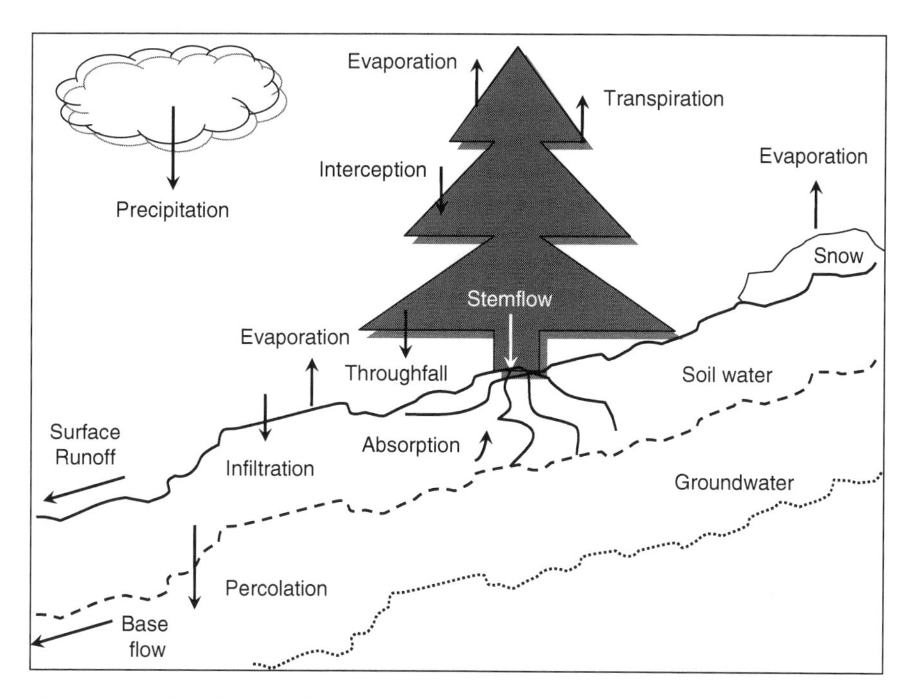

Fig. 5.3 Hydrologic balance of a forest ecosystem (modified from Chapin et al. 2002)

For a given forest cover, there exists a good relationship between long-term average evapotranspiration and rainfall (Zhang et al. 2001). The transfer of water from the plant canopy to the soil occurs via *SF* and *TF*. Tree characteristics relevant to *SF* and *TF* are crown size, leaf shape and orientation, branch angle, flow path obstruction, bark type and canopy gaps (Crockford and Richardson 2000). Rainfall characteristics such as continuity and proportion of dry periods, rainfall intensity, and the angle of the rain affect *SF*. However, knowledge about *SF* production in forests is particularly weak as accurate measurements are difficult (Levia and Frost 2003). For example, the values of *SF* between 0.07% and 56% of *PP* have been reported (Crockford and Richardson 2000; Johnson and Lehmann 2006). The understory may also be important to *SF* generation. However, a major difficulty is the reliable estimation of *TF* in forests.

Water enters the soil through infiltration and moves along pressure gradients associated with gravity and matric potential (Chapin et al. 2002). In comparison to other land uses, forests maintain or improve water infiltration into soils, tend to have a better infiltration capacity and a higher capacity to retain water (Bates et al. 2008; FAO 2008). Fluxes of *SF* are funneled preferentially belowground along tree roots and other preferential flow paths and, thus, bypass much of the bulk soil (Johnson and Lehmann 2006). Water is absorbed from the soil during the day by roots of transpiring C_3 and C_4 plants (inverse pattern for CAM plants; see Section 2.1.1.1), and flows from zones of higher water potential in the soil to those of a lower potential in the atmosphere. In roots, however, water can also move passively, usually at night from moister soil layers (higher water potential) to drier soil layers (lower water potential), and, thus, drier soil layers are hydrated (Richards and Caldwell 1987). The direction of this hydraulic redistribution can be upward, horizontal or downward (Caldwell et al. 1993). Water transfer downward by inverse hydraulic lift may occur especially in arid areas to facilitate roots growing at depth towards sources of moisture (Dawson 1998).

Hydraulic redistribution is not restricted to arid and semiarid environments, and has also been well documented for boreal, temperate and tropical trees (summarized in Liste and White 2008). The upward hydraulic lift probably buffers against water stress during seasonal water deficits and can also benefit neighboring plants (Caldwell and Richards 1989). For example, *Acer saccharum* Marshall recharges the upper soil layers with up to 25% of the total daily water use of the tree, and supplies up to 60% of the water used by neighboring shallow-rooted species (Dawson 1993, 1998). In addition, rhizosphere microorganisms and soil fauna may also benefit from hydraulically lifted water, and this may eventually increase nutrient availability to trees (Liste and White 2008). Relatively large amounts of hydraulically lifted water can be transported along and into the hyphae of mycorrhizal fungi in drought-prone forest ecosystems, and influence the fungal ability to fruit during summer drought (Lilleskov et al. 2009).

Within the plant, water flows along potential gradients through stems and branches to the leaves. In the leaves, transpiration through plant stomata causes water losses to the atmosphere because of a vapor-pressure gradient. Maximum rates of daily whole-tree water use range between 10 and 200 kg day^{-1} for trees that average 21 m in height (Wullschleger et al. 1998). Estimates for the vapor flow derived from boreal forests range between 293 and 339 mm year^{-1}, for temperate forests between 510 and 592 mm year^{-1}, and between 795 and 1,146 mm year^{-1} for tropical forests (Gordon et al. 2005). Deforestation has already significantly reduced vapor flows derived from forests.

Water movement through the soil-plant-atmosphere continuum is affected by the density and depth of root channels and the amount of organic residues incorporated into soil (Waring and Running 2007). Maximum rooting depth, thus, determines the soil depth than can be exploited by plants for evaporative demands and nutrient uptake by passive flow in water-limited ecosystems (Schenk 2008). Maximum rooting depths vary widely among biomes and species within biomes, ranging from 2.0 m for boreal, 2.9–3.9 m for temperate and 3.7–7.3 m for tropical forests (Canadell et al. 1996). However, complete tree root profiles are rarely studied (Schenk and Jackson 2005). Forests regulate hydrological flows (Calder et al. 2007). Water leaves forest ecosystems through surface runoff and ground water fluxes as not all water that enters forests is either lost by evapotranspiration or is held in the soil. Aside soil, tree roots and stems are other storage compartments for water (Meinzer et al. 2001). For example, up to 50% of the total water loss by transpiration during a 24-h cycle may be supplied by stem tissues. Otherwise, long-term storage of water occurs in the groundwater aquifers (Kimmins 2004).

5.3.2 Water and Carbon Sequestration in Forest Ecosystems

Photosynthesis is the major process of C input into forest ecosystems. In addition to CO_2, H_2O is one of the substrates in photosynthesis (Eq. (2.3)), and part of the fixed CO_2 and H_2O are released from photosynthetic products to the environment through respiration, during the production of energy in the form of ATP and NADPH (Eq. *(5.2)*); Kramer and Boyer 1995). Thus, precipitation is strongly linked to GPP and NPP.

$$C_6H_{12}O_6 + 6O_2 \rightarrow 6CO_2 + 6H_2O + \text{Energy} \tag{5.2}$$

The rate of photosynthesis depends on the stomatal conductance which is controlled primarily by the availability of H_2O to the plant and the CO_2 concentration inside the leaf. Both leaf area and the photosynthetic efficiency of the leaf area are determined by the availability of H_2O as well as by the availability of plant nutrients and light (Kimmins 2004). Stomatal responses play a key role in rapid responses to moderate drought, but tree genotypes differ in their levels of drought tolerance, as is indicated by differences in reduction of biomass accumulation among *Populus* genotypes (Monclus et al. 2006). Severe drought stress, on the other hand, may cause the inhibition of photosynthesis although it is unlikely that a moderate lack of the substrate H_2O causes a loss in photosynthesis (Kramer and Boyer 1995; Ericsson et al. 1996). Downy oak (*Quercus pubescens* Willd.), for example, withstands and survives extreme summer droughts as photoprotective mechanisms minimize drought stress induced damages to the photosynthetic apparatus (Gallé et al. 2007).

The process of C partitioning into foliage, stems, roots and reproductive organs is driven by the availability of H_2O aside from the availability of light and plant nutrients (Körner 2006). For example, the radial growth of Scots pine (*Pinus sylvestris* L.) growing on dry alluvial terraces is sensitive to water-level fluctuations as root growth responds to differences in soil water supply (Polacek et al. 2006). With increase in severity of drought stress, C partitioning to tree roots is favored causing an increase in root:shoot ratios (Ericsson et al. 1996). Losses of C in gaseous, liquid and particulate forms during surface litter and SOM decomposition depend on sufficient soil moisture availability for the activity of the decomposer community and mass transport (see Chapter 2). Pronounced soil water deficits, in particular, may compensate for the effect of warmer temperatures in reducing soil respiration (Ciais et al. 2005). Otherwise, excess soil moisture levels retard decomposition due to decrease in oxygen availability. By affecting fluxes other than that through C fixation and respiration (e.g., losses from forest ecosystems by surface runoff, percolation) precipitation is also linked to NECB (Chapin et al. 2006). Furthermore, a small C input into forest ecosystems is associated with bulk precipitation as it contains dissolved C (see Section 2.2.2.1). In comparison, larger DOC inputs occur through *TF* and, in particular, through *SF* (9.9–23.5 mg DOC L^{-1} and 13.7–129.9 mg DOC L^{-1}, respectively; Johnson and Lehmann 2006). *SF* water contains also high concentrations of POM.

The relationship between H_2O and NPP is extremely complex (Kimmins 2004). In drier and colder forests, NPP increases linearly with increase in mean annual precipitation (MAP). However, increase in NPP with increase in MAP diminishes in mesic and warm forests. Otherwise, NPP does not change with additional MAP in forest in wetter and warmer regions (Schuur 2003). In warm and wet tropical forests, however, NPP decreases with increase in precipitation. In particular, the influence of H_2O availability on tree growth can hardly be separated from the influences of other closely related factors. For example, the responses of tree growth to elevated CO_2 are amplified when water is limiting (McMurtrie et al. 2008). Lack or surplus of soil moisture restricts the nutrient release by decomposer activity, the uptake of nutrients by the roots and the exploitation of soil nutrients by fine-root growth.

The hydraulic-lift water can also promote tree growth and may have important implications for tree NPP (Horton and Hart 1998). For example, sugar maple (*Acer saccharum* Marshall) seedlings had 24–30% higher rates of root growth and 8–14% higher rates of shoot growth compared to seedlings not benefiting from the hydraulic lift (Dawson 1998). Neighboring trees with well designed hydraulic lift may also benefit through higher C gains. On the other hand, inverse hydraulic lift enables trees in arid regions to survive and continue to assimilate C by exploring water resources in the sub-soil layers. Hydraulically lifted water may provide moisture for enhancing nutrient availability, accentuating the decomposition process, increasing the acquisition of nutrients by roots, and prolonging or enhancing the fine-root activity (Dawson 1998). Thus, C sequestration in forest ecosystems may benefit from hydraulic redistribution of sub-soil water to the surface.

Increase in precipitation is linked with increase in cloudiness and decrease in solar radiation at the forest canopy level and, thus, may limit NPP (Schuur 2003). In the humid environments, however, nutrient availability seems to be the predominant control of soil moisture availability in the regulation and allocation of NPP (Kimmins 2004). *SF* may affect plant productivity by regulating leaching of nutrients such as NO_3^-, K and PO_4^{3-} near the plant stem (Levia and Frost 2003; Johnson and Lehmann 2006). On the other hand, maximum forest ecosystem productivity on a continental scale is ensured by the biotic pump of atmospheric moisture (Makarieva and Gorshkov 2007). Yet, high LAI of natural forests and high evaporation exceeding those from open water surface are capable of pumping atmospheric moisture form the oceans for the maintenance of optimal soil moisture stores in forest and for compensating the river runoff.

The boreal forest C sink, in particular, may be altered by changes in the surface water balance as recent summer drought periods may lead to marked decrease in NPP (see Chapter 4). Strong reductions in forest C uptake are also observed during the prolonged periods of summer drought in temperate forests in Europe and during ENSO episodes in moist tropical forests. In dry environments, available soil moisture often limits leaf area and NPP, and is probably the most dominant constraint in C allocation. However, the indirect effects of climate warming on C dynamics via changes in ecosystem-scale hydrological cycling are yet not fully known (Luo 2007).

Differences in evapotranspiration among forests contribute to differences in temperatures (Zhang et al. 2001). Otherwise, increased atmospheric CO_2 levels cause stomatal closure and suppress plant transpiration and, thus, leave more water at the land surface. Thus, a modified rate of evapotranspiration, and an increase in average continental runoff has been observed over the 20th century (Betts et al. 2007; Valladares 2008). This process may feedback on the land C sink as surface energy losses due to evaporation are reduced and surface warming is enhanced (Gedney et al. 2006). Otherwise, forest plantations may cause substantial reductions in stream flow, and either salinize or acidify some soils with possible adverse feedbacks on forest C sequestration (Jackson et al. 2005; van Dijk and Keenan 2007).

Predicting changes in forest ecosystem C fluxes through more extreme precipitation regimes relative to those from elevated atmospheric CO_2 and warming are difficult (Knapp et al. 2008). Larger individual precipitation events with longer intervening dry periods than at present are projected for the future. This trend may reduce NPP and total ecosystem respiration in more mesic forest ecosystems but increase DOC leaching through the soil profile. An opposite trend may occur in some xeric forests. On the other hand, in hydric or wetland forest ecosystems changes in net C fluxes may be dominated by potentially rapid rates of OM decomposition during oxic conditions. Theoretically, long-term changes in precipitation regimes have the potential to alter root distribution of forest biomes by affecting root growth of individual species and the species community composition. The consequences of these changes on the C transfer through tree roots and their associated mycorrhizal communities into the SOC pool are, however, unknown but important to C sequestration in forest ecosystems (see also Section 2.2.2.5).

Hydrological cycles in forest ecosystems may also affect C sequestration in the surrounding ecosystems. Specifically, entire ecosystems are supported through effects of forested watersheds on water production and quality, reduced salinization, erosion and sediment control, and the risks of avalanches (FAO 2008).

5.3.3 Conclusions

The availability of nutrients and water affect C sequestration in forest ecosystems. Lack of an adequate level of soil N supply, in particular, constraints the productivity of boreal and temperate forests whereas P deficit often limits growth in tropical forests. Anthropogenic transformations of global N and P cycles interact with C sequestration processes. However, net effects of increase in N and P availability on the forest ecosystem C balance is not well understood. Anthropogenic perturbations are responsible for the increase in atmospheric CO_2 concentrations but any tree growth stimulation may be only sustained in the long term when it is not constrained by N and P supply. Otherwise, smaller stomatal openings at elevated CO_2 concentrations may reduce the limitations of H_2O availability on NPP. Recent extreme drought events, however, indicate that strong reductions in forest C uptake may occur in a drier future climate. Furthermore, C losses through respiration are also affected by projected temperature increases. By affecting non-respiratory C fluxes, the altered precipitation regime interacts also with the NECB of forests. However, separating the influences of changes in H_2O availability from those of nutrient availability on C sequestration in forests ecosystems, and predicting how these processes may be affected by ACC remains to be scientific challenges.

5.4 Review Questions

1. Describe the major processes of the forest N and P cycles. What are the major differences among them?
2. How does N affect C sequestration in trees from the cellular to the whole plant scale? What are the differences between forests and agricultural ecosystems?
3. Discuss N effects on SOC sequestration processes in boreal, temperate and tropical forests.
4. Compare and contrast the effects of N with those of P on C sequestration in forest ecosystems.
5. Which other nutrients affect tree growth? Describe how these may limit C sequestration in forest ecosystems.
6. Schematically describe the water balance of forest ecosystems and the effect of anthropogenic perturbations.
7. List major functions of water from the cellular level to the whole tree and forest ecosystem level.
8. How would you set-up experiments to study ACC effects on water and nutrient limitations of C sequestration in forest ecosystems?

References

Aber J, McDowell W, Nadelhoffer K, Magill A, Berntson G, Kamakea M, McNulty S, Currie W, Rustad L, Fernandez I (1998) Nitrogen saturation in temperate forest ecosystems. BioScience 48:921–934

Allen CD (2009) Climate-induced forest dieback: an escalating global phenomenon? Unasylva 60:43–49

Alo CA, Wang G (2008) Potential future changes of the terrestrial ecosystem based on climate projections bei eight general circulation models. J Geophys Res 113:G01004. doi:10.1029/2007JG000528

Andréassian V (2004) Waters and forests: from historical controversy to scientific debate. J Hydrol 291:1–27

Attiwill PM (1981) Energy, nutrient flow, and biomass. In: Australian forest nutrition workshop, productivity in perpetuity, August 10–14, Canberra, Australia, pp 131–144

Attiwill PM, Adams MA (1993) Nutrient cycling in forests. New Phytol 124:561–582

Austin AT, Horwath RW, Baron JS, Stuart Chapin III F, Christensen TR, Holland EA, Ivanov MV, Lein AY, Martinelli LA, Melillo JM, Shang C (2003) Human disruption of element interactions:

drivers, consequences, and trends for the twenty-first century. In: Melillo JM, Field CB, Moldan B (eds) Interactions of the major biogeochemical cycles: global change and human impacts. Island Press, Washington, DC, pp 15–45

Barron AR, Wurzburger N, Bellenger JP, Wright SJ, Kraepiel AML, Hedin LO (2009) Molybdenum limitation of asymbiotic nitrogen fixation in tropical forest soils. Nat Geosci 2:42–45

Bates BC, Kundzewicz ZW, Wu S, Palutikof JP (eds) (2008) Climate change and water. Technical paper of the intergovernmental panel on climate change. IPCC Secretariat, Geneva, Switzerland

Betts RA, Boucher O, Collins M, Cox PM, Falloon PD, Gedney N, Hemming DL, Huntingford C, Jones CD, Sexton DMH, Webb MJ (2007) Projected increase in continental runoff due to plant responses to increasing carbon dioxide. Nature 448:1037–1042

Billings SA (2008) Nitrous oxide in flux. Nature 456:888–889

Binkley CS, Apps MJ, Dixon RK, Kauppi PE, Nilsson LO (1997) Sequestering carbon in natural forests. Crit Rev Env Sci Technol 27:S23–S45

Bowman WD, Cleveland CC, Halada Ĺ, Hreško J, Baron JS (2008) Negative impact of nitrogen deposition on soil buffering capacity. Nat Geosci 1:767–770

Brantley SL (2008) Understanding soil time. Science 321:1454–1455

Calder IR (1998) Water use by forests, limits and controls. Tree Physiol 18:625–631

Calder I, Hofer T, Vermont S, Warren P (2007) Towards a new understanding of forests and water. Unasylva 58:3–10

Caldwell MM, Dawson TE, Richards JH (1993) Hydraulic lift: consequences of water efflux from the roots of plants. Oecologia 113:151–161

Caldwell MM, Richards JH (1989) Hydraulic lift: water efflux from upper roots improves effectiveness of water uptake by deep roots. Oecologia 79:1–5

Canadell J, Jackson RB, Ehleringer JR, Mooney HA, Sala OE, Schulze E-D (1996) Maximum rooting depth of vegetation types at the global scale. Oecologia 108:583–595

Cannell MGR (1999) Environmental impacts of forest monocultures: water use, acidification, wildlife conservation, and carbon storage. New Forest 17:239–262

Chang M (2006) Forest hydrology: an introduction to water and forests. Taylor & Francis, Boca Raton, FL

Chapin FS III, Matson PA, Mooney HA (2002) Principles of terrestrial ecosystem ecology. Springer, New York

Chapin FS III, McFarland J, McGuire AD, Euskirchen ES, Ruess RW, Kielland K (2009) The changing global carbon cycle: linking plant-soil carbon dynamics to global consequences. J Ecol 97:840–850

Chapin FS III, Woodwell GM, Randerson JT, Rastetter EB, Lovett GM, Baldocchi DD, Clark DA, Harmon ME, Schimel DS, Valentini R, Wirth C, Aber JD, Cole JJ, Goulden ML, Harden JW, Heimann M, Howarth RW, Matson PA, McGuire AD, Melillo JM, Mooney HA, Neff JC, Houghton RA, Pace ML, Ryan MG, Running SW, Sala OE, Schlesinger WH, Schulze E-D (2006) Reconciling carbon-cycle concepts, terminology, and methods. Ecosystems 9:1041–1050

Chapuis-Lardy L, Wrage N, Metay A, Chotte J-L, Bernoux M (2007) Soils, a sink for N_2O? A review. Glob Change Biol 13:1–17

Chave J, Coomes D, Jansen S, Lewis SL, Swenson NG, Zanne AE (2009) Towards a worldwide wood economics spectrum. Ecol Lett 12:351–366

Churkina G, Trusilova K, Vetter M, Dentener F (2007) Contributions of nitrogen deposition and forest regrowth to terrestrial carbon uptake. Carbon Bal Manag 2:5

Ciais P, Reichstein M, Viovy N, Granier A, Ogée J, Allard V, Aubinet M, Buchmann N, Bernhofer C, Carrara A, Chevallier F, De Noblet N, Friend AD, Friedlingstein P, Grünwald T, Heinesch B, Keronen P, Knohl A, Krinner G, Loustau D, Manca G, Matteucci G, Miglietta F, Ourcival JM, Papale D, Pilegaard K, Rambal S, Seufert G, Soussana J-F, Sanz M-J, Schulze E-D, Vesala T, Valentini R (2005) Europe-wide reduction in primary productivity caused by the heat and drought in 2003. Nature 437:529–533

Claire Horner-Devine C, Martiny AC (2008) News about nitrogen. Science 320:757–758

Clark DA (2007) Detecting tropical forests' response to global climatic and atmospheric change: current challenges and a way forward. Biotropica 39:4–19

Colman BP, Fierer N, Schimel JP (2008) Abiotic nitrate incorporation, anaerobic microsites, and the ferrous wheel. Biogeochemistry 91:223–227

Cornwell WK, Cornelissen JHC, Allison SD, Bauhus J, Eggleton P, Preston CM, Scarff F, Weedon JT, Wirth C, Zanne AE (2009) Plant traits and wood fate across the globe-rotted, burned, or consumed? Glob Change Biol (in press) doi: 10.1111/j.1365-2486.2009.01916.x

Cornwell WK, Cornelissen JHC, Amatangelo K, Dorrepaal E, Eviner VT, Godoy O, Hobbie SE, Hoorens B, Kurokawa H, Pérez-Harguindeguy N, Quested HM, Santiago LS, Wardle DA, Wright IJ, Aerts R, Allison SD, van Bodegom P, Brovkin V, Chatain A, Callaghan TV, Díaz S, Garnier E, Gurvich DE, Kazakou E, Klein JA, Read J, Reich PB, Soudzilovskaia NA, Victoria VM, Westoby M (2008) Plant species traits are the predominant control on litter decomposition rates within biomes worldwide. Ecol Lett 11:1065–1071

Crockford RH, Richardson DP (2000) Partitioning of rainfall into throughfall, stemflow and interception: effect of forest type, ground cover and climate. Hydrol Process 14:2903–2920

Davidson EA (2008) Fixing forests. Nat Geosci 1:421–422

Davis SC, Hessl AE, Thomas RB (2008) A modified nitrogen budget for temperate deciduous forests in an advanced stage of nitrogen saturation. Global Biogeochem Cy 22, GB4006, doi:10.1029/2008GB003187

Dawson TE (1993) Hydraulic lift and water use by plants: implications for water balance, performance, and plant–plant interactions. Oecologia 95:565–574

Dawson TE (1998) Water loss from tree roots influences soil water and nutrient status and plant performance. In: Flores HE, Lynch JP (eds) Radical biology: advances and perspectives in the function of plant roots. Current topics in plant physiology, Vol 17. American society of plant physiologists, Rockville, MD, pp 195–210

DeLuca TH, Zackrisson O, Gundale MJ, Nilsson M-C (2008) Ecosystem feedbacks and nitrogen fixation in boreal forests. Science 320:1181

De Schrijver A, Verheyen K, Mertens J, Staelens J, Wuyts K, Muys B (2008) Nitrogen saturation and net ecosystem production. Nature 451:E1

Dessler AE, Sherwood SC (2009) A matter of humidity. Science 323:1020–1021

Dessler AE, Zhang Z, Yang P (2008) Water-vapor climate feedback inferred from climate fluctuations, 2003–2008. Geophys Res Lett 35:L20704. doi:10.1029/2008GL035333

De Vries W, Reinds GJ, Gundersen P, Sterba H (2006) The impact of nitrogen deposition on carbon sequestration in European forests and forest soils. Glob Change Biol 12:1151–1173

De Vries W, Solberg S, Dobbertin M, Sterba H, Laubhahn D, Jan Reinds G, Nabuurs G-J, Gundersen P, Sutton MA (2008) Ecologically implausible carbon response? Nature 451:E1–E3

De Vries W, Solberg S, Dobbertin MD, Sterba H, Laubhann D, van Oijen M, Evans C, Gundersen P, Kros J, Wamelink GWW, Reinds GJ, Sutton MA (2009) The impact of nitrogen deposition on carbon sequestration by European forests and heathlands. For Ecol Manag (in press) doi:10.1016/j.foreco.2009.02.034

Dewar RC, Franklin O, Mäkelä A, McMurtrie RE, Valentine HT (2009) Optimal function explains forest responses to global change. BioScience 59:127–139

Doney SC, Schimel DS (2007) Carbon and climate system coupling on timescales from the Precambrian to the Anthropocene. Annu Rev Environ Resour 32:14.1–14.36

Donner LJ, Large WG (2008) Climate modeling. Annu Rev Environ Resour 33:1–17

Doutriaux-Boucher M, Webb MJ, Gregory JM, Boucher O (2009) Carbon dioxide induced stomatal closure increases radiative forcing via a rapid reduction in low cloud. Geophys Res Lett 36:L02703. doi:10.1029/2008GL036273

Elbert W, Weber B, Büdel B, Andreae MO, Pöschl U (2009) Microbiotic crusts on soil, rock and plants: neglected major players in the global cycles of carbon and nitrogen? Biogeosciences Discuss 6:6983–7015

Elser JJ, Bracken MES, Cleland EE, Gruner DS, Stanley Harpole W, Hillebrand H, Ngai JT, Seabloom EW, Shurin JB, Smith JE (2007) Global analysis of nitrogen and phosphorus

limitation of primary producers in freshwater, marine and terrestrial ecosystems. Ecol Lett 10:1135–1142
Ericsson T (1994) Nutrient dynamics and requirements of forest crops. N Z J For Sci 24:133–168
Ericsson T, Rytter L, Vapaavuori E (1996) Physiology of carbon allocation in trees. Biomass Bioenerg 11:115–127
Evans CD, Goodale CL, Caporn SJM, Dise NB, Emmett BA, Fernandez IJ, Field CB, Findlay SEG, Lovett GM, Meesenburg H, Moldan F, Sheppard LJ (2008) Does elevated nitrogen deposition or ecosystem recovery from acidification drive increased dissolved organic carbon loss from upland soil? A review of evidence from field nitrogen addition experiments. Biogeochemistry 91:13–35
Falkowski PG, Fenchel T, Delong EF (2008) The microbial engines that drive Earth's biogeochemical cycles. Science 320:1034–1039
Falkowski P, Scholes RJ, Boyle E, Canadell J, Canfield D, Elser J, Gruber N, Hibbard K, Högberg P, Linder S, Mackenzie FT, Moore B III, Pedersen T, Rosenthal Y, Seitzinger S, Smetacek V, Steffen W (2000) The global carbon cycle: a test of our knowledge of earth as a system. Science 290:291–296
FAO (Food and Agricultural Organization of the United Nations) (2008) Forests and water. FAO Forestry paper 155. FAO, Rome
Farley KA, Jobbágy EG, Jackson RB (2005) Effects of afforestation on water yield: a global synthesis with implications for policy. Glob Change Biol 11:1565–1576
Fleischer S, Bouse I (2008) Nitrogen cycling drives a strong within-soil CO_2-sink. Tellus 60B:782–786
Franklin O, McMurtrie RE, Iversen CM, Crous KY, Finzi AC, Tissue DT, Ellsworth DS, Oren R, Norby RJ (2009) Forest fine-root production and nitrogen use under elevated CO_2: contrasting responses in evergreen and deciduous trees explained by a common principle. Glob Change Biol 15:132–144
Friedlingstein P, Cox P, Betts R, Bopp L, von Bloh W, Brovkin V, Cadule P, Doney S, Eby M, Fung I, Bala G, John J, Jones C, Joos F, Kato T, Kawamiya M, Knorr W, Lindsay K, Matthews HD, Raddatz T, Rayner P, Reick C, Roeckner E, Schnitzler K-G, Schnur R, Strassmann K, Weaver AJ, Yoshikawa C, Zeng N (2006) Climate-carbon cycle feedback analysis: results from the c4mip model intercomparison. J Clim 19:3337–3353
Gaige E, Dail DB, Hollinger DY, Davidson EA, Fernandez IJ, Sievering H, White A, Halteman W (2007) Changes in canopy processes following whole-forest canopy nitrogen fertilization of a mature spruce-hemlock forest. Ecosystems 10:1133–1147
Gallé A, Haldimann P, Feller U (2007) Photosynthetic performance and water relations in young pubescent oak (*Quercus pubescens*) trees during drought stress and recovery. New Phytol 174:799–810
Gallet-Budynek A, Brzostek E, Rodgers VL, Talbot IM, Hyzy S, Finci AC (2009) Intact amino acid uptake by northern hardwood and conifer trees. Oecologia 160:129–138
Galloway JN, Aber JD, Erisman JW, Seitzinger SP, Howarth RW, Cowling EB, Cosby BJ (2003) The nitrogen cascade. BioScience 53:341–356
Galloway JN, Townsend AR, Erisman JW, Bekunda M, Cai Z, Freney JR, Martinelli LA, Seitzinger SP, Sutton MA (2008) Transformation of the nitrogen cycle: recent trends, questions, and potential solutions. Science 320:889–892
Gedney N, Cox PM, Betts RA, Boucher O, Huntingford C, Stott PA (2006) Detection of a direct carbon dioxide effect in continental river runoff records. Nature 439:835–838
Giardina CP, Binkley D, Ryan MG, Fownes JH, Senock RS (2004) Belowground carbon cycling in a humid tropical forest decreases with fertilization. Oecologia 139:545–550
Goldberg SD, Gebauer G (2009) Drought turns a Central European Norway spruce forest soil from an N_2O source to a transient N_2O sink. Glob Change Biol 15:850–860
Gordon LJ, Steffen W, Jönsson BF, Folke C, Falkenmark M, Johanessen Å (2005) Human modification of global water vapor flows from the land surface. Proc Natl Acad Sci USA 102:7612–7617

de Graaff M-A, van Groenigen K-J, Six J, Hungate B, van Kessel C (2006) Interactions between plant growth and soil nutrient cycling under elevated CO_2: a meta-analysis. Glob Change Biol 12:2077–2091

Grandy AS, Sinsabaugh RL, Neff JC, Stursova M, Zak DR (2008) Nitrogen deposition effects on soil organic matter chemistry are linked to variation in enzymes, ecosystems and size fractions. Biogeochemistry 91:37–49

Gruber N, Galloway JN (2008) An earth-system perspective of the global nitrogen cycle. Nature 451:293–296

Harrington RA, Fownes JH, Vitousek PM (2001) Production and resource-use efficiencies in N- and P-limited tropical forest ecosystems. Ecosystems 4:646–657

Harrison A, Schulze E-D, Gebauer G, Bruckner G (2000) Canopy uptake and utilization of atmospheric pollutant nitrogen. In: Schulze E-D (ed) Carbon and nitrogen cycling in European forest ecosystems. Springer, Berlin, pp 171–188

van der Heijden MGA, Bardgett RD, van Straalen NM (2008) The unseen majority: soil microbes as drivers of plant diversity and productivity in terrestrial ecosystems. Ecol Lett 11:296–310

Heimann M, Reichstein M (2008) Terrestrial ecosystem carbon dynamics and climate feedbacks. Nature 451:289–292

Herbert DA, Fownes JH (1995) Phosphorus limitation of forest leaf area and net primary productivity on a weathered tropical soil. Biogeochemistry 29:223–235

Hessen DO, Ågren GI, Anderson TR, Elser JJ, De Ruiter PC (2004) Carbon sequestration in ecosystems: the role of stoichiometry. Ecology 85:1179–1192

Högberg P (2007) Nitrogen impacts on forest carbon. Nature 447:781–782

Holland EA, Carroll MA (2003) Atmospheric chemistry and the bioatmospheric carbon and nitrogen cycles. In: Melillo JM, Field CB, Moldan B (eds) Interactions of the major biogeochemical cycles: global change and human impacts. Island Press, Washington, DC, pp 273–292

Horton JL, Hart SC (1998) Hydraulic lift: a potentially important ecosystem process. Trends Ecol Evol 13:232–235

Houlton BZ, Wang Y-P, Vitousek PM, Field CB (2008) A unifying framework for dinitrogen fixation in the terrestrial biosphere. Nature 454:327–334

Hüttl RF, Schaaf W (1997) Magnesium deficiency in forest ecosystems. Kluwer, Dordrecht, The Netherlands

Hungate BA, Naiman RJ, Apps M, Cole JJ, Moldan B, Satake K, Stewart JWB, Victoria R, Vitousek PM (2003) Disturbance and element interactions. In: Melillo JM, Field CB, Moldan B (eds) Interactions of the major biogeochemical cycles: global change and human impacts. Island Press, Washington, DC, pp 47–62

Hyvönen R, Ågren GI, Linder S, Persson T, Cotrufo MF, Ekblad A, Freeman M, Grelle A, Janssens IA, Jarvis PG, Kellomäki S, Lindroth A, Loustau D, Lundmark T, Norby RJ, Oren R, Pilegaard K, Ryan MG, Sigurdsson BD, Strömgren M, Oijen M, Wallin G (2007) The likely impact of elevated [CO_2], nitrogen deposition, increased temperature and management on carbon sequestration in temperate and boreal forest ecosystems: a literature review. New Phytol 173:463–480

Hyvönen R, Persson T, Andersson S, Olsson B, Ågren GI, Linder S (2008) Impact of long-term nitrogen addition on carbon stocks in trees and soils in northern Europe. Biogeochemistry 89:121–137

Jackson RB, Jobbágy EG, Avissar R, Baidya Roy S, Barrett DJ, Cook CW, Farley KA, le Maitre DC, McCarl BA, Murray BC (2005) Trading water for carbon with biological carbon sequestration. Science 310:1944–1947

Jandl R, Lindner M, Vesterdahl L, Bauwens B, Baritz R, Hagedorn F, Johnson DW, Minkkinen K, Byrne KA (2007) How strongly can forest management influence soil carbon sequestration? Geoderma 137:253–268

Jarvis PG, Linder S (2007) Forests remove carbon dioxide from the atmosphere: spruce forest tales! In: Freer-Smith PH, Broadmeadow MSJ, Lynch JM (eds) Forestry and climate change. CAB International, Wallingford, UK, pp. 60–72

Jassal RS, Black TA, Chen B, Roy R, Nesic Z, Spittlehouse DL, Trofymow JA (2008) N_2O emissions and carbon sequestration in a nitrogen-fertilized Douglas fir stand. J Geophys Res 113:G04013. doi:10.1029/2008JG000764

Jetten MSM (2008) The microbial nitrogen cycle. Environ Microbiol 10:2903–2909

Johnson DW, Curtis PS (2001) Effects of forest management on soil C and N storage: meta analysis. For Ecol Manag 140:227–238

Johnson MS, Lehmann J (2006) Double-funneling of trees: stemflow and root-induced preferential flow. Ecoscience 13:324–333

Jones DL, Kielland K, Sinclair FL, Dahlgren RA, Newsham KK, Farrar JF, Murphy DV (2009) Soil organic nitrogen mineralization across a global latitudinal gradient. Global Biogeochem Cy 23, GB1016, doi:10.1029/2008GB003250

Kergoat L, Lafont S, Arneth A, Le Dantec V, Saugier B (2008) Nitrogen controls plant canopy light-use efficiency in temperate and boreal ecosystems. J Geophys Res 113:G04017. doi:10.1029/2007JG000676

Kimmins JP (2004) Forest ecology. Prentice Hall, Upper Saddle River, NJ

Knapp AK, Beier C, Briske DD, Classen AT, Luo Y, Reichstein M, Smith MD, Smith SD, Bell JE, Fay PA, Heisler JL, Leavitt SW, Sherry R, Smith B, Weng E (2008) Consequences of more extreme precipitation regimes for terrestrial ecosystems. BioScience 58:811–821

Knorr M, Frey SD, Curtis PS (2005) Nitrogen additions and litter decomposition: a meta-analysis. Ecology 86:3252–3257

Koehler B, Corre MD, Veldkamp E, Wullaert H, Wright SJ (2009) Immediate and long-term nitrogen oxide emissions from tropical forest soils exposed to elevated nitrogen input. Glob Change Biol 15:2049–2066

Körner C (2006) Plant CO_2 responses: an issue of definition, time and resource supply. New Phytol 172:393–411

Kramer PJ, Boyer JS (1995) Water relations of plants and soils. Academic, San Diego, CA

Kull O (2002) Acclimation of photosynthesis in canopies: models and limitations. Oecologia 133:267–279

Landsberg J (2003) Modelling forest ecosystems: state of the art, challenges, and future directions. Can J For Res 33:385–397

LeBauer DS, Treseder KK (2008) Nitrogen limitation of net primary productivity in terrestrial ecosystems is globally distributed. Ecology 89:371–379

Leininger S, Urich T, Schloter M, Schwark L, Qi J, Nicol GW, Prosser JI, Schuster SC, Schleper C (2006) Archaea predominate among ammonia-oxidizing prokaryotes in soils. Nature 442:806–809

Levia DF, Frost EE (2003) A review and evaluation of stemflow literature in the hydrologic and biogeochemical cycles of forested and agricultural ecosystems. J Hydrol 274:1–29

Lewis SL, Malhi Y, Phillips OL (2004) Fingerprinting the impacts of global change on tropical forests. Philos Trans R Soc Lond B 359:437–462

Lilleskov EA, Bruns TD, Dawson TE, Camacho FJ (2009) Water sources and controls on water-loss rates if epigeous ectomycorrhizal fungal sporocarps during summer drought. New Phytol 182:483–494

Limm EB, Simonin KA, Bothman AG, Dawson TE (2009) Foliar water uptake: a common water acquisition strategy for plants of the redwood forest. Oecologia 161:449–459

Liste H-H, White JC (2008) Plant hydraulic lift of soil water – implications for crop production and land restoration. Plant Soil 313:1–17

Litton CM, Raich JW, Ryan MG (2007) Review: carbon allocation in forest ecosystems. Glob Change Biol 13:2089–2109

Liu L, Greaver TL (2009) A review of nitrogen enrichment effects on three biogenic GHGs: the CO_2 sink may be largely offset by stimulated N_2O and CH_4 emission. Ecol Lett (in press), DOI: 10.1111/j.1461-0248.2009.01351.x

Long SP, Ainsworth EA, Rogers A, Ort DR (2004) Rising atmospheric carbon dioxide: plants FACE the future. Annu Rev Plant Biol 55:591–628

Lovelock CE, Feller IC, Ball MC, Engelbrecht BMJ, Ewe ML (2006) Differences in plant function in phosphorus- and nitrogen-limited ecosystems. New Phytol 172:514–522

Luo Y (2007) Terrestrial carbon-cycle feedback to climate warming. Annu Rev Ecol Evol Syst 38:683–712

Luo Y, Su B, Currie WS, Dukes JS, Finzi A, Hartwig U, Hungate B, McMurtrie RE, Oren R, Parton WJ, Pataki DE, Shaw MR, Zak DR, Field CB (2004) Progressive nitrogen limitation of ecosystem responses to rising atmospheric carbon dioxide. BioScience 54:731–739

Magnani F, Mencuccini M, Borghetti M, Berbigier P, Berninger F, Delzon S, Grelle A, Hari P, Jarvis PG, Kolari P, Kowalski AS, Lankreijer H, Law BE, Lindroth A, Loustau D, Manca G, Moncrieff JB, Rayment M, Tedeschi V, Valentini R, Grace J (2007) The human footprint in the carbon cycle of temperate and boreal forests. Nature 447:848–852

Mahowald N, Jickells TD, Baker AR, Artaxo P, Benitez-Nelson CR, Bergametti G, Bond TC, Chen Y, Cohen DD, Herut B, Kubilay N, Losno R, Luo C, Maenhaut W, McGee KA, Okin GS, Siefert RL, Tsukuda S (2009) Global distribution of atmospheric phosphorus sources, concentrations and deposition rates, and anthropogenic impacts. Global Biogeochem Cy 22, GB4026, doi:10.1029/2008GB003240

Makarieva AM, Gorshkov VG (2007) Biotic pump of atmospheric moisture as driver of the hydrological cycle on land. Hydrol Earth Syst Sci 11:1013–1033

Makarieva AM, Gorshkov VG, Li B-L (2009) Precipitation on land versus distance from the ocean: evidence for a forest pump of atmospheric moisture. Ecol Complex 6:302–307

Marschner H (1995) Mineral nutrition of higher plants. Academic, London

Matson PA, McDowell WH, Townsend AR, Vitousek PM (1999) The globalization of N deposition: ecosystem consequences in tropical environments. Biogeochemistry 46:67–83

McGroddy ME, Silver WL, de Oliveira RC, de Mello WZ, Keller M (2008) Retention of phosphorus in highly weathered soils under a lowland Amazonian forest ecosystem. J Geophys Res 113:G04012. doi:10.1029/2008JG000456

McMurtrie RE, Norby RJ, Medlyn BE, Dewar RC, Pepper DA, Reich PB, Barton CVM (2008) Why is plant-growth response to elevated CO_2 amplified when water is limiting, but reduced when nitrogen is limiting? A growth-optimisation hypothesis. Funct Plant Biol 35:521–534

Meehl GA, Stocker TF, Collins WD, Friedlingstein P, Gaye AT, Gregory JM, Kitoh A, Knutti R, Murphy JM, Noda A, Raper SCB, Watterson IG, Weaver AJ, Zhao Z-C (2007) Global climate projections. In: Solomon S, Qin D, Manning M, Chen Z, Marquis M, Averyt KB, Tignor M, Miller HL (eds) Climate change 2007: the physical science basis. Contribution of working group I to the fourth assessment report of the intergovernmental panel on climate change. Cambridge University Press, Cambridge, United Kingdom/New York, NY, pp 747–845

Meinzer FC, Clearwater MJ, Goldstein G (2001) Water transport in trees: current perspectives, new insights and some controversies. Environ Exp Bot 45:239–262

Melillo JM, Field CB, Moldan B (2003) Element interactions and the cycles of life: an overview. In: Melillo JM, Field CB, Moldan B (eds) Interactions of the major biogeochemical cycles: global change and human impacts. Island Press, Washington, DC, pp 1–12

Michalzik B, Kalbitz K, Park J-H, Solinger S, Matzner E (2001) Fluxes and concentrations of dissolved organic carbon and nitrogen – a synthesis for temperate forests. Biogeochemistry 52:173–205

Millard P, Sommerkorn M, Grelet G-A (2007) Environmental change and carbon limitation in trees: a biochemical, ecophysiological and ecosystem appraisal. New Phytol 175:11–28

Monclus R, Dreyer E, Villar M, Delmotte FM, Delay D, Petit JM, Barbaroux C, Thiec D, Brechet C, Brignolas F (2006) Impact of drought on productivity and water use efficiency in 29 genotypes of *Populus deltoides* × *Populus nigra*. New Phytol 169:765–777

Nadelhoffer KJ, Emmett BA, Gundersen P, Janne Kjønaas O, Koopmans CJ, Schleppi P, Tietema A, Wright RF (1999) Nitrogen deposition makes a minor contribution to carbon sequestration in temperate forests. Nature 398:145–148

Näsholm T, Kielland K, Ganeteg U (2009) Uptake of organic nitrogen by plants. New Phytol 182:31–48

Neary DG, Ice GG, Jackson CR (2009) Linkages between forest soils and water quality and quantity. For Ecol Manage (in press) doi:10.1016/j.foreco.2009.05.027

Novaes E, Osorio L, Drost DR, Miles BL, Boaventura-Novaes CRD, Benedict C, Dervinis C, Yu Q, Sykes R, Davis M, Martin TA, Peter GF, Kirst M (2009) Quantitative genetic analysis of biomass and wood chemistry of Populus under different nitrogen levels. New Phytol 182:878–890

Oliver CD, Larson BC (1996) Forest stand dynamics. Wiley, New York

Ollinger SV, Richardson AD, Martin ME, Hollinger DY, Frolking SE, Reich PB, Plourde LC, Katul GG, Munger JW, Oren R, Smith M-L, Paw KT, Bolstad PV, Cook BD, Day MC, Martin TA, Monson RK, Schmid HP (2008) Canopy nitrogen, carbon assimilation, and albedo in temperate and boreal forests: functional relations and potential climate feedbacks. P Natl Acad Sci USA 105:19336–19341

Oren R, Ellsworth DS, Johnsen KH, Phillips N, Ewers BE, Maier C, Schäfer KVR, McCarthy H, Hendrey G, McNulty SG, Katul GG (2001) Soil fertility limits carbon sequestration by forest ecosystems in a CO_2-enriched atmosphere. Nature 411:469–472

Ostertag R (2001) The effects of nitrogen and phosphorus availability on fine root dynamics in Hawaiian forests. Ecology 82:485–499

Ostle NJ, Smith P, Fisher R, Woodward FI, Fisher JB, Smith JU, Galbraith D, Levy P, Meir P, McNamara NP, Bardgett RD (2009) Integrating plant-soil interactions into global carbon cycle models. J Ecol 97:851–863

Pallardy SG (2008) Physiology of woody plants. Academic, Burlington, MA

Parton W, Silver WL, Burke IC, Grassens L, Harmon ME, Currie WS, King JY, Adair EC, Brandt LA, Hart SC, Fasth B (2007) Global-scale similarities in nitrogen release patterns during long-term decomposition. Science 315:361–364

Paungfoo-Lonhienne C, Lonhienne TGA, Rentsch D, Robinson N, Christie M, Webb RI, Gamage HK, Carroll BJ, Schenk PM, Schmidt S (2008) Plants can use protein as a nitrogen source without assistance from other organisms. Proc Natl Acad Sci USA 105:4524–4529

Perry DA (1994) Forest ecosystems. The John Hopkins University Press, Baltimore, MD

Plante AF (2007) Soil biogeochemical cycling of inorganic nutrients and metals. In: Paul EA (ed) Soil microbiology, ecology, and biochemistry. Academic, Burlington, MA, pp 389–432

Polacek D, Kofler W, Oberhuber W (2006) Radial growth of *Pinus sylvestris* growing on alluvial terraces is sensitive to water-level fluctuations. New Phytol 169:299–308

Pregitzer KS, Burton AJ, Zak DR, Talhelm AF (2008) Simulated chronic nitrogen deposition increases carbon storage in Northern Temperate forests. Glob Change Biol 14:142–153

Purves D, Pacala S (2008) Predictive models of forest dynamics. Science 320:1452–1453

Rasmussen J, Kuzyakov Y (2009) Carbon isotopes as proof for plant uptake of organic nitrogen: relevance of inorganic carbon uptake. Soil Biol Biochem 41:1586–1587

Rauch JN, Pacyna JM (2009) Earth's global Ag, Al, Cr, Cu, Fe, Ni, Pb, and Zn cycles. Global Biogeochem Cy 23, GB2001, doi:10.1029/2008GB003376

Ravishankara AR, Daniel JS, Portmann RW (2009) Nitrous oxide (N_2O): the dominant ozone-depleting substance emitted in the 21st century. Science (in press), doi:10.1126/science.1176985

Reay DS, Dentener F, Smith P, Grace J, Feely RA (2008) Global nitrogen deposition and carbon sinks. Nat Geosci 1:430–437

Reich PB, Hungate BA, Luo Y (2006) Carbon-nitrogen interactions in terrestrial ecosystems in response to rising atmospheric carbon dioxide. Annu Rev Ecol Evol Syst 37:611–636

Reich PB, Oleksyn J, Wright IJ (2009) Leaf phosphorus influences the photosynthesis-nitrogen relation: a cross-biome analysis of 314 species. Oecologia 160:207–212

Reich PB, Tjoelker MG, Pregitzer KS, Wright IJ, Oleksyn J, Machado J-L (2008) Scaling of respiration to nitrogen in leaves, stems and roots of higher land plants. Ecol Lett 11:793–801

Reichstein M (2007) Impacts of climate change on forest soil carbon: principles, factors, models, uncertainties. In: Freer-Smith PH, Broadmeadow MSJ, Lynch JM (eds) Forestry and climate change. CAB International, Wallingford, U.K., pp 127–135

Richards JH, Caldwell MM (1987) Hydraulic lift: substantial nocturnal water transport between soil layers by *Artemisia tridentata* roots. Oecologia 73:486–489

Robertson GP, Groffman PM (2007) Nitrogen transformations. In: Paul EA (ed) Soil microbiology, ecology, and biochemistry. Academic, Burlington, MA, pp 341–364
Schenk HJ (2008) The shallowest possible water extraction profile: a null model for global root distributions. Vadose Zone J 7:1119–1124
Schenk HJ, Jackson RB (2005) Mapping the global distribution of deep roots in relation to climate and soil characteristics. Geoderma 126:129–140
Schimel JP, Bennett J (2004) Nitrogen mineralization: challenges of a changing paradigm. Ecology 85:591–602
Schlesinger WH (1997) Biogeochemistry – an analysis of global change. Academic, San Diego, CA
Schlesinger WH (2009) On the fate of anthropogenic nitrogen. Proc Natl Acad Sci USA 106:203–208
Schuur EAG (2003) Productivity and global climate revisited: the sensitivity of tropical forest growth to precipitation. Ecology 84:1165–1170
Sheil D, Murdiyarso D (2009) How forests attract rain: an examination of a new hypothesis. BioScience 59:341–347
Sievering H, Tomaszewski T, Torizzo J (2007) Canopy uptake of atmospheric N deposition at a conifer forest: part I-canopy N budget, photosynthetic efficiency and net ecosystem exchange. Tellus 59B:483–492
Silver WL (1998) The potential effects of elevated CO_2 and climate change on tropical forest soils and biogeochemical cycling. Clim Change 39:337–361
Sparks JP (2009) Ecological ramifications of the direct foliar uptake of nitrogen. Oecologia 159:1–13
Sutton MA, Simpson D, Levy PE, Smith RI, Reis S, van Oijen M, de Vries W (2008) Uncertainties in the relationship between atmospheric nitrogen deposition and forest carbon sequestration. Glob Change Biol 14:2057–2063
Thomas PA, Packham JR (2007) Ecology of woodlands and forests – description, dynamics and diversity. Cambridge University Press, Cambridge, U.K
Townsend AR, Asner GP, Cleveland CC (2008) The biogeochemical heterogeneity of tropical forests. Trends Ecol Evol 23:424–431
Treseder KK (2004) A meta-analysis of mycorrhizal responses to nitrogen, phosphorus, and atmospheric CO_2 in field studies. New Phytol 164:347–355
Treseder KK (2008) Nitrogen additions and microbial biomass: a meta-analysis of ecosystem studies. Ecol Lett 11:1111–1120
Treseder KK, Turner KM, Mack MC (2007) Mycorrhizal responses to nitrogen fertilization in boreal ecosystems: potential consequences for soil carbon storage. Glob Change Biol 13:78–88
Valladares F (2008) A mechanistic view of the capacity of forests to cope with climate change. In: Bravo F, LeMay V, Jandl G, von Gadow K (eds) Managing forest ecosystems: the challenge of climate change. Springer, New York, pp 15–40
van Dijk AIJM, Keenan RJ (2007) Planted forests and water in perspective. For Ecol Manag 251:1–9
Vitousek P (2004) Nutrient cycling and limitation: Hawai'i as a model system. Princeton University Press, Princeton, NJ
Vitousek P, Aber J, Howarth R, Likens G, Matson P, Schindler D, Schlesinger W, Tilman D (1997) Human alteration of the global nitrogen cycle: sources and consequences. Ecol Appl 7:737–750
Vitousek PM, Fahey TJ, Johnson DW, Swift MJ (1988) Element interactions in forest ecosystems: succession, allometry and input–output budgets. Biogeochemistry 5:7–34
Vrede T, Dobberfuhl D, Kooijman SALM, Elser JJ (2004) Fundamental connections among organism C:N:P stoichiometry, macromolecular composition, and growth. Ecology 85:1217–1229
Walbridge MR (1991) Phosphorus availability in acid organic soils of the lower North Carolina coastal plain. Ecology 72:2083–2100
Walker TW, Syers JK (1976) The fate of phosphorus during pedogenesis. Geoderma 15:1–19
Wamelink GWW, van Dobben HF, Mol-Dijkstra JP, Schouwenberg EPAG, Kros J, de Vries W, Berendse F (2009) Effect of nitrogen deposition reduction on biodiversity and carbon sequestration. For Ecol Manag (in press) doi:10.1016/j.foreco.2008.10.024

Wang X (2007) Effects of species richness and elevated carbon dioxide on biomass accumulation: a synthesis using meta-analysis. Oecologia 152:595–605

Wang Y-P, Houlton BZ, Field CB (2007) A model of biogeochemical cycles of carbon, nitrogen, and phosphorus including symbiotic nitrogen fixation and phosphatase production. Global Biogeochem Cy 21, GB1018, doi:10.1029/2006GB002797

Wardle DA, Walker LR, Bardgett RD (2004) Ecosystem properties and forest decline in contrasting long-term chronosequences. Science 305:509–513

Waring RW, Running SW (2007) Forest ecosystems – analysis at multiple scales. Elsevier Academic, Burlington, MA

Weedon JT, Cornwell WK, Cornelissen JHC, Zanne AE, Wirth C, Coomes DA (2009) Global meta-analysis of wood decomposition rates: a role for trait variation among tree species? Ecol Lett 12:45–56

Whiteside MD, Treseder KK, Atsatt PR (2009) The brighter side of soils: quantum dots track organic nitrogen through fungi and plants. Ecology 90:100–108

Wild M (2009) Global dimming and brightening: a review. J Geophys Res 114, D00D16, doi:10.1029/2008JD011470

Wright IJ, Reich PB, Westoby M, Ackerly DD, Baruch Z, Bongers F, Cavender-Bares J, Chapin T, Cornelissen JHC, Diemer M, Flexas J, Garnier E, Groom PK, Gulias J, Hikosaka K, Lamont BB, Lee T, Lee W, Lusk C, Midgley JJ, Navas M-L, Niinemets Ü, Oleksyn J, Osada N, Poorter H, Poot P, Prior L, Pyankov VI, Roumet C, Thomas SC, Tjoelker MG, Veneklaas EJ, Villar R (2004) The worldwide leaf economics spectrum. Nature 428:821–827

Wullschleger SD, Meinzer FC, Vertessy RA (1998) A review of whole-plant water use studies in trees. Tree Physiol 18:499–512

Xia J, Wan S (2008) Global response patterns of terrestrial plant species to nitrogen addition. New Phytol 179:428–439

Zhang L, Dawes WR, Walker GR (2001) Response of mean annual evapotranspiration to vegetation changes at catchment scale. Water Resour Res 37:701–708

Chapter 6
The Importance of Carbon Sequestration in Forest Ecosystems

Forests are important components of the terrestrial C cycle, and store large amounts of C in vegetation, detritus, and soil. Thus, C sequestration in sustainably managed forests can contribute to the drawdown of atmospheric CO_2. Substitution of fossil fuel based products with forest biomass for heat and power generation, and liquid transport fuels can contribute to abrupt climate change (ACC) mitigation through sustainable forest management. Forest biomass for bioenergy can be derived from forest residues and dedicated energy plantations of fast-growing, high-yielding tree species. However, harvest of forest residues must not result in the depletion of the soil organic carbon (SOC) pool due to decrease in C input and decrease in net primary production (NPP) associated with nutrient export. Judicious soil management is essential to sustainable productivity in short-rotation woody plantations dedicated to bioenergy production. Furthermore, negative impacts of large-scale plantations on the hydrological cycle must be minimized through conservation of water resources. Tree improvement by traditional breeding techniques and biotechnology can contribute to meet the increasing demand for renewable energy production from woody biomass.

International agreements on ACC such as those developed under the United Nations Framework Convention on Climate Change (UNFCCC) under the auspices of IPCC include forests as a significant component due to their large C pools and fluxes (Article 3.3 of the Kyoto Protocol). However, the Kyoto Protocol has been ineffective in enhancing forest C sequestration. Specifically, C accounting by forestry activities in Annex I (industrialized) countries is limited to changes associated with afforestation, reforestation and deforestation that are human-induced but it is unclear what specific changes are included. Furthermore, potentially beneficial projects including those that could avoid deforestation are excluded from forest management activities in non-Annex I countries. Thus, only one Clean Development Mechanism (CDM) forestry project has been certified globally by the end of 2008 (Basu 2009). Most important, approaches under the Kyoto Protocol may increase CO_2 emissions from forestry by accelerating deforestation. Yet, reduced emissions from deforestation and degradation (REDD) and compensated reduction (CR) in deforestation are improved mechanisms to include avoided tropical deforestation under the Kyoto Protocol or its successor. Also, the national inventory (NI)

K. Lorenz and R. Lal, *Carbon Sequestration in Forest Ecosystems*,
DOI 10.1007/978-90-481-3266-9_6, © Springer Science+Business Media B.V. 2010

approach may be an attractive alternative to the Kyoto Protocol approach. Both established and future international agreements rely particularly on forest C accounting and monitoring systems.

The importance of C sequestration in forest ecosystems depends on indirect human-induced effects on ecosystem functions through ACC. The forest management practices must be adjusted to adapt forests to ACC (Chapter 3). However, the importance of forest C sequestration depends also on direct effects of human activities on the large and most vulnerable forest C pools. In particular, tropical deforestation must be halted as it is the main contributor to global C emissions from forest ecosystems. Furthermore, exploitation of peatland forests must be reduced as disturbance of these ecosystems has the potential to release large amounts of C stored as peat. Old-growth forests steadily accumulate large amounts of C in biomass and soils for centuries, and, therefore, must be excluded from direct disturbance by human activities.

This chapter describes the use of biomass derived from forest residues and dedicated energy plantations of fast-growing, high-yielding tree species for bioenergy including a section about tree improvement through biotechnology. Then, the role of forests in current and proposed international agreements on ACC mitigation are compared. This section includes also a discussion about methods for forest C accounting and monitoring systems. Finally, the importance of the large and most vulnerable forest C pools in tropical, peatland and old-growth forests are discussed.

6.1 Bioenergy from Tree Plantations

Adverse effects of greenhouse gas (GHG) emissions on the environment together with declining petroleum reserves and rising energy costs enhance efforts for production and use of fuels from plants (Gomez et al. 2008). Fuel is any material that burns readily in air, and biofuels are materials of biological origin that are used for producing heat and other forms of energy (Seth 2004). Woody biomass is organic material from woody plants, especially trees, that is not utilized in conventional wood products (Nicholls et al. 2009). Woody biomass consists of small, stems, branches, twigs, residues of harvesting and wood-processing, and woody biomass from non-forest sources such as urban waste. Bioenergy is the energy available from biological sources and can be obtained from woody biomass in the form of charcoal, gas, ethanol, fire, heat, methanol, oil and producer gas. Bioenergy from forest biomass for heat and power generation, and for liquid transport fuels is renewable and can be a sustainable form of energy production (FAO 2008; Scharlemann and Laurance 2008). The potential sustainable use of woody biomass energy globally is about 30% of the global energy use (Nicholls et al. 2009). However, compared to other energy-related solutions to global warming, air pollution, and energy security biofuel options provide no specific benefit. On the contrary, biofuels may result in significant negative impacts with respect to climate, air pollution, land use, nutrient and water demand, wildlife damage, and chemical waste (Jacobson 2009). For example, biofuel combustion is a major contributor to global BC emissions and BC aerosols contribute to ACC by radiative forcing (Fernandes et al. 2007; Shindell and Faluvegi 2009).

Producing biomass for bioenergy contributes to the C mitigation options of the forestry sector (Nabuurs et al. 2007). Specifically, permanent savings in C emissions can be achieved by using sustainable biomass to substitute for fossil fuel (Schlamadinger and Marland 2000). In addition to maintaining the forest area, the use of forests as a source of renewable energy can have the most impact globally to mitigate ACC through forest management (Bravo et al. 2008). Biomass energy plantations can provide adaptation benefits such as shading, reduce water evaporation, and decrease vulnerability to heat stress (Barker et al. 2007). Increasing the productivity of plantation forests through traditional breeding techniques and application of biotechnology could simultaneously enhance C sequestration and meet greater demands for renewable wood and bioenergy products (Novaes et al. 2009).

Similar to natural forests, short-rotation tree plantations for biomass production are affected by increases in atmospheric CO_2 abundance. For example, no long-term photosynthetic acclimation to CO_2 has been observed in fast-growing plantation trees, but species-specific responses in wood composition (i.e., lignin content) and the energy content have been reported (Davey et al. 2006; Luo and Polle 2009). Otherwise, stomatal conductance of fast-growing *Populus* trees in short-rotation plantations is often reduced under elevated CO_2 but the trees can use more water with strong negative effects on the regional water use (Tricker et al. 2009). To reduce vulnerability to ACC, bioenergy plantations must include changes in rotation periods, salvage of dead timber, shift to tree species more productive under the new climatic conditions, mixed species forestry, mosaics of different species and ages, and fire protection measures (Nabuurs et al. 2007). However, policies and measures in the forestry sector must take adaptation and mitigation options into account (Bernier and Schoene 2009).

6.1.1 Bioenergy and Biofuels from the Forest Sector

Mitigation options are large when woodfuels are used to displace fossil fuels by providing sustained C benefits. Woodfuels consist of fuelwood, charcoal and black liquor, and currently supply about 7% of the global energy use (Mead 2005). Advanced wood combustion, for example, is being deployed throughout Europe and has a great potential for community-based energy production in the United States (Richter et al. 2009). Burning compacted biomass energy pellets as a heat source is probably the most efficient commercial use of biomass energy (Field et al. 2007; Nicholls et al. 2009). On the other hand, the successful industrialization of ethanol production from lignocellulosic or woody biomass remains to be demonstrated, and breakthrough technologies are still needed (Barker et al. 2007; Stephanopoulos

2007; Himmel and Picataggio 2008). For example, production in the first commercially operational cellulosic ethanol plant in the United States in southern Georgia is scheduled to commence in the second quarter of 2010 (http://www.rangefuels.com/construction-georgia.html). This plant will use thermo-chemical processes to convert woody biomass into cellulosic ethanol. Other processes convert a portion of the woody biomass into simple sugars that are fermented and converted into ethanol (Gomez et al. 2008). However, hydrocarbons derived from biomass rather than ethanol will likely be the dominant biofuel in the future (Regalbuto 2009). The global energy demand projected for the year 2030 can probably be met sustainably and economically from lignocellulosic biomass grown on areas which do not compete for food production, and are agriculturally marginal lands (Metzger and Hüttermann 2009). Thus, forests can become a major source of bioenergy without competing for increasing food demand provided by existing arable land, without endangering the supply of industrial roundwood and wood fuel, and without further deforestation (Smeets and Faaij 2007). However, how competing uses of forest residues for fiberboard production may affect the availability of forest biomass for bioenergy remains to be determined (Barker et al. 2007).

Forest residues can be used as feedstock for bioenergy (Nabuurs et al. 2007). Residues are available from additional stemwood fellings, from thinning salvage after natural disturbances, final harvest fellings, processing forest products, and after end use. Typically, between 25% and 50% of logging residues and between 33% and 80% of processing residues can be recovered and used for energy purposes. Other possible sources of woodfuel include the large plantations of cocoa (*Theobroma cacao* L.), coconut (*Cocos nucifera* L.), rubber and oil palm (Mead 2005). Municipal wood wastes, and biomass from urban trees and forests can also be used for energy. Up to 15% of the current primary energy consumption can be generated from forest biomass. However, harvested forest biomass for bioenergy must not exceed growth increment to maintain the CO_2 balance, and CO_2 emitted during production, transportation and processing must be taken into account. Indirect effects of dedicated bioenergy plantations on the CO_2 balance may arise through replacing natural forests harvested. Such a strategy would relieve pressure on the large above- and belowground C pool of old-growth forests (Fenning and Gershenzon 2002).

The forest sector can contribute to the mitigation of ACC by replacing fossil fuels and by sequestering C in forest biomass, soil, and in forests products (Lemus and Lal 2005; Hennigar et al. 2008). However, conversion of forest to plantations with conifers may reduce the SOC pool by 15% but changes are small after conversion to plantations with broadleaf trees (Guo and Gifford 2002). Soil C was, however, restored to the original levels in plantations more than 40 years old. Otherwise, establishment of tree plantations on degraded land can enhance SOC sequestration in the tropics (Lal 2005). Thus, whether short-rotation plantations, secondary forests or old-growth forests are more effective in sequestering atmospheric CO_2 is a matter of debate (Schulze et al. 2000). Yet, by slowing down the increase in atmospheric CO_2 concentration associated with fossil fuel burning, bioenergy from forest biomass can also diminish the effects of ACC on C sequestration in forest ecosystems. Forest plantations for C sequestration or biofuels may also affect ACC or its impact through land surface albedo change or changes in the surface moisture budget (Betts 2007). For example, the climate benefits of C storage reinforce biophysical effects of plantations in the tropics to cool the Earth (Jackson et al. 2008). In contrast, the climate benefits of C storage in forest plantations in the boreal region may be counteracted by warming through absorption of more sunlight. The greatest uncertainties about influences of biophysical changes on the ACC lie in temperate forests. Effects of large-scale bioenergy plantations on the NECB of forest ecosystems and the surrounding landscape are not thoroughly understood.

Plantations managed for growth and renewability are an increasing source of traditional industrial wood (Carle et al. 2002). Wood residues extracted from these plantations can also be source for energy production but dedicated bioenergy plantations are increasingly important (FAO 2008). In the long term, however, the recovery of wood residues left from harvesting operations may reduce C sequestration in forests. In particular, the SOC pool may decrease due to decrease in C inputs, NPP may decrease due to increase in nutrient export with residue removal, and C losses may increase through accelerated soil erosion (Powers et al. 2005). Also, forest ecosystem functions and biodiversity may be altered with possible negative effects on the C balance (Marland and Obersteiner 2008).

Forest biomass produced in short-rotation plantations is a source for bioenergy. High growth rates and yield and, thus, high C accumulation rates in the tree biomass are achieved in these plantations over a short period of time. As the major fraction of the aboveground nutrient pool is confined to the plant parts most frequently harvested, proper soil management with a focus on soil fertility enhancement is required to sustain productivity (Ericsson 1994; Powers 1999). Thus, intensified forest management in bioenergy plantations requires supplementary fertilizers. Increases in nutrient losses occur, for example, as short harvesting cycles are combined with enriched tissue nutrient concentrations. The availability of Ca and Mg, in particular, may be in short supply when they are continually depleted through harvest. Leaching and erosion losses may also exacerbate long-term decreases in productivity. More intensive soil preparation methods such as plowing, stump and harvesting residue removal, and herbicides may also accentuate nutrient losses.

Improved nutrition favors canopy over root development and thereby increase the demand for water in high-yielding energy plantations. As described in Section 5.3.2, increase in water consumption in plantations may cause water shortages and decrease biomass C sequestration, in particular, by degrading soil quality (Farley et al. 2005; van Dijk and Keenan 2007). Interception losses are more from those forests which have large LAI throughout the year (Bates et al. 2008). Thus, losses may be larger for fast-growing forests with high rates of C storage such as bioenergy plantations than for slow growing forests. Extensive tree plantations as feedstock for ethanol production, for example, can negatively affect water resources (Evans and Cohen 2009). However, many assumptions about forest plantations and water use have not been fully validated (Vanclay 2009). Well-designed experiments are scanty, especially, in the tropics. A few studies in the tropics on afforestation or planting trees in agricultural fields show increases in the infiltration capacity (Ilstedt et al. 2007), which influences groundwater recharge, and the partitioning of runoff into slow flow and quick flow. It is, however, less clear if replacing natural forests with plantations in the tropics increases water use when there are no changes in either the rooting depth or the stomatal behaviour of the tree species (Bates et al. 2008).

Environmental benefits of forest plantations for obtaining biofuels are compromised by increase in risks of soil degradation, water runoff, and downstream siltation and capture of polluting agricultural runoff (Bates et al. 2008). Plantations can also be designed and managed for increase in water use efficiency. In particular, the selection of species is important as, for example, eucalypts have a larger impact than pines on reducing runoff when grasslands are converted to energy plantations (Farley et al. 2005). Also, the rate of increase in evapotranspiration is more drastic under eucalypts because of their rapid early growth and canopy closure. Thus, in drier regions eucalypts cause more severe reductions in runoff whereas pines may cause more severe reductions in runoff in wetter regions. In addition, areas surrounding bioenergy plantations may also suffer from decrease in water availability (Ericsson 1994). For example, plantations of species with high water demand in water-limited areas can reduce water flow to other ecosystems and affect aquifers recharge (Bates et al. 2008).

Dedicated biomass tree plantations may serve as feedstock for biofuel production (Gomez et al. 2008). Native and introduced species are either planted or seeded. However, insect pests and pathogens can limit the sites on which introduced species can be grown successfully in plantations outside their natural range (FAO 2009). For example, the red bank needle blight (*Mycosphaerella pini*) and western gall rust (*Endocronartium harknessii*) may infect *Pinus radiata*. Fast growing tree species suitable for bioenergy production in temperate climates are Black or Hickory Wattle (*Acacia mangium* Willd.), Gamhar (*Gmelina arborea* Roxb.) and several *Eucalyptus*, *Salix* and *Populus* species (FAO 2008). *Salix* and *Populus* species are highly productive as they are characterized by relatively high CO_2-exchange rates, light-use efficiencies and photosynthetic capacity among woody species (Karp and Shield 2008). Fast growing tree species produce up to 20 Mg ha^{-1} wood annually in temperate regions and up to >30 Mg ha^{-1} annually in tropical dry forests (Metzger and Hüttermann 2009). Further enhancement in productivity can be achieved by silvicultural practices and genetic innovations through traditional breeding techniques. Fast-growing genotypes can be selected without compromising wood density (Weber and Sotelo Montes 2008). Traditional tree breeding is, however, constrained by long reproductive cycles and complex reproductive characteristics (van Frankenhuyzen and Beardmore 2004). Thus, biotechnology is increasingly used for genetic modification of trees (Strauss and Bradshaw 2004).

6.1.2 Genetic Modification of Dedicated Biomass Trees by Biotechnology

Biotechnology is defined as the use of the whole or targeted portions of organisms to provide quantitative information and/or desired products, including the isolation and/or manipulation of specific genetic components of that organism (El-Kassaby 2004).

It utilizes fundamental discoveries in the field of plant tissue culture for clonal forestry, gene transfer techniques, molecular biology, and genomics (Nehra et al. 2005). Thus, innovative technologies provide the basis for acceleration in the improvement of forestry through biotechnology. Possible target for tree improvement are economically important and more widespread planted species. Important among these are tropical/subtropical regions species such as Black or Hickory Wattle, Thorn mimosa or Lekkerruikpeul (*Acacia nilotica* (L.) Willd. *ex* Delile), *Eucalytpus* spp. (especially Flooded gum or Rose gum (*E. grandis*) and Blue Gum (*E. globulus* Labill.)), Common Teak (*Tectona* grandis), *Populus* spp. (and hybrids) and Monterey Pine or Radiata Pine (*Pinus radiata* D.Don) (FAO 2006). Also, the rubber tree and other tropical plantation species providing non-forest products can be improved by breeding for wood or fibre production (Kendurkar et al. 2006). Such energy plantations may also include oilseed plants of the genus *Jatropha* which includes shrubs and trees such as *Jatropha curcas* Linnaeus 1753. (FAO 2008). Economically important conifers for timber and pulpwood, primarily *Pinus* spp., *Picea* spp., and *Pseudotsuga* spp. are also target for genetic modification research (Nehra et al. 2005). However, as forest research and tree-improvement schemes are time consuming and expensive only a handful of tree species are likely to be focus of major improvement through biotechnology (Fenning and Gershenzon 2002). Also, additional challenges specific to conifers are the reasons why the application of genetic engineering to conifers has lagged behind those of non-coniferous trees (Henderson and Walter 2006).

A wide range of plant traits can be targeted for enhanced tree biomass production and processing through biotechnology (Ragauskas et al. 2006). Specifically, genes associated with complex value-added traits have been identified in trees (Nehra et al. 2005). Specific genes can now be used to alter tree growth and characteristics governing wood development. This scientific innovation has been achieved specifically in *Populus* spp. as deciduous trees are easier to transform than conifers. Scientists have been successful in modification of wood and lignin. Although a number of genes involved in growth rates have been mapped, little progress has been made in testing these genes (Van Frankenhuyzen and Beardmore 2004). However, improvements in yield similar to those achieved in agricultural crops may be possible (Fenning and Gershenzon 2002). Biotic and abiotic stress resistance can also be enhanced in forest trees through genetic transformation. Establishment and choice of genetic engineered trees can strengthen C sequestration in forest ecosystems through faster growth, increased wood quantity, reduced juvenile phase, delayed flowering to divert energy used for reproduction into growth, and increased tolerance against abiotic and biotic stresses (Hoenicka and Fladung 2006; Karp and Shield 2008). Genetically improved trees may have short stature to increase light access and to enable dense growth, large stem diameter, and reduced branch numbers to maximize energy density for transport and processing (Rubin 2008).

The first bioenergy tree to have its genome sequenced is black cottonwood (*Populus trichocarpa* (Torr. & Gray)) (Tuskan et al. 2006). The availability of this genome sequence has significantly enhanced gene discovery in *Salix* spp. due to the collinearity of both genomes (Karp and Shield 2008). Wood and crown development, and disease resistance are now studied in detail and can be improved for bioenergy production through genetic engineering (Novaes et al. 2009). This technology may also soon be possible for *Eucalyptus* as the whole genomic sequence of this species will be available in the near future (http://www.fabinet.up.ac.za/eucagen). Maximizing biomass yield per unit land area is one of the goals of gene technology (Rubin 2008). Research challenge, however, is how to change (photo)thermal time sensitivity to lengthen the growing season without risking frost damage or limiting remobilization of nutrients following senescence (Karp and Shield 2008)? Additional challenges include: increasing aboveground biomass without depleting belowground reserves required for next year's growth to reduce the requirement for nutrient application, and increasing the aboveground biomass without increasing the water use.

Trees domesticated for bioenergy have less need for rigid cell walls as their lifetime is shorter than those of wild trees. Less rigidity can be achieved by altering the ratios and structures of cell wall macromolecules (e.g., cellulose, hemicellulose, lignin, pectin) to facilitate post-harvest deconstruction. Increasing cellulose contents and the disordered regions in the cellulose microfibrils would facilitate the conversion of tree biomass into sugar (saccharification) for liquid biofuel production (Gomez et al. 2008). In particular, increasing the proportion of lignin and hemicellulose-poor but cellulose-rich tension wood is a promising approach. Hemicelluose, on the other hand, is also a rich substrate for saccharification as it contains up to 50% of the polysaccharides in lignocellulose. Metabolizing pentose sugars in hemicellulose by fermentative organisms is, however, difficult. In general, recalcitrance of biomass to digestion is mainly related to lignin. Thus, modifying lignin content and structure would greatly enhance the suitability of lignocellulosic forest biomass for liquid biofuel production.

The genetic control linking higher growth rates to wood with less lignin has recently been identified in poplar and this discovery may facilitate engineering of poplar trees for cellulosic ethanol production (Novaes et al. 2009). In addition, improved burning can also be achieved by altered lignin properties (Fenning and Gershenzon 2002). Poplar and *Eucalyptus* modified in lignin content and structure have been developed for improved pulping in the process of paper making, but the data from long-term field trials are scanty (Pilate et al. 2002; Baucher et al. 2003). The next generation of bioenergy trees may even harbor the genes encoding enzymes necessary for self-degradation of cellulose and lignin, activated before harvest or at the end of the growth cycle to overcome the barrier of lignocellulosic recalcitrance (Himmel et al. 2007).

Large-scale afforestations of transgenic trees are scanty despite the promising aspects of genetic modifications (Hoenicka and Fladung 2006). Yet, the only exceptions are plantations of insect resistant *Bacillus thuringiensis* poplar in China (Hu et al. 2001). Field tests with transgenic *Populus*, *Pinus*, *Liquidambar*, and *Eucalyptus* trees are currently conducted in several countries (van Frankenhuyzen and Beardmore 2004). In the United States, for example, trials with genetically modified (GM) trees test herbicide tolerance, insect and disease resistance, and altered lignin.

Proponents of transgenic forest plantations expect that by 2010 the first transgenic forest tree product may be commercialized but others are pessimistic in terms of their commercial pursuit (van Frankenhuyzen and Beardmore 2004; Nehra et al. 2005). The biosafety of GM trees must be evaluated with respect to horizontal and vertical gene transfer, invasive potential, gene instability and impact on other organisms and ecosystem processes (Hoenicka and Fladung 2006). For example, ecological impacts in a 4-year study with reduced-lignin poplar were higher decomposition rates of roots (Pilate et al. 2002). The issue of the stability of transgene expression is of particular relevance for long-lived crops like trees (van Frankenhuyzen and Beardmore 2004). The most difficult hurdle, however, for the advance of genetic modification of trees is the lack of public confidence as is observed for GM food crops in Europe. In summary, for safe deployment of this technology relevant biological data are needed that only medium- to large-scale release of transgenic trees on full rotation can provide (van Frankenhuyzen and Beardmore 2004).

6.2 Forest Carbon Sequestration Under the United Nations Framework on Climate Change, the Kyoto Protocol and Post-Kyoto Agreements

Forests store large amounts of C in vegetation, detritus and soil (see Chapters 2 and 4). About half of the total terrestrial C sink is located in forests (Fig. 1.2). Forest C flows, on the other hand, comprise a significant part of the global GHG emissions (Plantinga and Richards 2008). Improved forestry practices are a natural way to draw down CO_2 while, for example, reforestation could absorb a substantial fraction of the 60 ppm CO_2 emitted over the past few hundred years due to deforestation (Hansen et al. 2008). Furthermore, the application of biochar produced by pyrolysis of forestry residues to soil may be C negative as atmospheric CO_2 is sequestered for centuries to millennia (Lehmann 2007; Glaser et al. 2009). Yet, a forest/soil draw-down scenario that reaches 50 ppm by 2150 appears to be feasible (Hansen et al. 2008). Thus, forest management activities play a key role in the context of global change and sustainable development through mitigation of climate change (Nabuurs et al. 2007). Specifically, the forest sector can reduce emissions and/or increase removals through sinks by maintaining or increasing the forest area, the stand-level C density, the landscape-level C density, and by increasing off-site C stocks in wood products and enhancing wood products and fuel substitution. Forestry projects can remove CO_2 from the atmosphere at lower costs than projects to reduce emissions (van Kooten et al. 2004). Forest C offset projects must address, in particular, the risk of reversal, the intentional or unintentional release of C back to the atmosphere due to storms, fire, pests, climate change, land use decisions, and many other factors (Galik and Jackson 2009). However, successful increasing C storage through changes in forest management practices while maintaining economic profitability is challenging (Boyland 2006). Furthermore, market interactions in forest C sequestration program analyses require considerable attention (Richards and Stokes 2004). Appropriate institutions must be put in place to include terrestrial ecosystem sink activities including those in forests in a C trading system (van Kooten 2009). Nevertheless, forests are a significant component of any international agreement on climate change.

6.2.1 Current Commitments for Forest Carbon Sequestration

Most of the observed increase in global average temperatures since the mid-twentieth century is attributed to observed increase in anthropogenic GHG concentrations (i.e., CO_2, CH_4, N_2O, F-gases; IPCC 2007). Thus, the United Nations Framework Convention on Climate Change (UNFCCC) addressed the increasing concentrations of GHG in 1992 and called to stabilize GHG concentrations at levels to prevent dangerous anthropogenic interference with the climate system (UN 1992). This convention was formulated in 1997 with a binding, quantitative agreement for reducing emissions of GHG, the Kyoto Protocol, and signed by 183 parties by the end of 2008 (UNFCCC 2008). All GHGs other than CO_2 combined cause climate forcing comparable to that of CO_2 but their rate of enrichment is falling and, thus, total GHG climate forcing is determined mainly by CO_2 (Hansen and Sato 2009). To preserve the life-support processes of the planet Earth, atmospheric CO_2 concentration must be reduced <350 ppm (Hansen et al. 2008). The primary option to reduce emissions is increasing the negative growth rate in C intensity of the global economy (Raupach et al. 2008). However, as current levels already exceed 350 ppm net CO_2 emissions must approach zero because of the long lifetime of CO_2 in the atmosphere (Matthews and Caldeira 2008).

The Kyoto Protocol sets binding targets for emissions of six GHGs (CO_2, CH_4, N_2O, SF_6, hydrofluorocarbons and perfluorocarbons) from Annex I (industrialized) nations. Industrialized countries agreed to reduce their collective GHG emissions by 5.2% compared to the year 1990. Removing GHG from the atmosphere is also recognized as an important strategy. The rate of build-up of atmospheric CO_2, in particular, can be reduced through land management activities. By Land Use, Land-Use Change, and Forestry (LULUCF) activities, C losses from plants and soil can be reduced while return of atmospheric CO_2 into plants and soils can be enhanced (Schlamadinger and Marland 2000). Countries can, thus, create C sinks to offset emissions by planting trees (afforestation and reforestation) instead of deforestation (Kyoto Protocol Article 3.3: ARD). Furthermore, improving management of forests can create new or enhance existing C pools (Art. 3.4: Additional activities). These new C pools can be a tradable resource (Schulze et al. 2002). However, only ARD activities that are directly human-induced in Annex B countries and initiated after January 1, 1990 can be accounted for as C credits.

Trading of emission credits and mitigation projects in other countries are other components of the Kyoto Protocol. Achieved C units can be traded between Annex I countries. Clean Development Mechanism (CDM) projects, otherwise, involve developed and developing countries and result in an increase in global total of assigned amounts of emission reduction or enhancement of C sinks (Schlamadinger and Marland 2000). Offset credits are rewarded to individual sequestration projects. CDM allow industrialized countries with a GHG reduction commitment (Annex B countries) to invest in projects that reduce emissions in developing countries as an alternative to more expensive emission reductions in their own countries. Forestry-related C gains in non-Annex I countries, however, are limited to afforestation and reforestation projects (Plantinga and Richards 2008). Projects that reduce deforestation are, thus, excluded. Most important, deforestation may actually accelerate by shifting timber harvesting from Annex I to non-Annex I countries. However, relatively immediate and dramatic actions are needed to curb large-scale deforestation (Mann 2009). Thus, deforestation will be included in the successor to the Kyoto Protocol, but details about market-based approaches to reduce deforestation are under discussion (Tollefson 2008b).

Only one forestry project, in China's Pearl River basin, has been certified by the CDM mechanism by the end of 2008 compared to 1,270 schemes worldwide to offset C in other sectors (Basu 2009). Furthermore, around 25 forest-based CDM projects are set to be approved by UNFCCC. However, the UNFCCC's approval process is very stringent. Some of the biggest difficulties lie, in particular, in documenting how much C the growing trees will absorb (Basu 2009). The required methodologies and calculations are too complex and the costs are too high to justify. In addition to accounting for leakage (i.e., GHGs released into the atmosphere in the process of setting up the project) (Sathaye and Andrasko 2007a), demonstrating that the project meets the crucial additionality concept (i.e., GHG reductions are more than those that would have occurred anyway) are challenging (Basu 2009). In summary, the current system for including forestry activities in accounts for GHG mitigation efforts is not perfect (Schlamadinger et al. 2007). Mitigation has been given a dominant role as a policy goal rather than also addressing climate-change adaptation options (Seppälä et al. 2009). Thus, the Kyoto Protocol has proven ineffective in enhancing forest C sequestration (Streck et al. 2008).

6.2.2 *Future Forest-Based Systems for Carbon Sequestration*

National governments that have signed the UNFCCC and the Kyoto Protocol must report the results of periodic inventories of GHG emissions and removals. This includes also national forest C inventories as an integral part with reporting intervals of approximately 3–4 years. However, conventional forest inventories typically do not include C stocks in litter, detritus and soil (LeMay and Kurz 2008). A range of measures are required to obtain estimates of past, current, and future forest C stocks. First of all, the current extent of managed forest land needs to be determined, and the changes, additions and deletions to the forest land base during the monitoring period needs to be known. The C stocks and stock changes in above- and belowground biomass, including stems, foliage, tops, branches, fine and coarse roots, and those in litter, coarse woody debris (CWD) and soil need to be estimated. For forecasting future forest C storage, the natural ecosystem dynamics and disturbances, and human-caused disturbances such as harvesting, reforestation, afforestation, conversion to other land uses, and insect and disease management need to be known (LeMay and Kurz 2008). Forest C accounting and monitoring systems are comprised of combinations of remote sensing data acquisition systems, spatial databases on forest cover, management activities and natural disturbances, parameter databases, and simulation models to do the numerical analyses (Sánchez-Azofeifa et al. 2009). Integrating multiple data sources across spatial scales can be done either by a 'bottom-up' approach based on an extensive network of repeatedly measured plots or by a 'top-down' approach based on remotely-sensed data (LeMay and Kurz 2008). Such approaches have been used more in boreal and temperate forests to develop relatively accurate C budgets and less so in tropical forests.

Accuracy and credibility in reporting forest C pool and fluxes a national scale is a voluntary process (Andersson et al. 2009). In contrast to CDM, the national approach of forest C inventories have few problems of leakage and baseline definition, a wide range of C sequestration options, a shift to appropriate aggregate effects from local project effects, and manageable program at the international level (Andersson and Richards 2001). This is a promising approach to address problems associated with additionality, leakage, and permanence in which nations conduct periodic inventories of their entire forest C pool (Plantinga and Richards 2008). The measured pool is compared to a negotiated pool to determine the number of tradeable credits (or debits to cover) in the permit market. In contrast to project-specific approaches for estimating baseline emissions of climate mitigation projects, regional approaches explicitly acknowledge the heterogeneity of C density, land use change, and other key baseline driver variables across a landscape (Sathaye and Andrasko 2007b).

6.2.2.1 Carbon Monitoring in Forestry Projects

Forestry projects are easier to quantify and monitor than national inventories (Brown 2002b). For example, several potential candidate variables for measurements at landscape-scale C monitoring sites with recommended measurement intervals have been identified for the North American Carbon Program (Hoover 2008). Variables differ in their representation of a process or property to quantify and understand C cycling at the forest site, usefulness as a model variable, and practicality of implementation. Thus, the highest priority is given to measurements of the aboveground biomass. In decreasing order of priority, this is followed by measurements of CWD, litterfall, site history, foliar N, forest floor mass, temperature, stand (site) age and soil CO_2 flux. Otherwise, the often largest and most stable forest C pool in the soil is given a medium priority but to what depth SOC must be measured is unclear. Labile C, root allometry and litter moisture are variables with the lowest priority for measurement at landscape-scale C monitoring sites (Hoover 2008). However, developing a sampling strategy for assessing landscape-scale C dynamics requires an objective assessment of the costs and benefits of a range of potential measurements (Bradford et al. 2008). Specifically, relating the benefit of a given measurement to the magnitude of the C pool or flux that it measures provides an objective quantification of importance vis-à-vis the cost of the measurement. For example, small C pools contribute little to overall estimates of forest ecosystem C storage and errors in those pools are only of minor consequences. Furthermore, the benefit of measurement of an individual C flux for the net forest ecosystem C balance is equivalent to the flux magnitude. The cost of measuring a process is often determined by the sampling requirements for characterizing variability and, to a lesser extent, by differences in measurement difficulty. For example, soil CO_2 efflux and litterfall require multiple plot visits per year and are, therefore, labor intensive and costly. Finally, landscape-scale C measurements must be compared with other approaches of assessing forest C dynamics, notably eddy covariance techniques and ecological simulation models that use the remote sensing data (Bradford et al. 2008).

The future effectiveness of international climate mitigation efforts depends on the political will to invest in accurate GHG accounting methods including forest C inventories. Field-based inventories can serve as the basis for the development of allometric models to estimate timber yield based on easy to measure characteristics. For example, LAI, canopy height, canopy cover, dominant species and ground litter amounts may be used to model biomass and C levels in forest ecosystems as high correlation exist between timber yield and C content. Such allometric equations are available for many global forests, but knowledge gaps exist for tropical areas and wetlands (Brown 2002a).

6.2.2.2 Monitoring the Soil Organic Carbon Pool

Critical factors in forest C inventories are precise measurements of the SOC pool and verification of the amount of C sequestered in the soil for implementing C trading programs. However, forest SOC pools are spatially variable due to soil heterogeneity (Schöning et al. 2006). This large within stand variability affects the detection of changes in SOC on a landscape, watershed or regional scale (Conant et al. 2003). Estimates of forest SOC pools for larger spatial scales across different soil depth intervals are limited, in particular, by time and cost constraints. However, spatial extrapolation from smaller to larger scales can be applied for SOC mapping (Mishra and Lal 2010). A common method is the 'measure and multiply approach' (MMA) for coarse predictions of SOC pool at regional to global scales. Specifically, the study area is divided into different strata and the point measurements of SOC concentration within each stratum are multiplied by its area. However, this approach does not account for the spatial variability due to soil heterogeneity within each stratum. Furthermore, assigning mean SOC concentrations from a small number of samples to a mapping area may introduce large errors. Another approach is 'soil landscape modeling' (SLM) by which the variability of soils is analyzed with respect to changes in environmental variables known to impact variations in soil properties. Data of multiple environmental factors are input variables to calibrate a model which is then used to make predictions over the entire study area. Estimates of the SOC pool obtained by this approach have lower estimation errors compared to the MMA approach because changes in environmental variables are taken into account. Otherwise, the SLM approach assumes that the relationship between the environmental variables and the SOC pool is constant over the study area. Thus, estimation errors may increase in large-scale studies (Mishra and Lal 2010). Nevertheless, the combination of digitized data on SOC profile depth distribution with geostatistics can quantify SOC pool variability at different depth intervals and spatial scales (Mishra et al. 2009). However, the detection of temporal changes in SOC pools is difficult because SOC sequestration rates are generally small in comparison to the amount of C stored in forest soils (Watson et al. 2000).

6.2.2.3 Remote Sensing of Forest Carbon Pools

Costly fieldwork is required to collate data for independent model variables such as rate of growth (C sequestration in biomass) and the dynamics of detritus and SOC pools. Thus, remote sensing is a primary source of data for the assessment of forest C storage and fluxes (Waring and Running 2007). Remote sensing provides fast and consistent access to information required to assess forest C pools and fluxes (Sánchez-Azofeifa et al. 2009). However, accurate information on what is present below the forest canopy such as detritus, CWD and SOC cannot be provided by remote sensing. Information required for forest C flux accounting include the forest area, data on direct human-induced alterations and changes in C pools by natural disturbances such as wind throw and fire can be accessed through remote sensing at different spatial and spectral resolutions. Instruments with higher resolution (<5 m) are better adapted to measuring forest variable inputs for allometric models (Andersson et al. 2009). At moderate resolution, Landsat remains one of the most used and most applied remote-sensing tools. Moderate Resolution Imaging Spectroradiometer (MODIS) is explicitly designed to produce measures of GPP, and has been validated for boreal and temperate deciduous forests, and to produce land-cover data for tropical forests. In contrast to optical satellite-borne sensors, C estimation in forests covered by clouds is possible by radar-based approaches such as Synthetic Aperture Radar (SAR) (Andersson et al. 2009). However, one serious limitation of SAR for estimating national inventories of aboveground biomass and C is its sensitivity to topographical changes. Specifically for mapping aboveground biomass stocks, using data from satellite remote sensing can reduce uncertainty for C monitoring (Goetz et al. 2009).

6.2.2.4 Modeling Forest Carbon Pools and Fluxes

Accurate methods are required to monitor and report forest C pools or fluxes under current and future international agreements. To achieve the highest degree of estimation certainty the inventory change method must be applied (Kurz et al.

2009). From two detailed forest inventories at different points in time C pools are calculated by using models such as FORCARB (Heath and Birdsey 1993). However, to provide estimates for inter-annual variation the one inventory plus change method need to be applied (IPCC 2003). This requires inventory and other data such as changes in land-use, forest management activities, natural disturbances, and detailed models of forest C dynamics. Forest C growth in models may either be driven by empirical yield curves (e.g., EFISCEN, Nabuurs et al. 2000; CO2FIX, Masera et al. 2003) or growth may be driven by simulating photosynthesis (e.g., 3-PG, Landsberg and Waring 1997; BIOME-BGC, Running and Gower 1991; CENTURY, Metherall et al. 1993; TEM, Tian et al. 1999). A yield data driven model with explicit simulation of the detritus dynamics is CBM-CFS3 (Kurz et al. 2009). This model, originally developed for Canadian forests, has been adapted for other countries and can also be used for future forest C balances to assess policy and management alternatives. However, the fate of harvested forest products and global change effects on growth and decomposition processes must be added to lower the uncertainty in predictions of CBM-CFS3 (Kurz et al. 2009). Otherwise, all forest C cycle models consist of imperfect representations of reality (Larocque et al. 2008). To increase the certainty of predictions of forest C dynamics, model results must be evaluated for the degree of uncertainty to enhance their importance for decision-making processes in science and policy. However, applications of uncertainty analyses in models of forest C dynamics are no common procedures. Major obstacles for a more widespread application of uncertainty analyses are a lack of data on parameter variability and of appropriate uncertainty analysis software (Larocque et al. 2008).

6.2.2.5 Reduced Emissions from Deforestation and Degradation

The C trading has been advocated as a way to slow, in particular, tropical deforestation (Laurance 2008). In this respect the REDD concept is promising, and is encouraged by the Bali Action Plan addressing the role of forests in climate change and how forests can be included in long-term cooperative action up to 2012 and beyond (UNFCCC 2007). Reducing emissions from deforestation can be defined as avoiding emissions associated with burning or natural degradation of stored forest biomass on the site as it is converted to another land use that maintains or stores a lower quantity of C in biomass (Martin 2008). On the other hand, no fixed definition exists for forest degradation but the reduced capacity of the degraded forest to produce goods and services is emphasized. The most efficient policy measures may be the removal of agricultural subsidies that encourage deforestation. The greatest opportunities for forest-based mitigation of C emissions are in the developing world (Obersteiner 2009). The Forest Carbon Partnership Facility of the World Bank assists developing countries in their REDD efforts by providing value to standing forests (http://wbcarbonfinance.org/Router.cfm?Page=FCPF&ItemID=34267&FID=34267). Most important, the world's third largest GHG emitter Indonesia has applied to join this programme to trade C credits for protecting forests (Anonymous 2009). However, not only in developing countries but in all climate zones and also in industrialized countries are forests at risk of loosing C (Mollicone et al. 2007). Thus, a future climate change agreement is more effective if it includes all C losses and gains from land use in all countries and climate zones.

Emissions of CO_2 are reduced by avoiding a net decrease in forest area or volume (McKinsey & Company 2009). The estimated abatement potential is large, and most of the potential is at low cost. Nations which cannot meet their emission targets can buy C credits from other nations, especially, those in the tropics which have no emission target or produce fewer emissions than allowed (Gullison et al. 2007). However, this also implies being paid for doing nothing, i.e., not deforesting or not degrading, and may face antipathy (Martin 2008). Thus, reversing forest degradation, i.e., doing something additional (storing new C), has probably the most promising future in the REDD complex although less C is saved compared to avoiding deforestation, and monitoring forest degradation is difficult. Otherwise, tropical deforestation must also be monitored at the national level as described in the previous section.

Crucial to success of REDD is access to free or low-cost satellite data. Massive expansion of biofuels (e.g., oil palm, jatropha and rubber plantations) may, however, accelerate tropical deforestation as the tropics are favorable ecosystems for growing biofuel crops. This may drive up land prices and reduce the attractiveness of REDD (Laurance 2008). In addition to C trading, the potential of biodiversity offsets from tropical forests compared to, for example, rubber plantations may help to conserve the land and reduce deforestation (Qiu 2009). The cost of long rotations and thinning treatments in managed forests to enhance old forest conditions (e.g., biodiversity) may be reduced if the increased value of the C stored in the forest is credited but increased if the loss in C across all pools is included (Lippke and Perez-Garcia 2008). Many C pools are affected by management decisions and how the wood is used as a substitute for other construction materials. Thus, CO_2 abundance in the atmosphere can be reduced when forest landowners are credited for the increased rate of C growth that they can produce, wood processors for the C stored that they can use, and contractors for the fossil energy savings they create through substitution (Lippke and Perez-Garcia 2008).

6.2.2.6 Accounting for Forest Carbon Offsets

Forest climate activities in a post-Kyoto agreement must focus on avoiding deforestation and devegetation, on sustainable forest management, and afforestation (Dutschke 2007). Furthermore, climate forcing from concurrent changes in albedo, evapotranspiration, surface roughness and aerosols caused by forest management activities must be accounted for in a future policy framework as these may have larger impacts on the regional and local climate than the net effects of GHGs (Jackson et al. 2008). The effects of age-class legacy on C emissions and removals can be strong and must also be accounted for in a future climate protocol (Böttcher et al. 2008). The global C market must be a market for CO_2 rather than for all six Kyoto Protocol gases to facilitate emission reductions that involve CO_2 or energy production associated with CDM projects (Wara 2007). Although CH_4 or N_2O are significant to climate change in the next few decades or century, they do not persist over time in the same way as does CO_2 (Forster et al. 2007).

For a full accounting of forest C offsets, forest management actions that reduce the risk of C loss through stand-replacing fire must also be recognized (Hurteau et al. 2008). Thus, C trading mechanisms must account for the reduction in value associated with disturbance risk such as wildfires and insect outbreaks (Hurteau et al. 2009). How deforestation is included in the successor to the Kyoto Protocol is an important issue (Tollefson 2008b). For example, the Coalition for Rainforest Nations supports a market-based approach which would allow developed nations to offset their emissions by paying for forest conservation in the developing world. Other forest goods such as water, wildlife or other ecosystem services can also be included in C banking as an alternative system for C sequestration (Bigsby 2009). Carbon banking is aimed at creating a C market that is analogous to a capital market and with functions similar to a financial institution. Thus, a market for C rental is created to accommodate C sequestered in biological assets such as forests. Carbon banking has several advantages compared to the current broker-based system of trading C. In particular, a significant degree of flexibility is involved in allowing any forest owner to potentially be involved in the C market. In contrast, current C trading schemes involve only large scale forest owners with a normally structured forest with constant annual harvest or owners who will never harvest. Carbon banking, on the other hand, provides opportunities for small forest owners with different types, age classes and management strategies to participate in C markets as payments are based on the amount of C presently sequestered. However, the current schemes creates uncertainty as the price paid for C is a single payment for permanent ownership but forests are at risk of catastrophic disturbances, and the future price of C remains uncertain (Bigsby 2009).

6.3 Major Constraints on the Importance of Forest Carbon Sequestration: Tropical Deforestation, Perturbations in Peatlands and in Old-Growth Forests

The future of C sequestration in global forest ecosystems depends on the impacts of climate change on the vulnerable forest C pools in vegetation, detritus and soil, and on the land uses in tropical forest (i.e., deforestation), and perturbations in peatland and old-growth forests. The accelerated release of C from vulnerable C pools, especially peatlands, tundra frozen loess ('Yedoma sediments'), permafrost soils, and soils of boreal and tropical forests is an important issue (Parry et al. 2007).

6.3.1 Tropical Deforestation

Tropical forests are the major C sinks among global forest biomes (Chapter 4). The C sink in tropical forest trees is estimated at 1.3 Pg C $year^{-1}$, which is about 50% of the global terrestrial C sink (Lewis et al. 2009). The NPP of tropical forests may be as much as 2,170 g C m^{-2} $year^{-1}$ compared with merely 539 and 801 g C m^{-2} $year^{-1}$ in boreal and temperate forests, respectively. Otherwise, a recent compilation indicates that annual NPP in tropical forests is not different than annual NPP in temperate forests (Huston and Wolverton 2009). Most important, up to 63% of global forest plant C and 45% of global forest soil C to 1-m depth occur in the tropics. Yet, tropical deforestation results in the largest global CO_2 flux from land use change, and is responsible for 88% of deforestation emissions and a major disturbance of the C balance of tropical forests (Gullison et al. 2007; McKinsey & Company 2009). Aside from climate change, deforestation is also a driver of the dieback of the Amazon rainforest (Kriegler et al. 2009). In particular, deforestation contributes to dieback by the reduction in rainfall due to reduced water vapor flows (Gordon et al. 2005).

The annual C release from deforestation may be more than twice as much as the annual C sink in tropical forests. Brazil and Indonesia each account for one-third of 2005 deforestation emissions, and Africa accounts for an additional 16% (McKinsey & Company 2009). The large C pools in the soil under tropical forests may decrease after deforestation as fine root turnover and related C transfer into the soil are drastically altered (Hertel et al. 2009). Thus, significant SOC losses may occur by tropical deforestation in addition to the huge C losses from the aboveground biomass (Milne et al. 2007). Over and above C emissions directly associated with land clearing, deforestation also increases the fire risk and C emission from surrounding un-cleared forests by increasing the impacts of ACC (Gullison et al. 2007; van der Werf et al. 2008). The magnitude of CO_2 emitted through burning, clearing, and decay during tropical deforestation is not known (Kintisch 2007). Furthermore, the impacts of recent trends towards rapid urbanization in the tropics on the rate of deforestation are also not known (Stokstad 2009). Similarly, whether regrowth of natural forests on degraded tropical lands can incure C sequestration in comparison to un-cleared tropical forests is also uncertain (Chazdon 2008). While the rate of deforestation remains high (13 million ha $year^{-1}$), there are numerous unknowns (Grainger 2008).

Principal causes of tropical deforestation include industrial logging, mining, oil and gas development and, in particular, large-scale agriculture (Butler and Laurance 2008). Surging demands for beef and soy beans, for example, together with the increasing emphasis on biofuels from corn in the United States are among the major causes of deforestation in the Brazilian Amazon (Laurance 2007; Malhi et al. 2008; Scharlemann and Laurance 2008). Similarly, population increase in Africa increases the demand for fuel wood or charcoal and industrial logging. Yet, deforestation rates in Africa are smaller than those in the Amazon (Koenig 2008; Laporte et al. 2007). Massive expansion of oil-palm plantations in Indonesia for biofuel prdocution is a primary cause of deforestation in this region (Tollefson 2008a). Much of the global tropical deforestation in the future, however, may be caused by the projected drastic increase in global industrial activity (MEA 2005).

Strategies for reducing C emissions from tropical deforestation are debatable (Butler and Laurance 2008). For example, eco-certification of forest products by the Forest Stewardship Council (FSC) may reduce the rate of deforestation but logging as an alternative to clear-cutting old-growth forests may also reduce the C pool (Asner et al. 2005). Government policies such as United States subsidies for corn ethanol must be adjusted to reduce market distortions that promote tropical deforestation. Future international agreements on ACC discussed in the previous section eventually slow rapid deforestation by promoting international C trading (Laurance 2008; Malhi et al. 2008). Global deforestation may be markedly affected by a well-designed and focused program. More than 50% of the emissions from deforestation come from two states in the Brazilian Amazon and one province in Indonesia (Tollefson 2008a). In Amazonia, for example, areas for forest conservation can be prioritized based on their vulnerability to develop into savanna due to precipitation reduction and soil nutrient stress by deforestation (Senna et al. 2009). Furthermore, Indonesia recently joined a World Bank programme aimed at reducing deforestation by selling tradeable C credits (Anonymous 2009).

Emission reductions from reduced deforestation is among the least-expensive mitigation options (McKinsey & Company 2009). In particular, the removal of agricultural subsidies that encourage tropical deforestation is probably the most efficient policy measures (Martin 2008). Deforestation can be avoided by reducing slash-and-burn and other forms of subsistence farming with compensation payments and income support to rural poor and forest people. Slash-and-burn agriculture currently accounts for 53% of deforestation emissions in Africa, 44% in Asia, and 31% in Latin America. Other approaches include reducing the conversion to pastureland and cattle ranching or intensive agriculture with compensating land holders for the lost revenue from on-time timber extraction and future cash flow from ranching or agriculture. Pastureland and cattle ranching currently account for 65% of deforestation emissions in Latin America, 6% in Asia and 1% in Africa. Furthermore, intensive agriculture accounts for 44% of emissions in Asia, 35% in Africa, and 1% in Latin America. Finally, compensating land holders for lost timber revenue is an approach to reducing unsustainable timber extraction (McKinsey & Company 2009).

6.3.2 Perturbations in Peatland Forests

Peatland forests store large amounts of C as peat, more so than the surrounding upland forest soils on an area basis (Section 3.4.1). Exchanges of C with the atmosphere through fluxes of CO_2 and CH_4, and loss of DOC to aquatic ecosystems are the principal losses of forest C (Limpens et al. 2008). However, knowledge about the crucial components of C input to peatland forest soils (i.e., production and turnover rates of moss and root biomass) is scanty (Laiho et al. 2008). Peatlands are not explicitly included in global climate models and predictions of future climate change (Denman et al. 2007). Thus, the impact of peatlands on the global warming potential is not known (Nieder and Benbi 2008). Yet, any land use in peatland forests is important to GHG emissions. For example, peatland forests in Finnland dominate the C balance of the forestry sector (Laiho et al. 2008). When Finnish peatland forests lose C, it overrides any sink function of Finnish upland forests, which largely depends on changes in the biomass C pool. Also, in countries with smaller peatland forest areas, their contribution to GHG exchange may be significant.

Perturbations in peatland forests by land use activities such as drainage and fire have the potential to release large amounts of C. Many temperate peatland forests lost C upon intensive drainage (Laine et al. 2006). However, moderate drainage of peatland forest soils may increase peat C accumulation (Minkkinen et al. 2002). Land use changes in boreal peatland forests may contribute to thawing of C frozen at depth as permafrost, and expose OM to microbial decomposition thereby causing a large C release (Schuur et al. 2008). Major perturbations occur also in tropical peatland forests in Southeast Asia and in Amazonian peatlands (Lähteenoja et al. 2009). Drainage and fire associated with oil palm and other plantations in Indonesia, for example, can release CO_2 equal to 19–60% of the global C emissions from fossil fuels between 1997 and 2006 (Jaenicke et al. 2008). The GHG emissions from peatland fires are more than double as high as emissions from peat decomposition (McKinsey & Company 2009). Thus, drainage and burning of peatlands are major contributors to GHG emissions by the forestry sector, and responsible for 27% of emissions from forest land use and land use change. Drainage and burning of tropical peatland forests has been identified as major contributor to emissions by the forestry sector.

Indirect human induced threats by global warming will continue to affect the C balance of peatland forests. Direct human induced threats to boreal and temperate peatland forests, on the other hand, are concentrated in northern Europe, parts of Russia, southern Canada, and in parts of the northern USA (Moore 2002). However, the rate of destruction of peatlands in these regions is already declining as governments increasingly realize their importance to the local and global C balance. Furthermore, as temporarily net C losses may occur after harvest, sustainable forest management practices are important to reduce perturbations of the C balance in boreal peatland forests (Schils et al. 2008). In contrast, losses of peatland forests are likely to continue to occur in Asia. Thus, exploitation of tropical peatland forests must be reduced by shifting, in particular, oil-palm cultivation to areas of less importance to the global C balance.

6.3.3 Perturbations in Old-Growth Forests

Old-growth forests can be defined on the basis of forest structure and composition, including a wide range of tree sizes and the presence of some old trees approaching their maximum longevity (Bauhus et al. 2009). These forests provide numerous benefits and habitats unavailable in managed forests (Wirth et al. 2009). Specifically, the global terrestrial C budget strongly depends on pristine old-growth forests (Schulze 2006). Most important, contrary to the long-standing view based on data from a single site, old forests continue to accumulate C and remain C sinks (Kira and Sihdei 1967; Odum 1967). The common accepted pattern of stabilization or decline of live biomass accumulation and NPP with stand age is primarily based on small scale-studies in homogenous stands (Hudiburg et al. 2009). However, a forest area may consist of heterogeneous stands such as transitional forests, uneven-aged stands, and stands that have experienced partial disturbances. The hypothesis that old-growth forests are at equilibrium with respect to C balance must be validated (Suchanek et al. 2004). For example, continued C accumulation based on C flux data has been reported for boreal and temperate old-growth forests (Luyssaert et al. 2008). Forest inventory data indicate that large amounts of C are stored in biomass and soil in old-growth boreal and temperate forests, and they steadily accumulate atmospheric CO_2 for centuries (Section 3.2.4). Tropical old-growth forests in Africa, America and Asia also increase in tree biomass and C storage according to inventory data (Lewis et al. 2009).

Maintaining mature and old forests that already store large amounts of C is, therefore, one of the important options of mitigating ACC (IPCC 2007). Furthermore, managing forests recovering from historical disturbances to allow the development of old-growth characteristics similar to the original primary forests would increase ecosystem C pools (Rhemtulla et al. 2009). Possible explanations for continued C uptake in old-growth tropical forests, for example, include the recovery process of succession which may take hundreds to thousands of years (Muller-Landau 2009). Another explanation is global climate and/or atmospheric change. The continued formation of stabilized SOC fractions may also contribute to the C sink in old-growth forests (Schulze et al. 2002). However, catastrophic natural disturbances potentially release much of the C stored in old-growth forests (Section 3.1). Any human-induced disturbance, either indirectly through climate change or directly through exploration, has also the potential of C emission to the atmosphere. Thus, international agreements on ACC must address the significance of old-growth forests for the global C balance and ensure their protection from human exploration (Schulze et al. 2002).

6.4 Conclusions

Woody biomass can substitute fossil fuel based products for heat and power generation, and contribute to mitigation of ACC by forest land use. Source for woody biomass are forest residues, and dedicated bioenergy tree plantations. Conventional breeding techniques and biotechnology are used to improve the properties of high-yielding bioenergy trees. A major challenge for cellulosic ethanol production from woody biomass is to overcome the barrier of lignocellulosic recalcitrance. Negative impacts of large-scale plantations on soil nutrients and on the hydrological cycle must be minimized to ensure sustainable use of woody biomass as renewable energy source. Enhancing forest C sequestration has not been achieved under the Kyoto Protocol. Several approaches are discussed for reducing C emissions from deforestation and degradation which will be included in future international agreements on ACC. Forest C accounting and monitoring systems are required to assess the success of C cap-and-trade mechanisms by forest land-use and deforestation, especially, in the tropics. Future C sequestration in global forest ecosystems may be determined by human activities such as tropical deforestation, and exploration of peatland and old-growth forests.

6.5 Review Questions

1. Compare and contrast the environmental impact of using forest residues and biomass from dedicated tree plantations for bioenergy.
2. Where are major global land areas suitable for tree plantations? What are major obstacles in establishing tree plantations on these areas?
3. Compare and contrast the contribution of the forestry sector to bioenergy production with those of the agricultural sector.
4. With regard to the definition of C sequestration in Chapter 1 – what properties characterize a 'superficial' tree to increase the stable forest ecosystem C pools in vegetation, detritus and soil?
5. Contrast and compare how the role of forests in the global C cycle is appreciated in current and future agreements on climate change.
6. What is your definition of C sequestration in forest ecosystems? Contrast and compare it with other definitions.
7. How can changes in the vegetation, detritus and soil C pools of forests be monitored, estimated and modeled?
8. What measures are needed to protect the most important forest C pools in tropical, peatland and old-growth forests against natural and anthropogenic disturbances? How does climate change impact these pools?
9. The decarbonization of the global economy is a slow process. Can C sequestration in forest ecosystems and the forestry sector buy time and advert dangerous consequences of global climate change?

References

Andersson K, Richards K (2001) Implementing an international carbon sequestration program: can the leaky sink be fixed? Clim Policy 1:173–188

Andersson K, Evans TP, Richards KR (2009) National forest carbon inventories: policy needs and assessment capacity. Clim Change 93:69–101

Anonymous (2009) Indonesia to sell carbon credits to conserve forests. Nature 458:137. doi:10.1038/458137b

Asner GP, Knapp DE, Broadbent EN, Oliveira PJC, Keller M, Silva JN (2005) Selective logging in the Brazilian amazon. Science 310:480–482

Barker T, Bashmakov I, Alharthi A, Amann M, Cifuentes L, Drexhage J, Duan M, Edenhofer O, Flannery B, Grubb M, Hoogwijk M, Ibitoye FI, Jepma CJ, Pizer WA, Yamaji K (2007) Mitigation from a cross-sectoral perspective. In: Metz B, Davidson OR, Bosch PR, Dave R, Meyer LA (eds) Climate Change 2007: Mitigation Contribution of working group III to the fourth assessment report of the intergovernmental panel on climate change. Cambridge University Press, Cambridge/New York, pp 620–690

Basu P (2009) A green investment. Nature 457:144–146

Bates BC, Kundzewicz ZW, Wu S, Palutikof JP (eds) (2008) Climate change and water. Technical paper of the Intergovernmental Panel on Climate Change. IPCC Secretariat, Geneva, Switzerland

Baucher M, Halpin C, Petit-Concil M, Boerjan W (2003) Lignin: genetic engineering and impact on pulping. Crit Rev Biochem Mol 38:305–350

Bauhus J, Puettmann K, Messier C (2009) Silviculture for old-growth attributes. For Ecol Manag 258:525–531

Bernier P, Schoene D (2009) Adapting forests and their management to climate change: an overview. Unasylva 60:5–11

Betts R (2007) Implications of land ecosystem-atmosphere interactions for strategies for climate change adaptation and mitigation. Tellus B 59:602–615

Bigsby H (2009) Carbon banking: creating flexibility for forest owners. For Ecol Manag 257:378–383

Böttcher H, Kurz WA, Freibauer A (2008) Accounting of forest carbon sinks and sources under a future climate protocol-factoring out past disturbance and management effects on age-class structure. Environ Sci Policy 11:669–686

Boyland M (2006) The economics of using forests to increase carbon storage. Can J For Res 36:2223–2234

Bradford JB, Weishampel P, Smith M-L, Kolka R, Hollinger DY, Birdsey RA, Ollinger S, Ryan MG (2008) Landscape-scale carbon sampling strategy-lessons learned. In: Hoover CM (ed) Field measurements for forest carbon monitoring. Springer, New York, pp 227–238

Bravo F, del Río M, Bravo-Oviedo A, Del Peso C, Montero G (2008) Forest management strategies and carbon sequestration. In: Bravo F, LeMay V, Jandl G, von Gadow K (eds) Managing forest ecosystems: the challenge of climate change. Springer, New York, pp 179–194

Brown S (2002a) Measuring carbon in forests: current status and future challenges. Environ Pollut 116:363–372

Brown S (2002b) Measuring, monitoring, and verification of carbon benefits for forest-based projects. Phil Trans R Soc Lond A 360:1669–1683

Carle J, Vuorinen P, Del Lungo A (2002) Status and trends in global forest plantation development. For Prod J 52:12–23

Chazdon RL (2008) Beyond deforestation: restoring forests and ecosystem services on degraded lands. Science 320:1458–1460

Conant RT, Smith GR, Paustian K (2003) Spatial variability of soil carbon in forested and cultivated sites: implications for change detection. J Environ Qual 32:278–286

Davey PA, Olcer H, Zakhleniuk O, Bernacchi CJ, Calfapietra C, Long SP, Raines CA (2006) Can fast-growing plantation trees escape biochemical down-regulation of photosynthesis when grown throughout their complete production cycle in the open air under elevated carbon dioxide? Plant Cell Environ 29:1235–1244

Denman KL, Brasseur G, Chidthaisong A, Ciais P, Cox PM, Dickinson RE, Hauglustaine D, Heinze C, Holland E, Jacob D, Lohmann U, Ramachandran S, da Silva Dias PL, Wofsy SC, Zhang X (2007) Couplings between changes in the climate system and biogeochemistry. In: intergovernmental panel on climate change (ed) climate change 2007: the physical science basis, Chapter 7. Cambridge University Press, Cambridge

Dutschke M (2007) CDM forestry and the ultimate objective of the climate convention. Mitig Adapt Strat Glob Change 12:275–302

El-Kassaby Y (2004) Feasibility and proposed outline of a global review of forest biotechnology. Forest genetic resources working paper FGR/77E. Forest resources development service, forest resources division. FAO, Rome

Ericsson T (1994) Nutrient dynamics and requirements of forest crops. N Z J For Sci 24:133–168
Evans JM, Cohen MJ (2009) Regional water resource implications of bioethanol production in the Southeastern United States. Glob Change Biol 15:2261–2273
FAO (Food and Agricultural Organization of the United Nations) (2006) Global planted forests thematic study: results and analysis. In: Del Lungo A, Ball, Carle J (eds) Planted forests and trees working paper 38. FAO, Rome
FAO (Food and Agricultural Organization of the United Nations) (2008) Forests and energy. FAO Forestry paper 154. FAO, Rome
FAO (Food and Agricultural Organization of the United Nations) (2009) Global review of forest pests and diseases. FAO Forestry paper 156. FAO, Rome
Farley KA, Jobbágy EG, Jackson RB (2005) Effects of afforestation on water yield: a global synthesis with implications for policy. Glob Change Biol 11:1565–1576
Fenning TM, Gershenzon J (2002) Where will the wood come from? Plantation forests and the role of biotechnology. Trend Biotechnol 20:291–296
Fernandes SD, Trautmann NM, Streets DG, Roden CA, Bond TC (2007) Global biofuel use, 1850–2000. Global Biogeochem Cy 21, GB2019. doi:10.1029/2006GB002836
Field CB, Campbell JL, Lobell DB (2007) Biomass energy: the scale of the potential resource. Trends Ecol Evol 23:65–72
Forster P, Ramaswamy V, Artaxo P, Berntsen T, Betts R, Fahey DW, Haywood J, Lean J, Lowe DC, Myhre G, Nganga J, Prinn R, Raga G, Schulz M, Van Dorland R (2007) Changes in atmospheric constituents and in radiative forcing. In: Intergovernmental panel on climate change (ed) Climate change 2007: the physical science basis, Chapter 2. Cambridge University Press, Cambridge
Galik CS, Jackson RB (2009) Risks to forest carbon offset projects in a changing climate. For Ecol Manag 257:2209–2216
Glaser B, Parr M, Braun C, Kopolo G (2009) Biochar is carbon negative. Nat Geosci 2:2
Goetz SJ, Baccini A, Laporte NT, Johns T, Walker W, Kellndorfer J, Houghton RA, Sun M (2009) Mapping and monitoring carbon stocks with satellite observations: a comparison of methods. Carbon Bal Manage 4:2
Gomez LD, Steele-King CG, McQueen-Mason SJ (2008) Sustainable liquid biofuels from biomass: the writing's on the walls. New Phytol 178:473–485
Gordon LJ, Steffen W, Jönsson BF, Folke C, Falkenmark M, Johanessen Å (2005) Human modification of global water vapor flows from the land surface. P Natl Acad Sci USA 102:7612–7617
Grainger A (2008) Difficulties in tracking the long-term global trend in tropical forest area. Proc Natl Acad Sci USA 105:818–823
Gullison RE, Frumhoff PC, Canadell JG, Field CB, Nepstad DC, Hayhoe K, Avissar R, Curran LM, Friedlingstein P, Jones CD, Nobre C (2007) Tropical forests and climate policy. Science 316:985–986
Guo LB, Gifford RM (2002) Soil carbon stocks and land use change: a meta analysis. Glob Change Biol 8:345–360
Hansen J, Sato M (2009) Global warming: east–west connections. 'http://www.columbia.edu/~jeh1/2007/EastWest_20070925.pdf' accessed August 26, 2009
Hansen J, Sato M, Kharecha P, Beerling D, Berner R, Masson-Delmotte V, Pagani M, Raymo M, Royer DL, Zachos JC (2008) Target atmospheric CO_2: where should humanity aim? Open Atmos Sci J 2:217–231
Heath LS, Birdsey RA (1993) Carbon trends of productive temperate forests of the coterminous United States. Water Air Soil Pollut 70:279–293
Henderson AR, Walter C (2006) Genetic engineering in conifer plantation forestry. Silvae Genet 55:253–262
Hennigar CR, MacLean DA, Amos-Binks LJ (2008) A novel approach to optimize management strategies for carbon stored in both forests and wood products. For Ecol Manag 256:786–797
Hertel D, Harteveld MA, Leuschner C (2009) Conversion of a tropical forest into agroforest alters the fine root-related carbon flux to the soil. Soil Biol Biochem 41:481–490

Himmel ME, Picataggio SK (2008) Our challenge is to acquire deeper understanding of biomass recalcitrance and conversion. In: Himmel ME (ed) Biomass recalcitrance: deconstruction the plant cell wall for bioenergy. Blackwell, Oxford, pp 1–6

Himmel ME, Ding S-Y, Johnson DK, Adney WS, Nimlos MR, Brady JW, Foust TD (2007) Biomass recalcitrance: engineering plants and enzymes for biofuels production. Science 315:804–807

Hoenicka H, Fladung M (2006) Biosafety in *Populus* spp. and other forest trees: from non-native species to taxa derived from traditional breeding and genetic engineering. Trees 20:131–144

Hoover CM (ed) (2008) Field measurements for forest carbon monitoring. Springer, New York

Hu JJ, Tian YC, Han YF, Li L, Zhang BE (2001) Field evaluation of insect resistant transgenic *Populus nigra* trees. Euphytica 121:123–127

Hudiburg T, Law B, Turner DP, Campbell J, Donato D, Duane M (2009) Carbon dynamics of Oregon and Northern California forests and potential land-based carbon storage. Ecol Appl 19:163–180

Hurteau MD, Koch GW, Hungate BA (2008) Carbon protection and fire risk reduction: toward a full accounting of forest carbon offsets. Front Ecol Environ 6:493–498

Hurteau MD, Hungate BA, Koch GW (2009) Accounting for risk in valuing forest carbon offsets. Carbon Balance Manage 4:1

Huston MA, Wolverton S (2009) The global distribution of net primary production: resolving the paradox. Ecol Monogr 79:343–377

Ilstedt U, Malmer A, Verbeeten E, Murdiyarso D (2007) The effect of afforestation on water infiltration in the tropics: a systematic review and meta-analysis. For Ecol Manage 251:45–51

IPCC (2003) Penman J, Gytarsky M, Hiraishi T, Krug T, Kruger D, Pipatti R, Buendia L, Miwa K, Ngara T, Tanabe K, Wagner F (eds) Good practice guidance for land use, land-use change and forestry. Institute for Global Environmental Strategies. Hayama, Japan

IPCC (2007) Climate change 2007: synthesis report. Contribution of working groups I, II, and III to the fourth assessment report of the intergovernmental panel on climate change. IPCC, Geneva, Switzerland

Jackson RB, Randerson JT, Canadell JG, Anderson RG, Avissar R, Baldocchi DD, Bonan GB, Caldeira K, Diffenbaugh NS, Field CB, Hungate BA, Jobbágy EG, Kueppers LM, Nosetto MD, Pataki DE (2008) Protecting climate with forests. Environ Res Lett 3. doi:10.1088/1748-9326/3/4/044006

Jacobson MZ (2009) Review of solutions to global warming, air pollution, and energy security Energ Environ Sci 2:148–173

Jaenicke J, Rieley JO, Mott C, Kimman P, Siegert F (2008) Determination of the amount of carbon stored in Indonesian peatlands. Geoderma 147:151–158

Karp A, Shield I (2008) Bioenergy from plants and the sustainable yield challenge. New Phytol 179:15–32

Kendurkar SV, Naik VB, Nadgauda RS (2006) Genetic transformation of some tropical trees, shrubs, and tree-like plants. In: Fladung M, Ewald D (eds) Tree transgenesis: recent developments. Springer, Berlin, pp 67–102

Kintisch E (2007) Improved monitoring of rainforests helps pierce haze of deforestation. Science 316:536–537

Kira T, Sihdei T (1967) Primary production and turnover of organic matter in different forest ecosystems of the western pacific. Jpn J Ecol 17:70–87

Koenig R (2008) Critical time for African rainforests. Science 320:1439–1441

Kriegler E, Hall JW, Held H, Dawson R, Schellnhuber HJ (2009) Imprecise probability assessment of tipping points in the climate system. Proc Natl Acad Sci USA 106:5041–5046

Kurz WA, Dymond CC, White TM, Stinson G, Shaw CH, Rampley GJ, Smyth C, Simpson BN, Neilson ET, Trofymow JA, Metsaranta J, Apps MJ (2009) CBM-CFS3: a model of carbon-dynamics in forestry and land-use change implementing IPCC standards. Ecol Model 220:480–504

Lähteenoja O, Ruokolainen K, Schulman L, Oinonen M (2009) Amazonian peatlands: an ignored C sink and potential source. Glob Change Biol 15:2311–2320

Laiho R, Minkkinen K, Antilla J, Vávřová P, Pentillä T (2008) Dynamics of litterfall and decomposition in peatland forests: towards reliable carbon balance estimation? In: Vymazal J (ed) Wastewater treatment, plant dynamics and management in constructed and natural wetlands. Springer, Berlin, pp 53–64

Laine J, Laiho R, Minkkinen K, Vasander H (2006) Forestry and boreal peatlands. In: Wieder RK, Vitt DH (eds) Boreal peatland ecosystems. Ecological studies, vol. 188. Springer, Berlin, pp. 331–357

Lal R (2005) Soil carbon sequestration in natural and managed tropical forest ecosystems. J Sustain For 21:1–30

Landsberg JJ, Waring RH (1997) A generalized model of forest productivity using simplified concepts of radiation-use efficiency, carbon balance and partitioning. For Ecol Manag 95:209–228

Laporte NT, Stabach JA, Grosch R, Lin TS, Goetz SJ (2007) Expansion of industrial logging in Central Africa. Science 316:1451

Larocque GR, Bhatti JS, Boutin R, Chertov O (2008) Uncertainty analysis in carbon cycle models of forest ecosystems: research needs and development of a theoretical framework to estimate error propagation. Ecol Model 219:400–412

Laurance WF (2007) Switch to corn promotes Amazon deforestation. Science 318:1721

Laurance WF (2008) Can carbon trading save vanishing forests? BioScience 58:286–287

Lehmann J (2007) A handful of carbon. Nature 447:143–144

LeMay V, Kurz WA (2008) Estimating carbon stocks and stock changes in forests: linking models and data across scales. In: Bravo F, LeMay V, Jandl G, von Gadow K (eds) Managing forest ecosystems: the challenge of climate change. Springer, New York, pp 63–81

Lemus R, Lal R (2005) Bioenergy crops and carbon sequestration. Crit Rev Plant Sci 24:1–21

Lewis SL, Lopez-Gonzalez G, Sonké B, Affum-Baffoe K, Baker TR, Ojo LO, Phillips OL, Reitsma JM, White L, Comiskey JA, M-N DK, Ewango CEN, Feldpausch TR, Hamilton AC, Gloor M, Hart T, Hladik A, Lloyd J, Lovett JC, Makana J-R, Malhi Y, Mbago FM, Ndangalasi HJ, Peacock J, S-H PK, Sheil D, Sunderland T, Swaine MD, Taplin J, Taylor D, Thomas SC, Votere R, Wöll H (2009) Increasing carbon storage in intact African tropical forests. Nature 457:1003–1007

Limpens J, Berendse F, Blodau C, Canadell JG, Freeman C, Holden J, Roulet N, Rydin H, Schaepman-Strub G (2008) Peatlands and the carbon cycle: from local processes to global implications – a synthesis. Biogeosci Discuss 5:1475–1491

Lippke B, Perez-Garcia J (2008) Will either cap and trade or a carbon emissions tax be effective in monetizing carbon as an ecosystem service. For Ecol Manag 256:2160–2165

Luo Z-B, Polle A (2009) Wood composition and energy content in a poplar short rotation plantation on fertilized agricultural land in a future CO_2 atmosphere. Glob Change Biol 15:38–47

Luyssaert S, Schulze E-D, Börner A, Knohl A, Hessenmöller D, Law BE, Ciais P, Grace J (2008) Old-growth forests as global carbon sinks. Nature 455:213–215

Malhi Y, Roberts JT, Betts RA, Killeen TJ, Li W, Nobre CA (2008) Climate change, deforestation, and the fate of the Amazon. Science 319:169–172

Mann ME (2009) Defining dangerous anthropogenic interference. Proc Natl Acad Sci USA 106:4065–4066

Marland G, Obersteiner M (2008) Large-scale biomass for energy, with considerations and cautions: an editorial comment. Climatic Change 87:335–342

Martin RM (2008) Deforestation, land-use change and REDD. Unasylva 230:3–11

Masera OR, Garza-Caligaris JF, Kanninen M, Karjalainen T, Liski J, Nabuurs GJ, Pussinen A, de Jong BHJ, Mohrenf GMJ (2003) Modelling carbon sequestration in afforestation, agroforestry and forest management projects: the CO2FIX.V2 approach. Ecol Model 164:177–199

Matthews HD, Caldeira K (2008) Stabilizing climate requires near-zero emissions. Geophys Res Lett 35:L04705. doi:10.1029/2007GL032388

McKinsey&Company (2009) Pathways to a low-carbon economy – Version 2 of the global greenhouse gas abatement cost curve. http://globalghgcostcurve.bymckinsey.com/

Mead DJ (2005) Forests for energy and the role of planted trees. Crit Rev Plant Sci 24:407–421

Metherall AK, Harding LA, Cole CV, Parton WJ (1993) CENTURY soil organic matter model environment technical documentation, agroecosystem version 4.0, Great Plains System Research Unit, Tech Rep No. 4, USDA-ARS, Ft. Collins

Metzger JO, Hüttermann A (2009) Sustainable global energy supply based on lignocelluosic biomass from afforestation of degraded areas. Naturwissenschaften 96:279–288

Millenium Ecosystem Assessment (MEA) (2005) Ecosystems and human well-being: opportunities and challenges for businesses and industry. Island Press, Washington, DC

Milne E, Paustian K, Easter M, Sessay M, Al-Adamat R, Batjes NH, Bernoux M, Bhattacharyya T, Cerri CC, Cerri CEP, Coleman K, Falloon P, Feller C, Gicheru P, Kamoni P, Killian K, Pal DK, Powlson DS, Williams S; Rawajfh Z (2007) An increased understanding of soil organic carbon stocks and changes in non-temperate areas: national and global implications. Agri Ecosys Environ 122:125–136

Minkkinen K, Korhonen R, Savolainen I, Laine J (2002) Carbon balance and radiative forcing of Finnish peatland 1900–2100, the impacts of drainage. Glob Change Biol 8:785–799

Mishra U, Lal R (2010) Predictive mapping of soil organic carbon: a case study using geographic weighted regression approach. In: Clay D, Shanahan J (eds) GIS applications in agriculture-nutrient management for improved energy efficiency. CRC, Boca Raton, FL in press

Mishra U, Lal R, Slater B, Calhoun F, Liu D, Van Meirvenne M (2009) Predicting soil organic carbon stock using profile depth distribution functions and ordinary kriging. Soil Sci Soc Am J 73:614–621

Mollicone D, Freibauer A, Schulze E-D, Braatz S, Grassi G (2007) Elements for the expected mechanisms on 'reduced emissions from deforestation and degradation, REDD' under UNFCCC. Environ Res Lett 2. doi:10.1088/1748-9326/2/4/045024

Moore PD (2002) The future of cool temperate bogs. Environ Conserv 29:3–20

Muller-Landau HC (2009) Sink in the African jungle. Nature 457:969–970

Nabuurs GJ, Schelhaas MJ, Pussinen A (2000) Validation of the European Forest Scenario Model (EFISCEN) and a projection of Finish forests. Silva Fenn 34:167–179

Nabuurs GJ, Masera O, Andrasko K, Benitez-Ponce P, Boer R, Dutschke M, Elsiddig E, Ford-Robertson J, Frumhoff P, Karjalainen T, Krankina O, Kurz WA, Matsumoto M, Oyhantcabal W, Ravindranath NH, Sanz Sanchez MJ, Zhang X (2007): Forestry. In: Metz B, Davidson OR, Bosch PR, Dave R, Meyer LA (eds) Climate change 2007: mitigation. Contribution of working group III to the fourth assessment report of the intergovernmental panel on climate change. Cambridge University Press, Cambridge, UK and New York, NY, USA, pp. 541–584.

Nehra NS, Becwar MR, Rottmann WH, Pearson L, Chowdhury K, Chang S, Dayton Wilde H, Kodrzycki RJ, Zhang C, Gause KC, Parks DW, Hinchee MA (2005) Invited review: forest biotechnology: innovative methods, emerging opportunities. In Vitro Cell Dev Biol Plant 41:701–717

Nicholls D, Monserud RA, Dykstra DP (2009) International bioenergy synthesis-lessons learned and opportunities for the western United States. For Ecol Manage 257:1647–1655

Nieder R, Benbi DK (2008) Carbon and nitrogen in the terrestrial environment. Springer, Berlin, Germany

Novaes E, Osorio L, Drost DR, Miles BL, Boaventura-Novaes CRD, Benedict C, Dervinis C, Yu Q, Sykes R, Davis M, Martin TA, Peter GF, Kirst M (2009) Quantitative genetic analysis of biomass and wood chemistry of Populus under different nitrogen levels. New Phytol 182:878–890

Obersteiner M (2009) Storing carbon in forests. Nature 458:151

Odum EP (1967) The strategy of ecosystem development. Science 164:262–270

Parry ML, Canziani OF, Palutikof JP, Co-authors (2007) Technical summary. Climate change 2007: impacts, adaptation and vulnerability. In: Parry ML, Canziani OF, Palutikof JP, van der Linden PJ, Hanson CE (eds) Contribution of Working Group II to the Fourth Assessment Report of the Intergovernmental Panel on Climate Change. Cambridge University Press, Cambridge, UK, pp. 23–78

Pilate G, Guiney E, Holt K, Petit-Concil M, Lapierre C, Leplé J-C, Pollet B, Mila I, Webster EA, Marstorp HG, Hopkins DW, Jouanin L, Boerjan W, Schuch W, Cornu D, Hapin C (2002) Field and pulping performances of transgenic trees with altered lignification. Nat Biotechnol 20:607–612

Plantinga AJ, Richards KR (2008) International forest carbon sequestration in a post-Kyoto agreement. Discussion paper 2008-11. Harvard Project on International Climate Agreements. Cambridge, MA

Powers RF (1999) On the sustainable productivity of planted forests. New For 17:263–306

Powers RF, Scott DA, Sanchez FG, Voldseth RA, Page-Dumroese DS, Elioff JD, Stone DM (2005) The North American long-term soil productivity experiment: findings from the first decade of research. For Ecol Manage 220:31–50

Qiu J (2009) Where the rubber meets the garden. Nature 457:246–247

Ragauskas AJ, Williams CK, Davison BH, Britovsek G, Cairney J, Eckert CA, Frederick WJ Jr, Hallett JP, Leak DJ, Liotta CL, Mielenz JR, Murphy R, Templer R, Tschaplinski T (2006) The path forward for biofuels and biomaterials. Science 311:484–489

Raupach MR, Canadell JG, Le Quéré C (2008) Anthropogenic and biophysical contributions to increasing atmospheric CO_2 growth rate and airborne fraction. Biogeosciences 5:1601–1613

Regalbuto JR (2009) Cellulosic biofuels-got gasoline? Science 325:822–824

Rhemtulla JM, Mladenoff DJ, Clayton MK (2009) Historical forest baselines reveal potential for continued carbon sequestration. Proc Natl Acad Sci USA 106:6082–6087

Richards KR, Stokes C (2004) A review of forest carbon sequestration cost studies: a dozen years of research. Clim Change 63:1–48

Richter deB D Jr, Jenkins DH, Karakash JT, Knight J, McCreery LR, Nemestothy KP (2009) Wood energy in America. Science 323:1432–1433

Rubin EM (2008) Genomics of cellulosic biofuels. Nature 454:841–845

Running SW, Gower ST (1991) FOREST-BGC, a general model of forest ecosystem processes for regional applications. II. Dynamic carbon allocation and nitrogen budgets. Tree Physiol 9:147–160

Sánchez-Azofeifa GA, Castro-Esau KL, Kurz WA, Joyce A (2009) Monitoring carbon stocks in the tropics and the remote sensing operational limitations: from local to regional projects. Ecol Appl 19:480–494

Sathaye JA, Andrasko K (2007a) Special issue on estimation of baselines and leakage in carbon mitigation forestry projects. Mitig Adapt Strat Glob Change 12:963–970

Sathaye JA, Andrasko K (2007b) Land use change and forestry climate project regional baselines: a review. Mitig Adapt Strat Glob Change 12:971–1000

Scharlemann JPW, Laurance WF (2008) How green are biofuels? Science 319:43–44

Schils R, Kuikman P, Liski J, van Oijen M, Smith P, Webb J, Alm J, Somogyi Z, van den Akker J, Billett M, Emmett B, Evans C, Lindner M, Palosuo T, Bellamy P, Jandl R, Hiederer R (2008) Review of existing information on the interrelations between soil and climate change (CLIMSOIL). http://ec.europa.eu/environment/soil/pdf/climsoil_report_dec_2008.pdf

Schlamadinger B, Marland G (2000) Forests, land management, and the Kyoto Protocol. Report for the Pew Center on Global Climate Change. Arlington, VA

Schlamadinger B, Bird N, Johns T, Brown S, Canadell J, Ciccarese L, Dutschke M, Fiedler J, Fischlin A, Fearnside P, Forner C, Freibauer A, Frumhoff P, Hoehne N, Kirschbaum MUF, Labat A, Marland G, Michaelowa A, Montanarella L, Moutinho P, Murdiyarso D, Pena N, Pingoud K, Rakonczay Z, Rametsteiner E, Rock J, Sanz MJ, Schneider UA, Shvidenko A, Skutsch M, Smith P, Somogyi Z, Trines E, Ward M, Yamagata Y (2007) A synopsis of land-use, land-use change and forestry (LULUCF) under the Kyoto Protocol and Marrakech Accords. Environ Sci Policy 10:271–282

Schöning I, Totsche KU, Kögel-Knabner I (2006) Small scale spatial variability of organic carbon stocks in litter and solum of a forested Luvisol. Geoderma 136:631–642

Schulze E-D (2006) Biological control of the terrestrial carbon sink. Biogeosciences 3:147–166

Schulze E-D, Wirth C, Heimann M (2000) Managing forests after Kyoto. Science 289: 2058–2059

Schulze E-D, Valentini R, Sanz M-J (2002) The long way from Kyoto to Marrakesh: implications of the Kyoto Protocol negotiations for global ecology. Glob Change Biol 8:505–518

Schuur EAG, Bockheim J, Canadell JG, Euskirchen E, Field CB, Goryachkin SV, Hagemann S, Kuhry P, Lafleur PM, Lee H, Mazhitova G, Nelson FE, Rinke A, Romanovsky VE, Shiklomanov N, Tarnocai C, Venevsky S, Vogel JG, Zimov SA (2008) Vulnerability of permafrost carbon to climate change: implications for the global carbon cycle. BioScience 58:701–714

Senna MCA, Costa MH, Pires GF (2009) Vegetation-atmosphere-soil nutrient feedbacks in the Amazon for different deforestation scenarios. J Geophys Res 114:D04104. doi:10.1029/2008JD010401

Seppälä R, Buck A, Katila P (eds) (2009) Adaptation of forests and people to climate change. A Global Assessment Report. International Union of Forest Research Organizations (IUFRO), Helsinki, Finnland

Seth MK (2004) Trees and their economic importance. Bot Rev 69:321–376

Shindell D, Faluvegi G (2009) Climate response to regional radiative forcing during the twentieth century. Nat Geosci 2:294–300

Smeets EMW, Faaij APC (2007) Bioenergy potentials from forestry in 2050. Clim Change 81:353–390

Stephanopoulos G (2007) Challenges in engineering microbes for biofuels production. Science 315:801–804

Stokstad E (2009) Debate continues over rainforest fate - with a climate twist. Science 323:448

Strauss SH, Bradshaw HD (eds) (2004) The bioengineered forest: challenges for science and society. Resources for the Future, Washington, DC

Streck C, O'Sullivan R, Janson-Smith T, Tarasofsky RG (eds) (2008) Climate change and forests: emerging policy and market opportunities. Chatham House, London

Suchanek TH, Mooney HA, Franklin JF, Gucinski H, Ustin SL (2004) Carbon dynamics of an old-growth forest. Ecosystems 7:421–426

Tian H, Melillo JM, Kicklighter DW, McGuire AD, Helfrich J (1999) The sensitivity of terrestrial carbon storage to historical climate variability and atmospheric CO_2 in the United States. Tellus B 51:414–452

Tollefson J (2008a) Save the trees. Nature 452:8–9

Tollefson J (2008b) Climate talks defer major challenges. Nature 456:846–847

Tricker PJ, Pecchiari M, Bunn SM, Vaccari FP, Peressotti A, Miglietta F, Taylor G (2009) Water use of a bioenergy plantation increases in a future high CO_2 world. Biomass Bioenerg 33:200–208

Tuskan GA, DiFazio S, Jansson S, Bohlmann J, Grigoriev I, Hellsten U, Putnam N, Ralph S, Rombauts S, Salamov A, Schein J, Sterck L, Aerts A, Bhalerao RR, Bhalerao RP, Blaudez D, Boerjan W, Brun A, Brunner A, Busov V, Campbell M, Carlson J, Chalot M, Chapman J, Chen G-L, Cooper D, Coutinho PM, Couturier J, Covert S, Cronk Q, Cunningham R, Davis J, Degroeve S, Déjardin A, dePamphilis C, Detter J, Dirks B, Dubchak I, Duplessis S, Ehlting J, Ellis B, Gendler K, Goodstein D, Gribskov M, Grimwood J, Groover A, Gunter L, Hamberger B, Heinze B, Helariutta Y, Henrissat B, Holligan D, Holt R, Huang W, Islam-Faridi N, Jones S, Jones-Rhoades M, Jorgensen R, Joshi C, Kangasjärvi J, Karlsson J, Kelleher C, Kirkpatrick R, Kirst M, Kohler A, Kalluri U, Larimer F, Leebens-Mack J, Leplé J-C, Locascio P, Lou Y, Lucas S, Martin F, Montanini B, Napoli C, Nelson DR, Nelson C, Nieminen K, Nilsson O, Pereda V, Peter G, Philippe R, Pilate G, Poliakov A, Razumovskaya J, Richardson P, Rinaldi C, Ritland K, Rouzé P, Ryaboy D, Schmutz J, Schrader J, Segerman B, Shin H, Siddiqui A, Sterky F, Terry A, Tsai C-J, Uberbacher E, Unneberg P, Vahala J, Wall K, Wessler S, Yang G, Yin T, Douglas C, Marra M, Sandberg G, Van de Peer Y, Rokhsar D (2006) The genome of black cottonwood, *Populus trichocarpa* (Torr. & Gray). Science 313:1596–1604

UN (United Nations) (1992) United nations framework convention on climate change. http://unfccc.int/resource/docs/convkp/conveng.pdf

UNFCCC (United nations framework convention on climate change) (2007) Report of the conference of parties on its thirteenth session, Bali, Indonesia, 3-15 December 2007. Geneva, Switzerland, UN

UNFCCC (United nations framework convention on climate change) (2008) Kyoto Protocol: status of ratification. http://unfccc.int/files/kyoto_protocol/status_of_ratification/application/pdf/kp_ratification.pdf

van der Werf GR, Dempewolf J, Trigg SN, Randerson JT, Kasibhatla PS, Giglio L, Murdiyarso D, Peters W, Morton DC, Collatz GJ, Dolman AJ, DeFries RS (2008) Climate regulation of fire emissions and deforestation in equatorial Asia. Proc Natl Acad Sci USA 105:20350–20355

van Dijk AIJM, Keenan RJ (2007) Planted forests and water in perspective. For Ecol Manage 251:1–9

Van Frankenhuyzen K, Beardmore T (2004) Current status and environmental impact of transgenic forest trees. Can J For Res 34:1163–1180

van Kooten GC (2009) Biological carbon sequestration and carbon trading re-visited. Climatic Change 95:449–463

van Kooten GC, Eagle AJ, Manley J, Smolak T (2004) How costly are carbon offsets? A meta-analysis of carbon forest sinks. Environ Sci Policy 7:239–251

Vanclay JK (2009) Managing water use from forest plantations. For Ecol Manage 257:385–389

Wara M (2007) Is the global carbon market working? Nature 445:595–596

Waring RW, Running SW (2007) Forest ecosystems – analysis at multiple scales. Elsevier Academic, Burlington, MA

Watson RT, Noble IR, Bolin B, Ravindranath NH, Verardo DJ, Dokken DJ (eds) (2000) Land use, land-use change, and forestry. Cambridge University Press, Cambridge

Weber J, Sotelo Montes C (2008) Geographic variation in tree growth and wood density of Guazuma crinita Mart. in the Peruvian Amazon. New For 36:29–52

Wirth C, Gleixner G, Heimann M (2009) Old-growth forests: function, fate and value. Ecological Studies. Springer, New York

Index

D

E

Color Plates

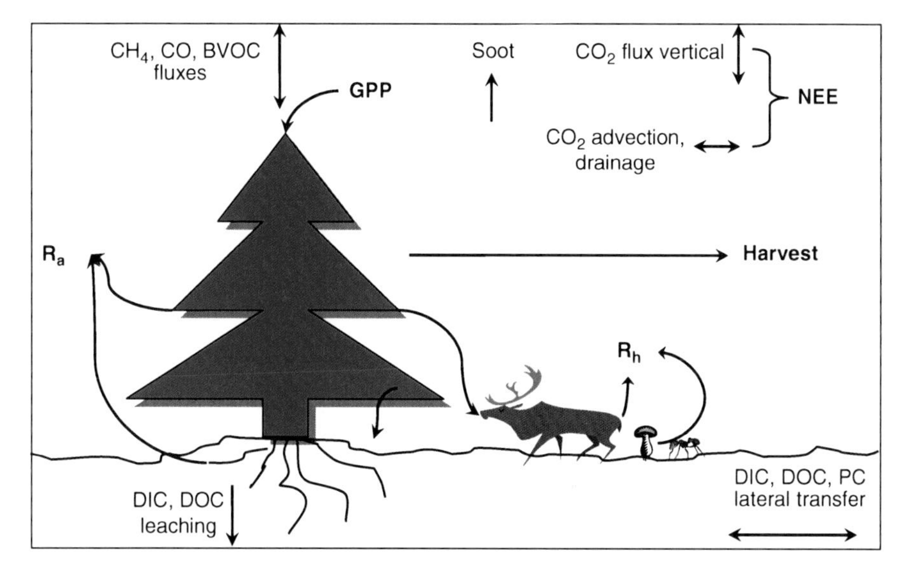

Fig. 2.1 Carbon fluxes associated with the net ecosystem C balance (GPP = gross primary production, NEE = net CO_2 exchange, R_a = autotrophic respiration, R_h = heterotrophic respiration; CH_4 = methane, CO = carbon monoxide, BVOC = biogenic volatile organic compounds, CO_2 = carbon dioxide, DIC = dissolved inorganic carbon, DOC = dissolved organic carbon, PC = particulate carbon) (Modified from Chapin et al. 2006)

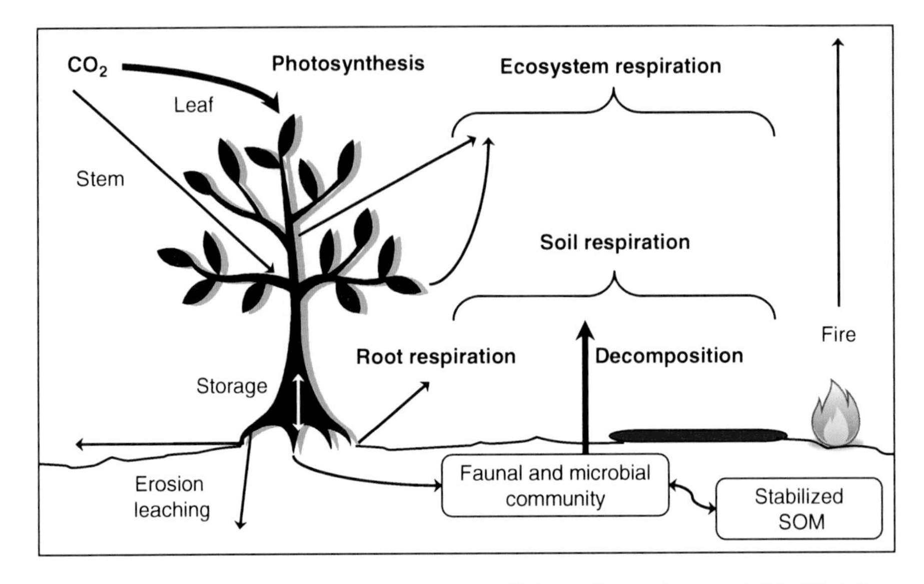

Fig. 2.2 Carbon flow through forest ecosystems (SOM = soil organic matter) (Modified from Trumbore 2006)

Fig. 3.1 Forest wildfire (Dave Powell, USDA Forest Service, Bugwood.org, http://creativecommons.org/licenses/by/3.0/us/)

Fig. 3.2 Forest disturbance by strong winds (Red and white pine, Steven Katovich, USDA Forest Service, Bugwood.org, http://creativecommons.org/licenses/by/3.0/us/)

Fig. 3.3 Forest damage by insect outbreak (Mountain pine beetle, Jerald E. Dewey, USDA Forest Service, Bugwood.org, http://creativecommons.org/licenses/by/3.0/us/)

Fig. 3.4 Trees uprooted by wind (European beech, Haruta Ovidiu, University of Oradea, Bugwood.org, http://creativecommons.org/licenses/by/3.0/us/)

Fig. 3.6 Secondary succession (Trees colonizing uncultivated fields and meadows, Tomasz Kuran, http://en.wikipedia.org/wiki/GNU_Free_Documentation_License)

Fig. 3.7 Old-growth forest (European beech, Snežana Trifunović, http://en.wikipedia.org/wiki/GNU_Free_Documentation_License)

Fig. 3.8 Timber harvest (Poplar clearcut, Doug Page, USDI Bureau of Land Management, Bugwood.org, http://creativecommons.org/licenses/by/3.0/us/)

Fig. 3.10 Forest plantation (*Pinus radiata* and *Eucalyptus nitens*, photo credit: Michael Ryan)

Fig. 3.11 Peatland forest (pine trees in *Sphagnum spp.* L. bog, Paul Bolstad, University of Minnesota, Bugwood.org, http://creativecommons.org/licenses/by/3.0/us/)

Fig. 3.12 Forest regrowth after surface mining for coal (Photo credit: Johannes Fasolt)

Fig. 3.13 Urban forest park (Joseph LaForest, University of Georgia, Bugwood.org, http://creativecommons.org/licenses/by/3.0/us/)

Fig. 4.1 Boreal forest (*Picea mariana*, photo credit: L.B. Brubaker)

Fig. 4.2 Temperate forest (*Fagus sylvatica*, photo credit: Malene Thyssen, http://commons.wikimedia.org/wiki/User:Malene)

Fig. 4.3 Tropical forest (photo credit: H.-D. Viktor Boehm)

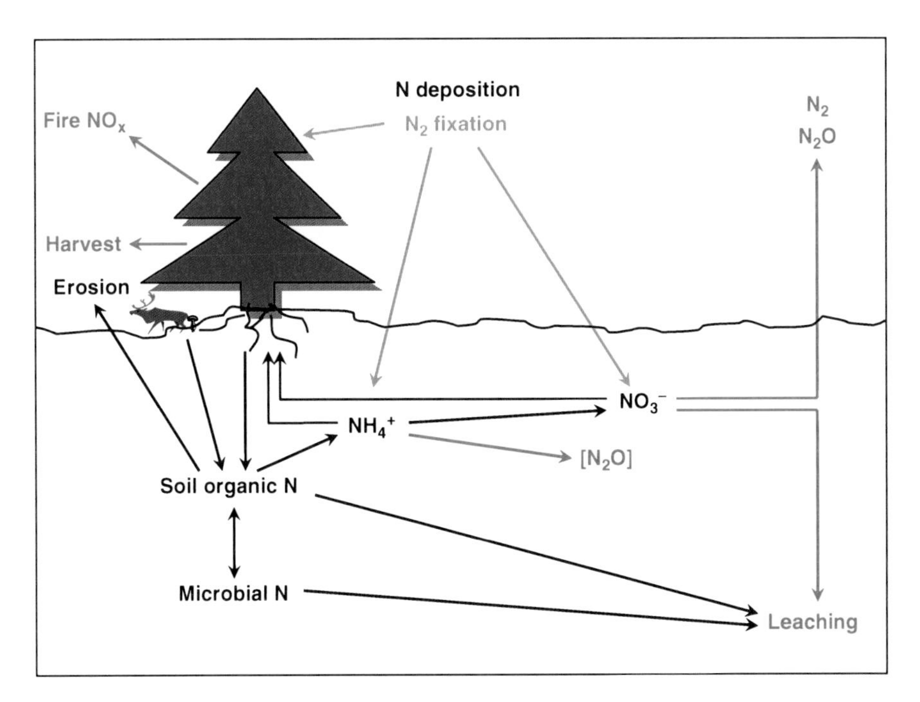

Fig. 5.1 Schematic representation of the major elements of the forest nitrogen cycle (N_2O = nitrous oxide; NO_x = nitrogen oxides; NO_3^- = nitrate; NH_4^+ = ammonium; major sources and losses underlined) (modified from Robertson and Groffman 2007)

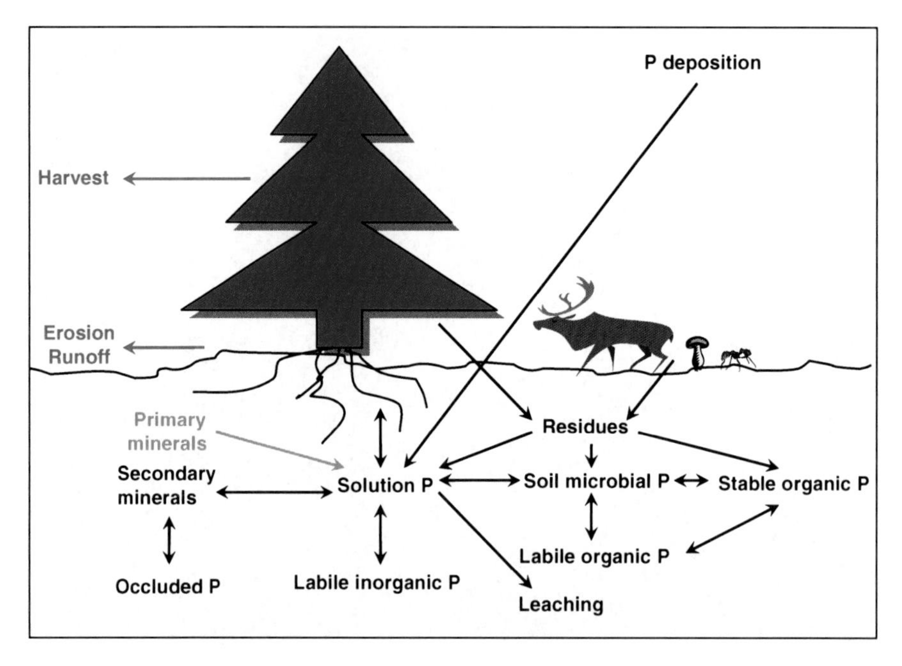

Fig. 5.2 Schematic representation of the major elements of the forest phosphorus cycle (major sources and losses underlined) (modified from Walbridge 1991)

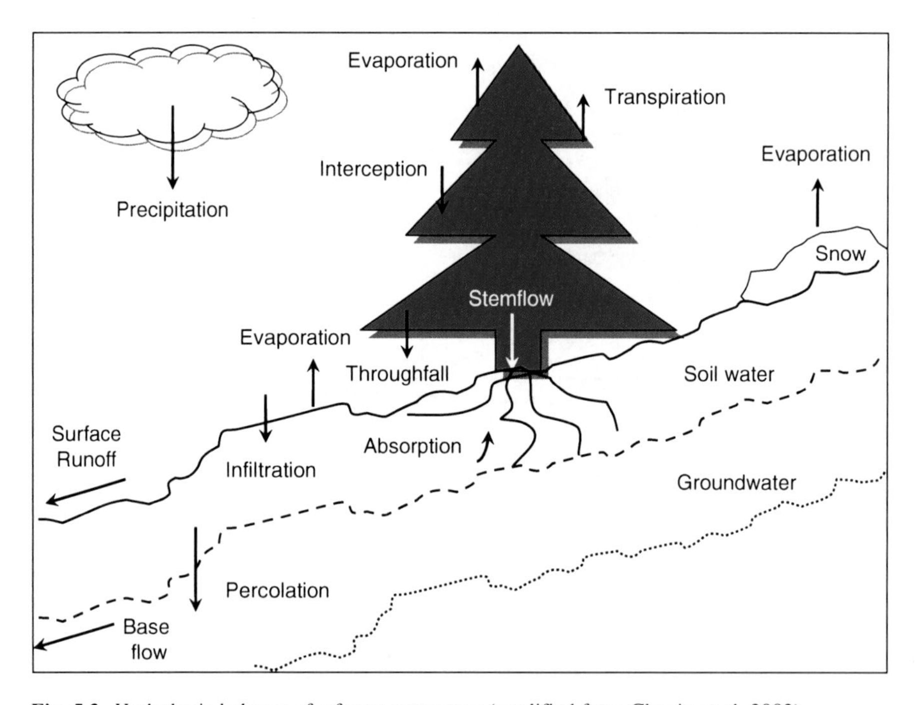

Fig. 5.3 Hydrologic balance of a forest ecosystem (modified from Chapin et al. 2002)